FLORICULTURE
Cultivation, Processing & Marketing

The Author

Dr. P.C. Bansil (1921–2016) belongs to first batch of Indian Economic Service. Worked with the FAO for about 15 years in Sri Lanka, Thailand, Rome, Zambia and Libya and for a similar period with the Government of India. Author of 46 books/Reports. Besides a large number of reports and papers prepared for various Governments, published over 400 papers in specialized technical journals. Hundreds of Technical Reports have been prepared on various aspects of Agriculture for Various Governments. Work experience relates to the formulation, implementation and evaluation of National and Regional Plans for Agriculture, Statistics, Livestock including poultry, Irrigation, Fisheries, Forestry, Wildlife, National Parks, Tourism, Agro Industries and Social Sectors like Demography, Health, Education and Housing *etc.* as well as project identification, formulation and evaluation in the above fields.

Dr. Bansil was basically an agri and livestock economist. He was appointed by the Planning Commission as chairman of the Tenth and Eleventh Plans Working Groups on Animal Husbandry Economics and Statistics. The Food and Agriculture Organization of the United Nations sponsored study – Livestock in India 2000-2030 – was completed by him in 2001. During his FAO assignments in Sri Lanka, Zambia, and Libya as Senior Planning advisor, he undertook micro level studies to improve the quality of slaughter houses, poultry, Agriculture and livestock development programmes – Agriculture in India, cattle in Sri Lanka and Zambia and mutton/chicken in Libya.

Over 50 years of experience in research on problems relating to agricultural and Livestock, economics & statistics, rural and regional economics and planning, demography, health and social sectors in India, Planning Commission, Ministry of Agriculture and FAO.

Some of the Papers were read at the Academic Conferences. Member: International Agricultural Economic Association. Life member: American Agricultural Economic Association, Indian Society of Agricultural Economics, Indian Society of Population Education, Indian Society of Medical Statistics. Indian Institute of Public Administration, Indian Society for Constitutional Studies, Indian Society of Agricultural Marketing and Indian Society of Training and Development, Life and Founder Member, North India Economic Association. He is preparing a **Thirteen Volume** Series – Economics of Agricultural Commodities to cover all aspects of agriculture. Six Volumes have already been published and the remaining are in the Press.

FLORICULTURE
Cultivation, Processing & Marketing

— *Author* —
P. C. Bansil

2018
Daya Publishing House®
A Division of
Astral International Pvt. Ltd.
New Delhi – 110 002

© 2018 AUTHOR

ISBN 9789387057623 (International Edition)

Published by   :   **Daya Publishing House®**
                                         A Division of
                                         Astral International Pvt. Ltd.
                                         – ISO 9001:2015 Certified Company –
                                         4736/23, Ansari Road, Darya Ganj
                                         New Delhi-110 002
                                         Ph. 011-43549197, 23278134
                                         E-mail: info@astralint.com
                                         Website: www.astralint.com

Preface

Government of India has identified floriculture as a sunrise industry and accorded it 100 per cent export oriented status. Owing to steady increase in demand of flower floriculture has become one of the important commercial trades in Agriculture. Hence commercial floriculture has emerged as hi-tech activity-taking place under controlled climatic conditions inside greenhouse. Floriculture in India, is being viewed as a high growth industry. Commercial floriculture is becoming important from the export angle. The liberalization of industrial and trade policies paved the way for development of export–oriented production of cut flowers. The new seed policy had already made if feasible to import planting material of international per unit area than most of the field corps and is therefore a lucrative business. Indian floriculture industry has been shifting from traditional flowers to cult flowers for export purposes. The liberalized economy has given an impetus to the Indian entrepreneurs for establishing export oriented floriculture units under controlled climatic conditions.

We have made extensive use of the data from the net. The study is the result of innumerable sources – published/unpublished, quite a large number of daily papers, journals, magazines and books. The final presentation, however, is in a specific order so as to give the reader the required information at one place. The entire work depends on secondary sources of data, including the net. Every possible source has been studied to provide the required data.

Our special thanks are due to Mr. Mukesh Kumar, Mr. Ravi Shankar and Dr. M.L. Srivastav, Institute Faculty who took upon themselves the arduous task of not only going through the first draft, but also giving their most valuable time to search through the net. The whole of the net data search is their contribution. Total responsibility of providing office support fell on our two diligent staff –Mrs.

Vasantha R. Pillai and Mrs. Ajitha Gangatharan - who cheerfully helped in printing/ reprinting of various drafts from time to time. We must, however, say that none of them is responsible for any errors/omissions from which this work may suffer as at present. I am personally responsible for all this.

P.C. Bansil

Contents

1

Floriculture: An Overview

Floriculture may be defined as the branch of Horticulture which deals with the culture and management of flowers and ornamental plants. It has been derived from two words 'flor' means flower and 'cultura' means cultivation. But now in addition to flower other ornamental plants are also included in floriculture. As Floriculture deals with the culture of flowers and ornamental plants, it has great importance in our daily life as well as national economy.

Floriculture comprises cultivation and trade of flowers (fresh and dried), pot plants, foliage and green like plants and their economic products like scent, oils and medicines. Due to diversity of climatic and soil conditions, flora of India is very rich, ranging from temperate to tropical flowering plants. Of late, floriculture has developed into a full-fledged scientific discipline. Further, with the declaration of floriculture as an extreme focus area by the Ministry of Commerce, Government of India, this industry has acquired a special status in the basket of India's export commodities.

India has a long tradition of floriculture. As it is an ancient creative skill with imagination and an advanced science that played a very importation role in the course of human civilization and its social development. In most part of the country, flower growing is carried out on small holdings and commercial floriculture has assumed importance only in the recent past.

Traditionally, flowers have been grown in India in the open fields, where they have been exposed to both biotic and abiotic stresses. Hence, the quality is not up to the standards. However, in the era of globalization, the produce has to be of International quality and globally competitive, as there is lot of demand for different floricultural products in the export market. The modern floriculture will meet the above demand of the present day's consumers.

Floriculture can be considered as "a specialized branch of horticulture which deals not only with the cultivation of flowers, foliage, climbers, trees, shrubs, cacti, succulents, *etc.*, but also with their marketing and production of value-added products from them"

Floriculture can also be considered as "a discipline of horticulture concerned with the cultivation of flowering and ornamental plants for gardens and floristry, comprising the floral industry. Floriculture includes bedding plants, flowering plants, foliage plants or houseplants, cut greens and cutflowers".

What is Floriculture?

Floriculture is the growing of cut flowers, potted flowering and foliage plants, and bedding plants in greenhouses and/or in fields. There are several thousand different species of flowers and plants that are grown as commercial crops. Cut flowers include such crops as roses, freesia, alstromeria and snapdragons. Some of the favourite flowering potted plants that are available year-round are African violets, orchids, cyclamen and potmums (potted Chrysanthemums). Some seasonal flowering plants are an important part of our traditions, for example, poinsettias for Christmas and Easter lilies for Easter.

Production of Floriculture Products

Growers who produce crops year-round rely on greenhouses to protect their crops Floriculture also involves a considerable amount of production that is not greenhouse based, such as field-grown specialty cut flowers. It includes such products as daffodils, tulips, gladiolus, snapdragons.

☆ Majority of floriculture crops are generally herbaceous.

☆ Bedding and garden plants consist of young flowering plants especially annuals and perennials.

☆ The floriculture business is growing in the world at around 6-10 per cent per annum.

☆ In spite of a long tradition of floriculture, India's share in the International market for these flowers is negligible (< 0.70 per cent).

☆ During the last few years, taking an advantage of the incentives offered by the Government of India, a number of Floriculture units were established in India for producing and exporting flowers to the developed countries. Most of them are located near Mumbai, Pune, Bengaluru, Hyderabad and Delhi and are obtaining the technical know-how from Dutch as well as Israeli Consultants.

☆ Tamil Nadu is the leader in floriculture followed by Karnataka, accounting for 75 per cent of India's total flower production and the state is having the highest area under both modern and traditional flowers.

☆ The country's first and the only Digital Flower Auction Centre is located in Bengaluru, running by Karnataka Agro Industrial Corporation (KAIC).

We all know, Flowers play an essential role in people's celebrations and every day lives. Weddings, graduations, funerals, Mother's Day, St. Valentine's Day, Easter and Christmas. Being a nature lover, we would like to share some of the interesting facts about the floriculture through this blog, with a hope that the information provided will be useful for those who are thinking of starting a business in floriculture which has tremendous opportunities in the present scenerio. And promise to keep posting about this interesting subject in future also with some more interesting facts of this business

Floriculture, or flower farming, is a discipline of horticulture concerned with the cultivation of flowering and ornamental plants for gardens and for floristry, comprising the floral industry. The development, via plant breeding, of new varieties is a major occupation of floriculturists. Floriculture crops include bedding plants, houseplants, flowering garden and pot plants, cut cultivated greens, and cut flowers. As distinguished from nursery crops, floriculture crops are generally herbaceous. Bedding and garden plants consist of young flowering plants (annuals and perennials) and vegetable plants. They are grown in cell packs (in flats or trays), in pots, or in hanging baskets, usually inside a controlled environment, and sold largely for gardens and landscaping. Pelargonium ("geraniums"), Impatiens ("busy lizzies"), and Petunia are the best-selling bedding plants. The many cultivars of Chrysanthemum are the major perennial garden plant in the United States.

Flowering plants are largely sold in pots for indoor use. The major flowering plants are poinsettias, orchids, florist chrysanthemums, and finished florist azaleas. Foliage plants are also sold in pots and hanging baskets for indoor and patio use, including larger specimens for office, hotel, and restaurant interiors.

Cut flowers are usually sold in bunches or as bouquets with cut foliage. The production of cut flowers is specifically known as the cut flower industry. Farming flowers and foliage employs special aspects of floriculture, such as spacing, training and pruning plants for optimal flower harvest; and postharvest treatment such as chemical treatments, storage, preservation and packaging. In Australia and the United States some species are harvested from the wild for the cut flower market.

Floriculture is increasingly regarded as a viable diversification from the traditional field crops due to increased per unit returns and the increasing habit of "saying it with flower" during all the occasions. Though the art of growing flowers is not new to India, protected cultivation in polyhouses is relatively new in India. Enormous genetic diversity, varied agro climatic conditions and versatile human resources offer India a unique scope for diversification into new avenues which have not been explored to a greater extent. With the opening up of world market in the WTO regime, there is a free movement of floriculture products worldwide. In this context, each and every country has equal opportunity for trade in each other's territory. Globally, more than 140 countries are involved in cultivation of floricultural crops. The USA continues to be the highest consumer with more than $ 10 billion per annum, followed by Japan with more than $ 7 billion. India has better scope in the future as there is a shift in trend towards tropical flowers and this can be gainfully exploited by India with enormous amount of diversity in indigenous flora.

What do Floriculture Products Look Like

Flowers play an essential role in people's celebrations and every day lives. Weddings, graduations, funerals, Mother's Day, St. Valentine's Day, Easter and Christmas are all peak periods of demand for flowers and plants. Cut flowers are combined into elaborate arrangements and bouquets, or several stems are packaged together for impulse cash-and-carry purchases. Flowering and foliage plants are combined together in baskets or planters, or sold individually with pot covers and sleeves to accent their beauty. Cut flowers, potted plants and bedding plants are available at florists, supermarkets, corner grocery stores, mass-market outlets and garden centers. More people are buying flowers at their supermarket as part of their weekly grocery shopping. Another shift in marketing is the move towards more direct farm marketing. Several growers have retail outlets on the farm where you can buy products such as longstem roses, potted orchids and bedding plants.

Floriculture Producers

Those involved in floriculture in floriculture are:

☆ Growers

☆ Greenhouse and field employees

☆ Wholesalers

☆ Florists

☆ Garden centres

☆ Supermarkets

☆ Corner stores

☆ Mass-market outlets

☆ Retail clerks

Importance of Floriculture

Besides food and nutritional security, the aesthetic value is also equally important for our daily lively hood as well as for environmental purity. Floriculture is important from the following point of view:

1. Economic point of view
2. Aesthetic point of view
3. Social point of view

1. Economic Point of View

☆ Floriculture is a fast emerging major venture in the world, especially as a potential money-spinner for many third-world countries.

☆ Many flowers and ornamental plants are being grown for domestic as well as for export market will provide more return/unit area than any other agricultural/horticultural crops.

☆ For example in markets such as Delhi and Mumbai and other metros a single spike of gladiolus and gerbera flower may sell up to ₹ 3-5 in Kharif and ₹ 5-10/spike in Rabi/Summer.

☆ Gestation period of flower crop is very less compared to other crops.

☆ Modern-day floriculture refers to the production of high-value cutflowers such as Rose, Gladiolus, Carnation, Orchids, Tuberose, Anthurium, Lilium, Gerbera *etc.*

☆ Now days, growing of these cutflower crops, suited for flower arrangements/decorations for bouquets preparation, and for floral boskets, have increased substantially and its share of the total trade has also improved.

☆ The sales of loose flowers of Jasmine, Crossandra, Marigold, China Aster, Chrysanthemums and Gaillardia *etc.*, are a roaring busyness in south India.

☆ The present trend in floriculture is for making dry flowers, extraction of natural colours and essential oils.

☆ There is lot of demand for good quality flower seeds and ornamental plant materials.

☆ At present the global ornamental crop industry is worth about US$ 70 billion.

☆ The global consumption of the flowers is about US$ 35 billion.

☆ More than three lakhs hectare are under flower production in different countries of the world.

☆ Floriculture generates self employment opportunities round the year. The employment opportunities in this field are as varied as the nature of work itself.

  a. One can join the floriculture field as farm/estate managers, plantation experts, supervisors and project coordinators *etc.*

  b. Research and teaching are some other avenues of employment in the field.

  c. Marketing of Floriculture products for different ventures is emerging as a potential segment of this field.

  d. Besides one can work as consultant, landscape architect etc with proper training.

  e. One can also work as entrepreneur and offer employment to others.

  f. In addition to these careers which involve research and actual growing of crops.

g. Floriculture also provides service career opportunities which include such jobs as floral designers, groundskeepers, landscape designers, architects and horticultural therapists.

h. Professional qualification combined with an inclination towards gardening and such other activities produces efficient floriculturists and landscaping professionals.

☆ Presently more than 145 countries are involved in flower production on commercial scale.

2. Aesthetic Point of View

☆ Lot of scope for landscaping and is considered as billion dollar earning industry in states which ultimately adds the monitory value of any building.

☆ To a Japanese flower arranger each flower expresses one or more meaning (*e.g.* Ikebana).

☆ The wealth of any nation is linked with the health of its people. Unless we can ensure the healthy development of our citizens, especially the younger generation, by providing for them open breathing places through bio-aesthetic planning and landscape gardening, we cannot expect to build up a healthy society and prosperous nation.

☆ Horticultural therapy - is the new dimension of horticultural science to heal the psychic debility and the science is to use garden, landscape plants, parts of plants, growing activity as tools to work.

☆ The bio-force of plants offer a permanent solution to the problems of bio force of human thus, bio aesthetic horticulture is emerging as a new occupational therapeutic tool to restore the lost rhythm and harmony back to human self or inner environment.

☆ It is being utilized in psychiatric hospital, general hospitals, physical rehabilitation center, homes for elderly, prisons and schools.

☆ The patients can achieve higher level of personal development and satisfaction.

3. Social Point of View

☆ Flowers symbolize the purity, beauty, peace, love, adoration, innocence and passion *etc.* Hence, many flowers are used to express the most sensitive, delicate and loving feelings eloquently what our words fail to express.

☆ In our society no social function is complete without the use of flowers, floral ornaments, bouquets or flower arrangements they are invariably used in all social functions.

☆ Used in social gatherings, birthday parties, welcoming friends or relatives and honoring dignitaries. The concept of Valentine's Day is fast catching up in India also.

☆ The arrival of new born is rejoiced with flowers.

☆ To an Indian, especially for Hindu's, flowers have a much greater significance in religions offerings. It has estimated that more than 30-40 per cent of the total flower productions are being consumed in Kolkata city alone used for worshiping purpose.

☆ Floral garlands, gajras and venis are required in marriage ceremonies for adornment of hairs by women of all ages, especially in the south India.

☆ In the present modern era sicks are wished for speedy recovery by offering beautiful cutflowers, while the deads are bidden farewell with flowers along with tear of sorrow.

☆ Flowers are very closely associated with mankind from the dawn of human civilization. There is increasing habit of 'saying with flowers.' Any Indians born with flowers live with flowers and finally dies with flowers.

Interesting Fact about Floriculture

Some of our important floriculture crops originate as weeds in other parts of the world. For example, gerberas (Transvaal Daisies) in South Africa and eustoma (Prairie Gentian) in Texas. Some countries grow dandelions commercially as a salad crop. Floriculture is a world-wide industry: the flowers you buy today could have been picked in South America, Europe or Israel two days ago. To compete with imports, local growers must be able to provide a fresh, high quality product for less money.

Floriculture and its Scope

We all know,Flowers play an essential role in people's celebrations and every day lives. Weddings, graduations, funerals, Mother's Day, St. Valentine's Day, Easter and Christmas. Being a nature lover, i would like to share some of the interesting facts about the floriculture through this blog, with a hope that the information provided will be useful for those who are thinking of starting a business in floriculture which has tremendous opportunities in the present scenario. And promise to keep posting about this interesting subject in future also with some more interesting facts of this business.

General Importance of Flower

☆ Flowers help getting out of illness through psychological enchantment.

☆ Flower gardens increase beauty of the house or an institution.

☆ Scent and perfumes are extracted from the fragrance of flowers *e.g.*, Rose water, Atar.

☆ It is the source of incentive to the poets and writers.

☆ Flower is a national symbol. Water lily (white) *Nymphaea nouchali* is the national flower of Bangladesh.

☆ Flower increases the aesthetic sense of human being and satisfies demand of heart feelings.

☆ Its beauty and fragrance give happiness to all irrespective of age and wealth.

☆ It is the principal component of public functions and anniversaries and also extensively used for decoration.

☆ Flower gardening is a good medium of passing leisure time and help keeping good health.

Flowers have always been an integral part of Indian culture. But, floriculture as a business proposition has not received the attention it deserves. This industry is still in a nascent stage and the potential remains untapped.

Floriculture includes not just cultivation of a wide variety of flowers but also extensive research and creation of healthier, stronger seed varieties and plugs. Globally, more than 140 countries are involved in the cultivation of floriculture crops.

India's floriculture business is worth ₹ 9 billion. With rich and varied climatic conditions and abundance of cheap labour, it is most suit- able for floriculture. Organised floriculture in India is 15 years old. Of the 110,000 hectares under floriculture in the country, only 500 hectares come under organised floriculture. Huge projects came up in the 1990s. However, not many of the older units have survived.

The Indian floriculture industry comprises the florist trade, nursery plants, potted plants, bulb and seed production, micro propagation material and extraction of essential oils from flowers.

Karnataka, Tamil Nadu, Andhra Pradesh, West Bengal, Maharashtra, Rajasthan, Uttar Pradesh, Delhi and Haryana are showing interest in floriculture. The southern states of Tamil Nadu, Karnataka and Andhra Pradesh have a major share. While Karnataka contributes a major portion of exports at ₹ 1 billion, Delhi accounts for over half of the domes- tic market at ₹ 4 billion.

Current Trends

Floriculture is a viable and profitable alternative for the new generation of farmers. For instance, the Punjabi farmers have taken up floriculture seriously. Sandeep Puri, a progressive farmer of Punjab, had gone to Holland to make himself familiar with the technical know-how of the business. He is now able to export Asiatic and Oriental Liliums to South-east countries.

It is interesting to note that those interested in the industry and mining are also showing interest in floriculture. For instance, the VSL group, which has interest in the mining sector, has developed a 25- acre floriculture facility in Hampi, which has only historical importance. The area is not suitable for agriculture. Yet, the VSL group wants to promote agrirelated activity in and around Hampi.

Economic Importance of Flower Production

☆ Perfume industries can be established in the country which can help improving national economy.

☆ Flowers can be a source of earning huge foreign currency by exporting them.

☆ Flowers can be considered as a commercial commodity. Commercial flower production may be helpful in increased earning of the grower.

☆ Establishment of flower production farms and perfume industries can help solving unemployment problem to a large extent.

☆ It provides scope to bring more unused land under flower cultivation.

Flowers to Cure Diseases

The famous homoeopath Dr Edward Bach had observed, "There is no healing unless there is peace of mind, inner happiness and change in our look." Dr Bach believed that nature takes care of most of the ailments from which human beings suffer. He spent six years roaming about in jungles and collected several plants and flowers. He produced 38 medicines which correspond to 38 different moods of human beings. These are called Bach flowers.

Dr Bach had given importance to psychological symptoms. Out of 38 medicines the one that suited the patient's mental state was selected to treat that person. The treatment is thus the simplest and the safest. This treatment can take care of both mental and physical ailments.

The medicines work well and have no side effect. For example, a Bach flow-ers remedy named 'Aspen' is given to the patient who has unspecified fears for which there is no cause.

Scope of Flower Cultivation

☆ Potted foliage and flowering plants are less perishable and has advantage over cut flowers. As most of the important foliage and flowering plants are native to the tropical region, they can successfully be grown in Bangladesh as well as for export from Bangladesh.

☆ Bangladesh has a very good potentiality to become an important supplier of flower and ornamental plants to different countries of Asia, Middle East and Europe.

☆ The agro climatic condition is well suited for growing many flowers and foliage plants in Bangladesh.

☆ Since Floriculture is a new industry in Bangladesh it has a tremendous scope to develop.

☆ Bonsai culture is a recent development in the field of floriculture in Bangladesh. In bonsai culture a large tree is transformed into a miniature form giving a lucrative look. Such plants have heavy demand in the society though they are expensive. To meet up local demand and for export many bonsai making farms can be established.

☆ Cactus is a high valued ornamental crop sold in plant nurseries and shops. Some progressive nurserymen and amateurs are trying to make special types of plants like grafted cactus and bonsai in small scale. This gives a hope for trade in Floriculture.

☆ Tissue culture technology can be exploited for developing quality seedling in many flower spp. in order to get local as well as international market.

☆ There is a high demand of fresh flower and pot plants in Europe, America, Japan, Holland and Middle East. Very recently Bangladesh Govt. has included flowers in the export policy. As such those flowers suitable for export such as rose, gladiolus, gerbera, tuberose, marigold, orchid *etc.* can easily be produced for this purpose.

Measures Needed

APEDA has suggested to the government some sops to strengthen floriculture business. These include subsidy to farmers for replanting, 50 per cent reduction in the freight charges for exporting flowers, and assistance to growers in transporting the flowers.

APEDA wants to set up integrated facilities for handling and storage of exportable perishable products at Kochi, Ahmedabad, Amritsar and Kolkata airports. Of course, integrated facilities are now available at international air-ports. To facilitate exports, APEDA has urged the government to exempt perishable products from sales tax.

The centre plans to set up a ₹260-million special floriculture fund to assist sick floricultural units. Also, there is a plan to set up a replanting subsidy fund of ₹ 470 million in the 11th plan and continue the Transport Assistance Subsidy to exporters to offset the high freight costs. The first International Flower Auction Centre was inaugurated in Bangalore on January 18, 2007.

The centre has initiated some measures to boost floriculture. There is a full-fledged department of floriculture in the Ministry of Agriculture.

APEDA has been playing an important role in the export pro- motion of floriculture products. There have been visits of Indian delegations to the Netherlands. A Dutch delegation also visited India for finalising joint ventures regarding growing plants in greenhouses, and also in the area of refrigeration marketing and propagation of planting.

Flowers and History

In the middle ages a Rose was suspended from the ceiling of a council chamber, pledging all present to secrecy, or sub Rosa, "under the Rose". Evidence of flowers dating back to the prehistoric period have been discovered through Flower Fossils. There are traces of association of flowers with humans during the paleolithic age.

Archaeologists discovered skeletons of a man, two women and an infant buried together in soil containing pollen of flowers in a cave in Iraq. This association of flowers with the cave dwelling Neanderthals of the Pleistocene epoch is indicative of the role of flowers in burial rituals. Analysis of the sediment pollen concentrated in batches, implied that possible bunches of flowers had been placed on the grave.

Closer examination of the flower pollen enabled scientists to identify many flowers that were present, all of which had some therapeutic properties. The use of flowers also testify abundant floral varieties available at that time.

People have used flowers to express their feelings, enhance their surroundings, and to commemorate important rituals and observances. All forms of art, depict the use of flowers: music, books, paintings, sculpture, ceramics, tapestries, *etc.* Some of the most opulent examples of source material are the flower pictures, produced by artists during the 17th, 18th and 19th centuries, which so accurately depict flowers in their incredible beauty. Scientists assert that there are over 270,000 species of flowers that have been documented and are existing in the 21st Century. The evolutionary history of flowers extends across some 125 years. During this time, an intricate assortment of more than 125,000 flower species has developed.

During the Victorian era, in England the language of flowers was as important to people as being "well dressed." For example, the recognizable scent of a particular flower, plant or perhaps a scented handkerchief sent its own unique message. But scientists have yet to answer basic questions about these marvels of beauty. What led to their amazing diversity? Are there flowers that have not changed much during the evolution of this planet?

Fossils of woody magnolia-like plants dating back 93 million years are the first evidence of plant life. More recently, tiny herb-like flower fossils dating back 120 million years have been uncovered by Paleobotanists have. Flowering plants, called angiosperms, were believed to be already diverse and found in most locations by the middle of the Cretaceous Period. 146 million years ago. A myriad of images of preserved flowers and flower parts have been found in fossils from Sweden, Portugal, England, and along the Eastern and Gulf coasts of the United States. Below are a few flowers which have a long history.

The history of flowers is older than of humans. Flowers are ubiquitous. Virtually, every non-meat food that we eat starts as a flowering plant somewhere. Even the cotton clothes we wear come from flowering plants. Flowers appeared on earth about 130 million years ago, during the Crustacean period. Humans are believed to have existed for a mere two hundred thousand years. Once flowers took firm root about 100 million years ago, they quickly diversified into some 250,000 species.

Flowering plants have the botanical name of 'angiosperm.' Angiosperms enclose their seed in fruit, and each fruit contains one or more carpels. Carpels are hollow chambers that protect and nourish the seeds. The oldest fossil of an angiosperm was discovered in China in sediments that date back to 130 million years. It was so primitive that it had not developed the lovely flower that is typical of a flowering plant. It is still considered a flowering plant since it had carpels enclosing seeds that grew into fruits.

The first angiosperms are believed to have evolved in areas where there were ecological disturbances like floodplains and volcanic regions. They were slow growing and had short life cycles. Thus, they matured and reproduced and multiplied much faster than trees. As they seeded quickly, they also evolved faster than other plants that were their competitors. Scientists believe that petals evolved 30 to 40 million years after the first angiosperms evolved. Once this happened, the angiosperms had the distinct advantage over all the other plants, as they attracted the birds and the bees, which, in turn, polinated them.

The world must have started bustling with activity with butterflies and bees and insects swarming everywhere. When this happened some 70 to 100 million years ago, the number of flowering species on Earth exploded. By the time the first flowering plant appeared, plant-eating dinosaurs had been around for a 100 million years. Dinosaurs were eating these and when the dinosaurs became extinct, another group of animals took their place—the mammals, which dispersed the angiosperm fruits, nuts and many vegetables. Now, the flower kingdom and the human race depended on each other.

Name of the Flower	History of the Flower
Alstroemeria	Alstroemeria is named after the Swedish botanist Baron Klas von Alstroemer.
	This South American flowerÂ´s seeds were among many collected by von Alstroemer on a trip to Spain in 1753.
Aster	These plants were believed to have healing properties.
	Asters were laid on the graves of French soldiers to symbolize the wish that things turned out differently.
Calendula	The Romans used Calendula mixed with vinegar to season their meat and salad dishes.
	Calendula blossoms in wine were purported to soothe indigestion, and the petals as ointments, cured skin irritations, jaundice, sore eyes, and toothaches.
Carnations	These flowers were used in Greek ceremonial crowns.
	Carnation, comes from Greece. carnis(flesh) refers to the original color of the flower, or perhaps the word incarnacyon(incarnation), which refers to the incarnation of God made flesh.
Chrysanthemums	Japanese emperors sat upon the Chrysanthemum throne.
	They put a single chrysanthemum petal on the bottom of a wine glass to sustain a long and healthy life.
	In Italy Chrysanthemums are associated with death.
Daisy	Beautiful gold hairpins, each ending in a daisy-like ornament were found when the Minoan palace on the Island of Crete was excavated.
	Daisy flowers are believed to be more than 4000 years old.
	Egyptian ceramics are also decorated with Daisies.
	"Marguerite", the French word for Daisy, is derived from a Greek word meaning "pearl".
	Francis I called his sister Marguerite of Marguerites and the lady used the Daisy as her device, so did Margaret of Anjou the wife of Henry IV and Margaret Beaufort, mother of Henry VII.
Dahlia	A herbal document written in Latin just sixty years after the coming of Columbus was discovered 1929.
	It noted that the Aztecs used dahlias as a treatment for epilepsy.
	Dahlias were late in coming to Europe.

Name of the Flower	History of the Flower
	European scientific specialists considered the dahlia as a possible source of food since a disease had destroyed the French potato crop in the 1840s.
	Between 1800 and 1805, Lord and Lady Holland lived in France and in Spain where Lady Holland first saw dahlias that had been introduced to Spain about 15 years before.
	She sent some home to England and introduced the Dahlia into England.
Delphinium	Delphinium comes from the Greek word delphis, meaning dolphin - the flower resembles the bottle-like nose of a dolphin.
	Delphiniums were used by West Coast Native Americans to make blue dye, and European settlers made ink from ground delphinium flowers.
	The most ancient use of Delphinium flowers was a strong external concoction thought to drive away scorpions.
Gladiolus	The Latin word gladius, meaning "sword," and this flower was named for the shape of its leaves.
	Gladiolus was also called "xiphium," from the Greek word xiphos, also meaning sword.
	This flower is said to have represented the Roman gladiators. British Gladiolus used the stem base (corms) as a poultice and for drawing out thorns and splinters.
	In the 18th century, African Gladioli were imported in large quantities to Europe from South Africa.
Holly	Medieval monks called this plant the Holy Tree.
	They believed Holly would keep evil spirits away, and protect their home from lightening.
	The early Romans decorated their hallways with garlands made from Holly for their mid-winter feast, Saturnalia.
	Later its pointed leaves represented the crown of thorns worn by Jesus, and the red berries his drops of blood.
Lily	Lilies have been associated with many ancient myths, and pictures of lilies were discovered in a villa in Crete, dating back to the Minoan Period, about 1580 B.C.
	Lilies are mentioned in the Old Testament, and in the New Testament, they symbolize chastity and virtue.
	In both the Christian and pagan traditions, the lily is a fertility symbol.
	In Greek marriage ceremonies the bride wears a crown of lilies and wheat implying purity and abundance.
Rose	The first cultivated roses appeared in Asian gardens more than 5,000 years ago.
	In ancient Mesopotamia, Sargon I, King of the Akkadians (2684-2630 B.C.) brought "vines, figs and rose trees" back from a military expedition beyond the River Tigris.
	Confucius wrote that during his life (551-479 B.C.), the Emperor of China owned over 600 books on the culture of Roses.
	Roses were introduced to Rome by the Greeks.
	During Roman public games all the streets were strewn with rose petals.

Name of the Flower	History of the Flower
	Egyptian wall paintings depicting roses have been found in tombs dating from the fifth century B.C. to CleopatraÂ´s time.
	Cleopatra had a passion for everything Roman, and she is said to have scattered rose petals before Mark AnthonyÂ´s feet.
	Roses were introduced to Europe during the Roman Empire, where they were mainly used for ornamental purposes.
	Early Christians saw the rose as a symbol of paganism, orgy, and lust.
	King Childebert I had a rose garden planted for his Queen in Paris.
	Charlemagne ordered the cultivation of Roses.
	Leo IX, elected Pope in 1084, sent a Golden Rose to favored monarchs.
Poinsettia	Dr. Joel Roberts-Poinsett, the US Ambassador to Mexico, brought the first poinsettia to the United States in 1928.
	Because Mexican legends say its bracts resemble the flower of Bethlehem, Poinsettias have the honor of decorating churches at Christmas time.
	Today, this flower is known worldwide as "the Christmas flower".
	This plant was used during the Medieval times as a purgative to rid the body of black bile and melancholy.
Queen Anne´s Lace	Queen AnneÂ´s Lace was named for Queen Anne, wife of King James I of England.
	The QueenÂ´s friends challenged her to create lace as beautiful as the flower.
	North African natives chewed it to protect themselves from the sun.
Snap Dragons	Snapdragons were common in the earliest gardens, but their actual origin is not known.
	Some botanists believe they grew wild in Spain and Italy.
Sun Flower	Sunflowers originated in Central and South America, and were grown for their usefulness, not their beauty.
	In 1532 Francisco Pizarro reported seeing the natives of the Inca Empire in Peru worshipping a giant sunflower.
	Incan priestesses wore large sunflower disks made of gold on their garments.
Tulip	Over a thousand years ago, Tulips grew wild in Persia, and near Kabul the Great Mogul Baber counted thirty-three different species.
	The word 'Tulip' is thought to be a corruption of the Turkish word for turbans.
	Persian poets sang its praises, and their artists drew and painted it so often, that all of Europe considered the Tulip to be the symbol of the Ottoman Empire.
	Wealthy people began to purchase tulip bulbs that were brought back from Turkey by Venetian merchants.
	In 1610, fashionable French ladies wore corsages of tulips, and many fabrics were decorated with tulip designs.
	In the seventeenth century, a small bed of tulips was valued at 15,000-20,000 francs.
	Tulipmania flourished between 1634-1637. just like the California Gold Rush, people abandoned jobs, businesses, wives, homes and lovers to become tulip growers.
	The frenzy spread from France, through Europe to the Low Countries.

Name of the Flower	History of the Flower
Violets	When Napoleon married Josephine, she wore Violets, and on each anniversary Josephine received a bouquet of violets.
	Following Napoleon´s lead, the French Bonapartists chose the violet as their emblem, and nicknamed Napoleon "Corporal Violet".
	In 1814, Napoleon asked to visit Josephine's tomb before being exiled to the Island of St. Helena.
	When he died, he wore a locket around his neck that contained violets he had picked from Josephine´s grave site.

Indian History of Gardening

India has a long history of flowering plants. In the era of Mahabharata, there was famous tree named Kadamba, which associated with Lord Krishna. Vatsyana (A.D. 300-400) described four kinds of gardens, which were made for the queens, kings, courtiers and ministers. Famous poet Bana Bhatta described the number of flowering plants in his famous book the Harsh Charita. These flowers plants were growing in the gardens. At that time, water pools built with red lotus and blue water lilies. Status of gardening had mentioned in Ramayana written by Valmiki and Tulsidas. At that time, Ayodhya city was having wide streets, large houses, noble palaces, richly decorated temples and gardens. These gardens were planted with fruit trees, flowers; lakes were full of lotuses and different kinds of birds.

In the period of Lord Buddha, the life of Buddha was associated with a number of trees from birth to his nirvana. He was born in 563 B.C. under the tree of Ashoka at Lumbini. Birth place of Buddha has been described by Hiuen Tsang who visited the place in 630 A.D., there was bathing tank of Sakya Muni filled with clear water, lotuses and lilies when Lord Buddha visited Vaishali, Amrapali presented a park known as Amrvana which was dominated by flowering trees. Buddhist was planting trees and flowering plants on a large scale for making surrounding peaceful, a place ideal for meditation.

When Aryans came in India about 1600 B.C., at that time, the country was called as Aryavrta, which means the country of lotus and sunshine because the lakes were studded with lotus flowers and there were wide-open spaces. Therefore, the lotus being a native of India found everywhere. Aryans started the use of flowers in religious and social ceremonies. They appreciated the beauty of flowering plants, lakes, mountains, flowers like Kamal, Champa, Madhavi, Bela, Chameli, Rukmani, *etc.*

The history of systematic gardening in India is as old as civilization of Indus of Harappa, which existed between 2400 B.C. and 1750 B.C. At that time, people were living in well-planned roads cut across one another almost at right angles. There are many evidences found that trees and ornamental plants were associated with the Harappa civilization. In Mogul era, Babar had founder of gardens. He made gardens at Panipat and Agra. Mogul gardens are synonymous of formal style of gardening. Grand Trunk Road from Lahore to Calcutta made by Sher Shah Suri and planted shady trees along both sides of the roads. Akbar has made Fatehpur Sikri (Agra) garden. There are so many gardens build by Mogul emperors.

During the British period, there was a lot of activity in gardening by Britishers and Indian kings. King Hyder Ali established most famous Lal Bag garden at Bangalore. In North India, Maharaja Ranjit Singh made garden at Amritsar. Britishers had managed well gardens in India. They imports plants from England. Britishers established Royal-Agri-Horticulture Societies and Botanical Gardens in India *i.e.* Royal Agri-Horticultural Society Garden, Calcutta., Lloyd botanical garden, Darjeeling, Botanical garden, Saharanpur, National Botanical Garden, Lucknow, Botanical garden of the forest research garden, Ootacmund; *etc.*

There have been changes in the field of gardening during post independence period. An effort has made for public gardens in big cities for improving environment. Several gardens in different cities have been providing recreational facilities. Some important gardens are Buddha Jayanti garden Delhi, Rose garden, Chandigarh *etc.* The gardens has made along with mega highway all over India recently. Because of these stages of gardening, have increased flowering habits among Indians. After globalization, modern trends are transferring from one nation to another, though the new trends arrive in India. Therefore, last two decades there has been raising commercial floriculture in India.

Floriculture in India

About two decades back or so, the floriculture was just a pastime of rich people and hobby of flower lovers, but now it has opened a new vista in agri-business i,e, commercial floriculture. With the increase in buying capacity of people, the flower lovers have now started buying them from the markets to beautify their home as well as to adore some one they love simply because they don't have time and enough space to grow flowers particularly in urban areas and in metropolitan cities. Flowers, it seems, is the most wanted item in any social occasions for conveying one's status and aesthetic sense. Flower is now so indispensable that one may cancel his her birthday celebration or Yama may postpone the death of a dying person in case flowers are not available at that time. No nuptial is performed and honeymoon of a young couple is not consummated till garden fresh rose and rajanigandha or tuberose with lingering and stupefying aroma are made available. Warm welcome cannot be offered to VIPs in the public functions without bouquet - flowers are so indispensable!

All these, no doubt, have set flower business on a top gear. One may wonder, the global market on flower is at present, carrying a business worth 2000 crores US dollar (1992) par annum. India is also having a business worth ₹280 crores in her domestic market (1992-93).

Indian Council of Agricultural Research (ICAR) conducted a survey of assessment on the possibilities of cut flowers trade in India during 1960-62. An important conclusion was that an internal sale as ₹9.26 Crores worth flower weighing 10,460 tones grown in an area of 4000 hector. Flowers like Rose, Gladiolus, Tuberose, Chrysanthemum, Aster, Carnation, Orchids, Marigold are most popular in cut flower market all over the World.

Introduction

Floriculture or flower farming is the study of growing and marketing flowers and foliage plants. Floriculture includes cultivation of flowering and ornamental plants for direct sale or for use as raw materials in cosmetic and perfume industry and in the pharmaceutical sector. It also includes production of planting materials through seeds, cuttings, budding and grafting. In simpler terms floriculture can be defined as the art and knowledge of growing flowers to perfection. The persons associated with this field are called floriculturists.

Worldwide more than 140 countries are involved in commercial Floriculture. The leading flower producing country in the world is Netherlands and Germany is the biggest importer of flowers. Countries involved in the import of flowers are Netherlands, Germany, France, Italy and Japan while those involved in export are Colombia, Israel, Spain and Kenya. USA and Japan continue to be the highest consumers.

Floriculture is an age old farming activity in India having immense potential for generating gainful self-employment among small and marginal farmers. In the recent years it has emerged as a profitable agri-business in India and worldwide as improved standards of living and growing consciousness among the citizens across the globe to live in environment friendly atmosphere has led to an increase in the demand of floriculture products in the developed as well as in the developing countries worldwide. The production and trade of floriculture has increased consistently over the last 10 years. In India, Floriculture industry comprises flower trade, production of nursery plants and potted plants, seed and bulb production, micro propagation and extraction of essential oils. Though the annual domestic demand for the flowers is growing at a rate of over 25 per cent and international demand at around ₹ 90,000 crore India's share in international market of flowers is negligible. However, India is having a better scope in the future as there is a shift in trend towards tropical flowers and this can be gainfully exploited by country like India with high amount of diversity in indigenous flora.

After liberalization the Government of India identified floriculture as a sunrise industry and accorded it 100 per cent export oriented status. The liberalization of industrial and trade policies paved the way for the development of export oriented production of cut flowers. The new seed policy has already made it feasible to import planting material of international varieties. Floriculture products mainly consist of cut flowers, pot plants, cut foliage, seeds bulbs, tubers, rooted cuttings and dried flowers or leaves. The important floricultural crops in the international cut flower trade are rose, carnation, chrysanthemum, gerbera, gladiolus, orchids, anthurium, tulip and lilies.

In recent decades there has been increasing in demand of floriculture products with increasing income. It is souring industry in Asian countries including India. Floriculture is an emerging area with great potential both in the domestic as well as export market. In India, commercial floriculture is ongoing development but have a long tradition of various types of flowers. Flowers have been representing in ancient painting, mural and coins. However, the social and economic aspect of flower growing recognized later. It is only in the last two three decades.

Since 1991, New Economic Policy has adopted in India. The main objective of this policy is to solve foreign currency crisis and remove the stagnancy through liberal economy. Thus, the major change has done in every sector of Indian economy. For the promotion of agriculture export, the Union Government gave incentives through the certain policies. NHB, APEDA and NABARD plays supporting role and recently NHM introduced as centrally sponsored scheme.

From 2001, there has been tremendous growth in floriculture production. In terms of area, production and export it can be seen extreme growth. All states in India have a tradition of growing flowers, commercial growing of flowers presently confined to Karnataka, Tamil Nadu, Andhra Pradesh, West Bengal, Maharashtra, Rajasthan, Delhi and Haryana. In India, marigold, aster, roses, tuberose, gladiolus, are grown in open field while gerbera, carnation, roses, anthorium, orchids, etc, are grown under greenhouse conditions.

The export of floricultural products has been increasing tremendously during 2001. However, the historical background, floriculture development and other related matter considered in this chapter with following sub heads.

Evolution in Progress

By tracing how a gene mutation over 100 million years ago led flowers to make male and female parts in different ways, research by University of Leeds plant scientists has uncovered a snapshot of evolution in progress. In a number of plants, the gene involved in making male and female organs has duplicated to create two, very similar, copies. In rockcress (Arabidopsis), one copy still makes male and female parts, but the other copy has taken on a completely new role: it makes seed pods shatter open.

In snapdragons (Antirrhinum), both genes are still linked to sex organs, but one copy makes mainly female parts, while still retaining a small role in male organs– but the other copy can only make male. The findings published in the *Proceedings of the National Academy of Sciences (PNAS)* Online Early Edition provide a perfect example of how diversity stems from such genetic 'mistakes'.

The research also opens the door to further investigation into how plants make flowers. "Snapdragons are on the cusp of splitting the job of making male and female organs between these two genes, a key moment in the evolutionary process," says lead researcher Professor of Plant Development, Brendan Davies.

Added Complexity

"More genes with different roles give an organism added complexity and open the door to diversification and the creation of new species." By tracing back through the evolutionary 'tree' for flowering plants, the researchers calculate the gene duplication took place around 120 million years ago. But the mutation which separates how snapdragons and rock cress use this extra gene happened around 20 million years later.

The researchers have discovered that the different behaviour of the gene in each plant is linked to one amino acid. Although the genes look very similar, the proteins they encode don't always have this amino acid. When it is present, the activity of the protein is limited to making only male parts. When the amino acid is not there, the protein is capable of interacting with a range of other proteins involved in flower production, enabling it to make both male and female parts, according to a University of Leeds press release.

Growth in India

In India, the floral enterprise is growing by about 25 per cent over the last decade. Some experts consider it to be a goldmine, with promises of many employment opportunities, and a steady inflow of foreign currency. It is likely to flourish in the future. The varied climatic zones, the extremely wide and colourful range of indigenous species, cheap labour and land costs—India does offer a very fertile ground for floriculture.

Currently, flowers are mainly grown in four zones here– Bangalore, Pune, Hyderabad and Delhi. Other regions where flowers are farmed are West Bengal, Tamil Nadu, North-east, Uttarakhand, Rajasthan and Jammu and Kashmir. The traditional varieties like rose, marigold, aster, jasmine are mainly consumed by buyers at home for devotional purposes, while the commercial blooms, including chrysanthemum, garberra, lily, tulip and orchids are produced for the high value global market. The total turnover of the traditional variety is around 105 crores and that of the commercial sector is around 100 crores. There is a growing demand for flowers in the fashion industry which uses the blossoms to make eco-friendly colours—the latest rage among the fashionistas. In the year 2008-09 flower production amounted to 700 crores and for the current year 2009-10 the Planning Commission has targeted 1000 crores. Floriculture is practiced over 1.4 lakh hectares of land, producing around 6, 70,000 metric tonnes of loose flowers and 13,009.3 million cut flowers.

Historical Background

It is interesting to study that, history of gardening so how long like civilization. The history of gardens connected with the history of civilization, the history of gardens and gardening has connected with the history of the people and their culture, which includes their science, art and literature. Hence, the account of gardens and gardening given as below.

World History of Gardening

The evidence about the existence of gardens in ancient Egypt is archaeological. Different portions of the plants recovered from tombs and a number of plants painted on the tombs. These painting show that the Egyptians had religious. There was Temple Garden at Karnak dating from the region of Tuth Mosis II, about 1500 B.C. Queen Hatsheput (1505-1483 B.C.) built a Temple Garden on a hill. It was a terrace garden with the temple on the uppermost terrace. It also indicate that the idea and technique of garden making from Mesopotamia to Egypt.

China is one of the centres of humans evolution. The great poet Li-Po (A.D.705-762) of Tang period has noted importance of garden and nature in his poem and we seen by this evidence China has the origin of the flora.

The Japanese are great lovers of plants and flowers. The first garden was constructed in Kyoto for holding garden parties dating A.D. 794 in the period of Heian.

Iran apart from love, war and hunting they were founder of gardening from 558- 323 B.C. Cyrusi who was the first Persian Gardner in that period. He planted a garden of Sandins in Lydia, had set the design of the Chahav Bagh.

The first park in Greece was dedicated to Diana in the period 434-356 B.C. In this park, fruit trees were planted around the temple. The king Xenophon got his inspiration from Persian garden in that period.

Italy, one of the heartland of the Rome Empire. Italy was the state of agriculture in third and second centuries B.C. At that time, few farmers were horticulturist and growers of gardens. In Arab countries, during A.D. 1058 to A.D.1259 that a galaxy of men of genius flourished that is Al Bakri and Idrisi.

In modern age Europe made several gardens, *i.e.* Dahlias and Zinnias are the gifts of Mexican gardeners to the world. Thus, it would be seen that the countries of new world has sophisticated gardens.

Gyangtak Show

The annual International Flower Festival is here again. Held every year in the month of May 2011 at Gangtok, the festival is a visual treat for all nature lovers and enthusiasts. More than 600 species of orchids, 240 species of trees and ferns, 150 varieties of gladioli, 46 types of rhododendrons along with a variety of magnolias and many other foliage plants are displayed which reflect the State's rich flora.

Dining options at The Royal Plaza include Orchids – the multi-cuisine restaurant, Round The Clock – the lobby lounge and Aura– the bar. A three nights/ four days stay during the festival starts from ₹23499 onwards. The offer is valid throughout the month of May. The package includes non-alcoholic welcome drink, half-day sightseeing to Tashi view, Ganesh Tok, Hanuman Tok and Burtuk Falls, half-day sightseeing to Handicraft Centre, Flower Show, Tibetology and Do Drul Chorten, complimentary entry to the Fitness Centre and Casino, special rate for Nature Trekking, River Rafting (subject to weather condition and advance booking).

Evolution of Greenhouse Technique

Man is social and thinking animal. He discovered so many things from nature. In modern age, he tries to solve the problems through science and technology. However, agriculture and floriculture made considerable progress in modern era. Greenhouse technique is one of the techniques, which is useful to give horticultural produce. Greenhouse is one of the framed structures covered with transparent material in which crops can be growing under the conditions of controlled environment.

In the 16th century, glass lanterns, bell jars and hot beds covered with glass were used to protect horticultural crops against cold. In 17th century, low portable wooden frames covered with an oiled transparent paper were used to warm the plan environment. In Japan, straw mats were used in combination with oilpaper to protect crops.

During 17th century greenhouse heated by manure and covered glass panes on one side in France and England. Then 18th century glass was used on both sides. After World War II, protected agriculture was fully established with the introduction of polythene. Thereafter, plastic greenhouse was established in 1948. In India, the cultivation in the plastic greenhouse is one of the recent origins.

All Types of Flowers Can be Placed into Two Main Groups: Monocots or Dicots.

Different types of flowers (the Angiosperms) can easily be identified by dividing them into Monocots or Dicots. By just looking at one flower we should soon be able to identify it as one or the other. This not only makes it easier to pick out the different flower types, but it is very interesting to know exactly what we are looking at.

There are other ways of classifying flowers, especially Annuals; Perennials, Biennials and Ephemerals; or even by the way they reproduce such as Sexual Reproduction and Asexual Reproduction *e.g.* Bulbs. These are the terms most useful to gardeners, whereas classifying by Monocots and Dicots is very basic but is very useful if you are really studying flowers or just want a general understanding of them. Of course the one other big way is by Botanical Names and Classes, *etc.* But that's for the Botanists.

When we try to group flowers into Monocots and Dicots, just remember that there are always exceptions when a flower doesn't seem to be one type or the other. Sometimes there are fusions of petals and leaves and other parts making it very confusing. If ever you find any types of flowers which just don't fit and you want to know their names, it's a good idea to either take a flower to your local Botanical

Gardens, or send them a photo. They are really helpful with ID in different types of flowers. Of course the local nursery usually has the ID we want.

This Portulaca has narrow leaves and the petals are not clearly defined. Because the leaves look like Monocots (not branching), I would have assumed that this was a Monocot. But it is definitely a Dicot Flower as you can see below with the broader leaved form. It is a good example of just how tricky Nature can be. The photo below is much clearer. If you click on it to enlarge it you can see the five petals distinctly. Sometimes, unless you have access to the original baby seed leaves of a flower, things can get pretty tricky.

Portulacas (above) are often called Moss Roses. They are Dicots because they have combinations of five petals. The singles have five petals and the doubles may have ten or more. But they do look so much like Wild or Single Roses that it's easy to see how they came to be considered Moss Roses. (The true Moss Rose is an actual

Centifolia Rose). See more about them on the Portulacas Page. Here you will see just why these hardy, drought tolerant little plants are so popular.

The Plant Kingdom

Besides the Angiosperms (flowering plants) and the Gymnosperms (cone bearing plants), in the Plant Kingdom there are the Bryophytes: Mosses, Liverworts and Hornworts; and the Pteridophytes (ferns, horsetails and club mosses). The last two groups still use spores to reproduce. Spores (brown) can be seen on the backs of fern fronds in Spring. This is a type of sexual reproduction. And finally Legumes are often classed as another member of the Plant Kingdom all by themselves. Examples are beans, peas, and lentils. So the Plant Kingdom can be classified into five main groups:

☆ Angiosperms - Flowers

☆ Gymnosperms - Cones

☆ Bryophytes - Mosses, Liverworts, Hornworts

☆ Pteridophytes - Ferns, Horsetails and Club Mosses

☆ Legumes - Beans, Peas and Lentils

The Angiosperms or Flowering Plants

The Angiosperms are the Flowering Plants and there are around 250,000 to 400,000 different flower types. Fortunately they can be divided into two groups: Monocots and Dicots. Note that it is only the Flowering Plants that are Monocots or Dicots. Other plants such as Mosses and Ferns are neither.

Monocots have one Cotyledon (the seed capsule where the seed develops) and dicots have two. The Cotyledon contains food for the growing Embryo/s. This means that when the seed develops and grows, it will send up either one (Monocot) or two (Dicot) leaves. I used to think that Dicots had two seeds. Wrong.

Monocots form one quarter of all the Angiosperms and Dicots the rest. Roses are Dicots. Click here to see the Wild Rose. There are several ways to distinguish between Monocots and Dicots. The leaves of Monocots have parallel veins that begin at the base of the leaf and end at the tip without any branching (Lily family). The Dicots' veins start at the bottom and branch out in an ordered network all over the leaf (as in a rose). And of course, if you are there at the beginning when the seedlings push up through the soil, you can tell just by looking at them. The Monocot seed has one leaf and the Dicot has two, as in the side photos:

The Difference Between Monocots and Dicots

Monocots	Dicots
One leaf emerges from the cotyledon	Two leaves emerge from the cotyledon
Leaves have parallel veins	Leaves form a branching network
Fibrous root system	Tap root system
Petals in combinations of 3	Petals in combinations of 4 or 5
Stamens in combinations of 3	Stamens in multiples of 4 or 5
Example: a Lily	Example: a Rose.

The Monocot Plant

A Monocot Has A Fibrous Root System. Monocots may have millions of individual Fibrous roots. There is no main central root. Their leaves are similar in that the veins do not branch out from a central vein, but run parallel to each other. Obviously, the roots aren't parallel but they can still cover a huge area.

The Dicot Plant

The Broad Bean Seedling Has A Main Tap Root. Dicots have a main Tap root with many smaller roots branching off it. Just like their leaves which have a central vein with other veins branching off it. The Dicot root system can cover a huge area to get the essential nutrients for the plant.

Lists of Monocots and Dicots

Monocots	Dicots
Lily: Oriental Lily, Day Lily	Rose
Tulip	Daisy
Orchid	Sweet Pea

Monocots	Dicots
Bluebell	Cosmos
Daffodil	Nasturtium
Jonquil	Hollyhock
Crocus	Foxglove
Freesia	Portulaca(Moss Rose or Purslane)
Amaryllis	Begonia
Lily of the Valley	Ranuculus
Day Flower	Marigold
Lesser Celandine (below)	Pansy
Agapanthus	

Lily of the Valley. Another plant to die for. I honestly would have thought that this plant was definitely a Dicot because its leaves are so big and fat. But don't be fooled. This is a great photo where you can definitely see that although the leaves are broad, the leaf veins are parallel. Definitely a Monocot. So many of our delightful Spring Bulbs are Monocots, usually with narrow blade-like leaves. The Rose Leaf, on the other hand, is a Dicot because it has a main vein with other veins branching off it.

A Snapshot of Evolution in Progress

By tracing how a gene mutation over 100 million years ago led flowers to make male and female parts in different ways, research by University of Leeds plant scientists has uncovered a snapshot of evolution in progress. In a number of plants, the gene involved in making male and female organs has duplicated to create two, very similar, copies. In rockcress (Arabidopsis), one copy still makes male and female parts, but the other copy has taken on a completely new role: it makes seed pods shatter open.

In snapdragons (Antirrhinum), both genes are still linked to sex organs, but one copy makes mainly female parts, while still retaining a small role in male organs – but the other copy can only make male. The findings published in the *Proceedings of the National Academy of Sciences* (*PNAS*) Online Early Edition provide a perfect example of how diversity stems from such genetic 'mistakes'.

The research also opens the door to further investigation into how plants make flowers. "Snapdragons are on the cusp of splitting the job of making male and female organs between these two genes, a key moment in the evolutionary process," says lead researcher Professor of Plant Development, Brendan Davies.

Added Complexity

"More genes with different roles give an organism added complexity and open the door to diversification and the creation of new species." By tracing back through the evolutionary 'tree' for flowering plants, the researchers calculate the gene duplication took place around 120 million years ago. But the mutation which separates how snapdragons and rock cress use this extra gene happened around 20 million years later.

The researchers have discovered that the different behaviour of the gene in each plant is linked to one amino acid. Although the genes look very similar, the proteins they encode don't always have this amino acid. When it is present, the activity of the protein is limited to making only male parts. When the amino acid is not there, the protein is capable of interacting with a range of other proteins involved in flower production, enabling it to make both male and female parts, according to a University of Leeds press release.

The Business of 'Blooming

Flowers. They spell colour, fragrance and festivity. They also spell money. Yes, the pretty blossoms are actually worth a pretty packet, and a global business of considerable proportions rests on their tender petals. Floriculture, say economists, is a 'sunshine' industry today. It has a bright future.

The flower route It is, quite literally, a blooming industry, with a global turnover of over US $ 60 billion. The floriculture market has a very well demarcated trade link. Europe, Japan and the US are the bulk importers of flowers, while the 'production', meaning the actual process of growing the ware is mainly carried out in the West African and Latin American countries like Kenya, Ethiopia, South Africa, Columbia and Ecuador.

So what makes a country a good grower? Conducive climate, availability of land and labour, sound knowledge base (on the art of flower production) are the most-in-demand conditions. Africa has the largest flower market in the world, accounting for 90 per cent of the global production. As of now, India contributes a meagre 0.38 per cent (US $47 million). Should it aspire to be a more powerful player in this arena?

Bouquet of Toxins

May be it is wise to look around a bit and find out how the bulk growers are faring before taking the plunge. Because artificial cultivation of flowers in greenhouses is branded as one of the most polluting agricultural activities. No wonder most of the growing activities have been gradually shifted out to the poorer nations, during the past two decades, even though the rich countries remain the prime buyers. The Result? Giant portion of the industry's profit is pumped back to the developed world, while the poor producers are left to deal with the damages... of the environmental kind.

Yes, the sweet smelling blossoms leave a severely toxic trail. Pesticide use in this industry is rampant. Flowers are sprayed with 50 times more chemicals than any other agri product. A worker in a greenhouse is exposed to about 127 different kinds of harmful chemicals in the form of fertilisers, insecticides, fungicides, and nematocides, which are mostly sprayed on flower crops without any proper standardization or precaution. These chemicals can cause cancer, birth defects, reproductive and nervous system damage, skin infections, miscarriages, serious respiratory diseases and even reproduction disorders Considering the fact that more than 190,000 people in Latin America, Africa and some parts of Asia are reportedly employed in the flower business, its impact on human health definitely requires attention.

Ravaging Resources?

Growing flowers require a lot of water. Ground water levels are recorded to have dropped drastically in regions where floriculture is practiced. The problem has scaled up more sharply because floriculture is being promoted as an alternative to traditional crops by the multinational companies. So flowers are taking over lands that grew food crops in the producer nations. This is not only taking a heavy toll on water and soil, but is triggering food shortages in some countries.

Doing Business with the Flowers is, however, Infested with many Thorny Issues

☆ There is severe shortage of the technology required to store, preserve and transport this highly perishable cargo from the field to the market. In Delhi's Connaught Place mandi, for instance, where small traders do business, lack of clod storage and display facilities wreck havoc in the profit margins of the dealers.

☆ Often farmers find it tough even to arrange for basic inputs like seeds and planting material. Mostly imported varieties are grown for export, so the costs soar.

☆ The economies of scale do not work here. The Indian farms lose their competitive edge by virtue of being smaller in size (4 hectare average) as compared to the huge African farms (40 hectare average). African units are able to capture more market share and offer more diverse varieties.

☆ Indian flower cultivators need a lot more hand holding to become as market savvy as their counter parts elsewhere in the world. It is only recently that government funding through banks like NABARD has been arranged, and quarterly journals dedicated to floriculture have been launched.

Flower Power is the New Weapon to Fight Wrinkles

A tropical flower has been hailed as the latest weapon against wrinkles, after scientists found that a sugar compound derived from it can actively regenerate skin, making it feel plumper and more elastic.

A team at cosmetic giant LO real tested the sugar compound called rhamnose, from Uncaria flower, nicknamed cat claw, on 400 women after using it in a cream. The chemical is thought to stimulate cells into producing collagen the main component of connective tissues such as skin. Full results have not been published because LOreal is waiting for patents to be granted. Julie McManus, the head of scientific affairs at LOreal, said, This is a breakthrough, as we have found a compound which can produce an effect on this very important group of cells. The new generation of treatments have been dubbed cosmeceuticals as they're created using research technology from the pharmaceutical industry. In this case, computer software was used to screen thousands of naturally occurring chemicals to find one that stimulates a specific type of skin cell that produces collagen. They were then further filtered to see if they were small enough to penetrate through the outer waterproof layer of the skin. It then underwent human trials involving 400 women aged 50 to 70.

Rose GYAN

Rose, undoubtedly is the queen of all flowers. It is one of the most sought after flowers for all occasions. In West the demand for its different varieties changes like fashion. However, a bloom with a long, straight and sturdy stem is always the most sought after. This comes from the class of roses that are called Hybrid Teas. Flowers in this class are borne singly on almost a meter long stem. The green-house roses are also the Hybrid Teas and command a good price in the market. In rare instances, some of the Hybrid Teas modify themselves to develop into climbers roses. Second important class of roses is Floribundas. This class of roses is valued for their bearing habit as the blooms appear in bunches at the terminals. They may be single-floret or multi-floret blooms. The life of these blooms is short but due to frequent and profuse blooming the bed never wears an empty look.

Miniatures are another form of roses. Leaves of this variety are the miniature replicas of the original. The blooms are also small but due to the free flowering habit of the bushes, the plant looks full and attractive at all times. While planting these distances between individual plants should be kept the minimum to get that "full" look. The plant hardly grows more than a foot high. There is another interesting variety is - climbing roses. The climbing roses spread and cover any surface in no time. The beauty of this variety lies in profuse flowering. When in bloom. One can find thousands of them on a single climber which may have spread to about 30 ft. Roses bloom only twice a year - once in mid-December and the next flush appears in March. Gardeners want to have the maximum blooms in this period.

A Georgina Shimmering in the Garden is the Sweetest Way to begin your day

English humourist Thomas Hood said, "A double dahlia delights the eye" and he was right. This corpulent flower grows so tall that it shows its bright purple and white, yellow, deep red and orange ruffles against the blue skies and is very regal. It has two names thanks to two men, who fell in love with it! It is called dahlia after Swedish botanist Andreas Dahl. In eastern Europe and Russia, it is called Georgina named for a botany professor, Johann Georgi, who was so thrilled when he saw it for the first time on a holiday in Paris, that he took several flowers home for his friends.

Europeans discovered the dahlias during the Spanish conquest of Mexico early in the 16th century. The Aztecs grew them for their tuberous roots, which were used in medicine. It was also used in cooking till people thankfully realised that it wasn't very tasty and stopped eating it. Georginas attract butterflies because of their stunning colours. They often get mildew in cold weather. If you give them good air circulation by pulling off the bottom row of leaves, with 8 to 12 inches of clear space, under the plant, you will solve the mildew problem.

Dahlias are water guzzlers and regularly watered produce more flowers. If these get pests, don't use garden chemicals. A study has shown that children of non-chemical gardeners were six times less likely to contract leukaemia.

Dahlias discourage nematodes when you grow them around a flower or vegetable affected by them. They also discourage moles. When we use pesticides, 60 to 90 per cent (by volume) of what we spray misses the intended target. It goes into the air or water table. To get rid of slugs affecting your dahlias and other flowers, keep a couple of boards in the garden. If you flip them open and scoop up the slugs and drown them in boiling water, your problem is solved.

Indian Flowers going to Flourish in Global Market

Cymbidiums orchids, the unique large sized orchids exclusive to altitudes of 4,500 to 5,500 meter is among the most expensive flowers produced here. The sturdy orchid, a stem of which costs between ₹ 100 and ₹ 200 in domestic markets depending upon the grade of the flower and length of the stick, is now headed for Gulf markets for the first time, adding new dimension to India's growing cut-flower industry. If properly handled, cymbidium can survive for almost 20 to 25 days, even a month in current weather, Sikkim Deputy Director Padam Subba, who was in Delhi for Horti Expo and Flora Expo that concluded.

Expo coordinator S Jafar Naqvi explains that as far as flowers are concerned, India has advantages of robust gene pool, climatic conditions and large agriculture lands. However, though Ooty, Kodai and even foothills of the Himalayas in north India have right weather conditions, floriculture operations are difficult there because flowers are highly perishable and air connectivity is a must for the business. "In the entire farm sector in India, there is a huge distance between growers and bulk buyers. The effect is more pronounced in floriculture. The distance can be reduced with better infrastructure and technology. Large corporate houses are taking to horticulture as a profitable business opportunity by bringing in investments and latest technologies. However, cold chain infrastructure is an issue that needs to be addressed.

India's growing flower power even the world's largest name in the business, the Netherlands, acknowledges. While India is among top producers of horticulture commodities and Holland is the world leader in cut flower exports and trade, explaining thereby the growing interest of the Netherlands in India's industry.

The growing importance of India in the supply of affordable high-tech propagation material and seedling for the world has made the Dutch companies take keen interest in being the natural partners with India. Our focus and areas of interest in India will be on production of vegetables, fruits, potatoes and vegetables seeds. Floriculture will be another major thrust area which includes flower production, greenhouse, greenhouse management, water efficiency management.

Unique Night-Flowering Orchid found

An orchid that unfurls its petals at night and loses its flowers by day has been found on an island off the coast of Papua New Guinea. The plant is the only known night-flowering orchid and was collected by botanists on a field trip to New Britain, an island in the Bismarck archipelago.

The flowers of the species, *Bulbophyllum nocturnum*, are thought to be pollinated by midges and last for only one night, according to a description of the plant published in the Botanical Journal of the Linnean Society. Orchid specialist Ed de Vogel, from the Netherlands, discovered the unusual flowering after he gathered some of the plants from trees in a logging area on the island and returned home to cultivate the orchids at the Hortus Botanicus in Leiden. Most orchids are epiphytes, which means they take root on trees.

The botanist was particularly eager to see the orchid's flowers because it was a member of the Epicrianthes group of orchids. This group contains several species that have bizarre flowers with strange appendages, which often resemble leggy insects, small hairy spiders or intricate sea-creatures.

Flowers that open only at night are seen in a small number of plant species, such as the queen of the night cactus, the midnight horror tree and night blooming jasmine. *Bulbophyllum nocturnum* is the only orchid among 25,000 species that is known to do so. Many orchids are pollinated by moths and other nocturnal insects, but have flowers that remain open during the day.

In 1862, Charles Darwin correctly predicted that the Christmas star orchid, which is endemic to Madagascar, was pollinated by a moth with a 30cm-long proboscis. The moth in question was not discovered until 20 years after his death. The small night-flowering orchid has yellow-green sepals that unfurl to reveal tiny petals adorned with dangling, greyish, thick and thin appendages. The flower, which is 2cm wide, has no noticeable smell, though some nocturnal species can time the release of their scents to attract night-time pollinating insects. Writing in the journal, the authors point out the resemblance between the flowers' appendages and the fruiting bodies of certain slime moulds found in the same part of the world. The similarity led the botanists to speculate that the orchids might be pollinated by midges that normally feed on slime moulds or small fungi.

Andre Schuiteman, an orchid specialist at the Royal Botanic Gardens in Kew said: "This is another reminder that surprising discoveries can still be made. But it is a race against time to find species like this that only occur in primeval tropical forests. which are disappearing fast.

India Blossoms as 2ⁿᵈ Largest Consumer of Flowers

More and more Indians are saying it with flowers and trade figures suggest that India is now the second largest consumer and fastest growing retail destination of flowers in the world. Among the biggest contributors to the country's growing flower power are Punjab and Haryana and the Himachal-Shivalik belt, while at its core are the 30-crore middle class and higher income Indians who are ensuring that the consumption grows at a whopping 30 per cent per annum - the fastest perhaps in any sector.

Growing in-house demand is also pushing India as a major importer. Flower imports are not encouraged officially following high import duties but traders are still importing to meet the growing demand, a clue perhaps for farmers to diversify to flowers in a major way. While Germany is the largest consumer, Holland is the largest importer and exporter of flowers in the world.

"Flowers are for all occasions - celebrations, marriage, New Year's Eve and Valentine's Day and even death," says S Jafar Naqvi, president of the Indian Flowers and Ornamental Plants Welfare Association (iFlora), and one of the organisers of the International Flora Expo beginning in Delhi from January 11. Rising incomes, globalisation and awareness are all contributors to this growth story which started with export oriented units around two decades ago in 1993.

Rosy Picture

☆ India is the fastest growing retail destination of flowers in the world

☆ Punjab-Haryana-Himachal belt among major flower growers

☆ Chandigarh, Amritsar among major consumers

☆ Germany is the largest consumer, while Holland is the largest importer and exporter of flowers in the world

The exports peaked in 2006-07 when India shipped flowers worth ₹ 750 crore in 2006-07, just a drop when compared to Holland which does flower business worth $ 17 billion. As of now, India's export market has dipped to ₹ 350 crore but the domestic market has grown to ₹ 13,000 crore and the reason is better prices within India than abroad. Punjab, Haryana and the Himachal-Shivalik belt are emerging as the fastest growing areas for cultivation of flowers, while Chandigarh and Amritsar are proving to be "good" consumers.

Reasons are quite predictable. "Punjabis have better exposure to western lifestyles, where per person per year consumption of flowers ranges between $10 and 15. Also Punjabis have stronger overseas connection so they adapt to the culture faster, therefore Chandigarh and Amritsar are emerging as major consumers. "The region is among the biggest utilisers of allocations under the National Horticulture Mission. Infrastructure like greenhouses and poly houses are selling the maximum in Punjab and area under flower cultivation is also increasing".

For Punjab farmers, looking for replacement to water-guzzling paddy, flowers may be too small an answer but it is emerging as one of the ways to make some extra cash in between the seasons. "Plus flowers are more environmentally friendly as they require much less water than other crops. Water used by flood irrigation method in one hour is sufficient to irrigate flowers for a month using the drip irrigation method.

Flowers that always Bloom

Gone are the days when artificial flowers meant cheap plastic replicas. Today the market is blooming with colourful stems and blossoms made of superior-quality, wrinkle-resistant material and look so real.

Flowers have held several philosophical connotations. Romantics have used these as a metaphor for love, while the sick have found hope in these. And the devout have used flowers in prayers. Indeed, flowers are a thing of joy but there is also a downside. Fresh flowers are seasonal, have a short life and need to be carefully nurtured and stored with their stems dunked in water. As décor accessories, these need to be replaced every week and as bouquets, flowers should give that just-plucked-from-the-garden look.

Bright and Colourful

The same explains the growing preference for faux flowers that have become a favourite with interior decorators. A decade or so ago, artificial blooms were made from cheap-moulded plastic. Displaying these was considered akin to wearing cheap clothes.

Today production refinements and new materials have given artificial flowers such a real look that the result is an array of beautiful and colourful stems and blossoms in wrinkle-resistant polyester, silk, latex, parchment and other sophisticated material. These retain the original shape and have a negative static charge, which keeps dust at bay. The hues are more life-like and subtle and colour gradations match that of real counterparts. Artificial flowers, faux flora, permanent botanicals, false blooms. the natural-looking flowers now enjoy a niche market.

From the Designer Houses

Walk into any high-end gift shop like Oma, Beyond Design or Home of the Traveller and you are likely to find designer artificial blooms of chrysanthemums, orchids, gardenias, roses, daisies, violets, lilies of the valley and many more. These are so beautifully crafted that these appear real to the eye and touch. Even leading global designer brands like John Lewis, Kate Spade and others, periodically launch their collections of exotic blooms like peonies, poinsettias, roses, English ivy, geraniums and a host of other faux florals.

Such is their appeal that artificial flowers created by some of the leading designers are treated as works of art. It is not surprising then that designers like Livia Cetti, Mary Delaney, Vladmir Kanevsky and Leopold Blaschka are put on the same pedestal as any accomplished artist of the world. The prices too have matched the sophistication in production. If you could get a bunch of cheap plastic flowers for under ₹ 100 some years ago, today these classy polyester flowers complete with silk foliage can be exorbitantly priced. Sometimes a single stem of a lily, gladioli or carnation can cost more than ₹ 4,500 even.

Interestingly these faux blooms do not compete with real flowers. A major part of the business comes from high-end hotels, restaurants, corporate offices, showrooms and hospitals, where real flowers are considered an expensive affair.

For a Better Display

Since these almost look like real flowers, people consider these dahlias, roses, lilacs and peonies as excellent dining-table centre pieces. Other flowers like orchids, purple roses, carnations and tulips are used as attractive display objects that last for years and don't need to be watered, re-potted, sprayed with insecticides and put out in the sun.

According to statistics indicated in the Handbook on Horticulture Statistics 2014, the total area under flower crops in 2012-13 was 232.70 thousand hectares. Total area under floriculture in India is second largest in the world and only next to China. Production of flowers was estimated to be 1729.2 MT of loose flowers and 76731.9 million (numbers) of cut flowers in 2012-13. Fresh and Dried cut flowers dominate floriculture exports from India.

Among states, Karnataka is the leader in floriculture with about 29,700 hectares under floriculture cultivation. Other major flower growing states are Tamil Nadu and Andhra Pradesh in the South, West Bengal in the East, Maharashtra in the West and Rajasthan, Delhi and Haryana in the North.

The expert committee set up by Govt. of India for promotion of export oriented floriculture units has identified Bangalore, Pune, New Delhi and Hyderabad as the major areas suitable for such activity especially for cut flowers. Of the four zones identified as potential centers for flower production namely Bangalore, Hyderabad, Pune and New Delhi, the area around Bangalore and Pune have got the advantage of ideal climatic conditions where the temperature ranges between 15 to 30°C. In view of this, the units established in these locations do not require either cooling or heating system. As a result maximum number of units has been established in these locations. There are more than 300 export oriented units in India. APEDA (Agricultural and Processed Food Products Export Development Authority) is the registering authority for such units.

Marketing

In India Marketing of cut flowers is much unorganized. In most of the Indian cities flowers are brought to wholesale markets, which mostly operate in open yards. From here the flowers are distributed to the local retail outlets which more often than not operate in the open on-road sides, with different flowers arranged in large buckets. In the metropolitan cities, however, there are some good florist show rooms, where flowers are kept under controlled temperature conditions, with considerable attention to value added service. The government is now investing in setting up of auction platforms, as well as organized florist shops with better storage facilities to prolong shelf life. The packaging and transportation of flowers from the farms to the retail markets at present is very unscientific. The flowers, depending on the kind, are packed in gunny bags, bamboo baskets, simple cartons or just wrapped in old newspapers and transported to markets by road, rail or by air. However, the government has provided some assistance for buying refrigerated cargos and built up a large number of export oriented units with excellent facilities of pre-cooling chambers, cold stores and reefer vans.

According to a study titled, 'Indian Floriculture Industry: The Way Ahead' released by the apex industry body ASSOCHAM, India's floriculture industry is growing at a compounded annual growth rate of about 30 per cent, and is likely to cross ₹ 8,000 crore mark by 2015. Currently, the floriculture industry in India is poised at about ₹ 3,700 crore with a share of a meagre 0.61 per cent in the global floriculture industry which is likely to reach 0.89 per cent by 2015.

Export Constraints

In spite of an abundant and varied production base, India's export of floricultural product is not encouraging. The low performance is attributed to many constraints like non-availability of air space in major airlines. The Indian floriculture industry is facing with a number of challenges mainly related to trade environment, infrastructure and marketing issues such as high import tariff, low availability of perishable carriers, higher freight rates and inadequate refrigerated and transport facilities. At the production level the industry is faced with challenges mostly related to availability of basic inputs including quality seeds and planting materials, efficient irrigation system and skilled manpower. In order to overcome these problems, steps must be taken to reduce import duty on planting material and equipment, reduce airfreight to a reasonable level, provide sufficient cargo space in major airlines and to establish model nurseries for supplying genuine planting material. Training centres should be established for training the personnel in floriculture and allied areas. Exporters should plan and monitor effective quality control measures right from production to postharvesting, storage, and transportation.

Government Programmes and Policies

Department of Agriculture and Cooperation under the Ministry of Agriculture is the nodal organization responsible for development of the floriculture sector. It is responsible for formulation and implementation of national policies and programmes aimed at achieving rapid agricultural growth through optimum utilization of land, water, soil and plant resources of the country. Production of cut flowers for exports is also a thrust area for support. The Agricultural and Processed Food Products Export Development Authority (APEDA), the nodal organization for promotion of agri exports including flowers, has introduced several schemes for promoting floriculture exports from the country. These relate to development of infrastructure, packaging, market development, subsidy on airfreight for export of cut flowers and tissue-cultured plants, database up-gradation *etc.* The 100 per cent Export Oriented Units are also given benefits like duty free imports of capital goods. Import duties have also been reduced on cut flowers, flower seeds, tissue-cultured plants, *etc.* Setting up of walk in type cold storage has been allowed at the International airports for storage of export produce.

Initiatives have also been launched for the benefit of exporters by providing cold storage and cargo handling facility for perishable products at various international airports. Direct subsidy up to 50 per cent is also available in cold storage units. Besides, subsidy is also provided by APEDA on improved packaging materials to promote their use. To attract entrepreneurship in floriculture sector, NABARD is providing financial assistance to hi-tech units at reasonable interest rates.

Several schemes have been initiated by the Government for promotion and development of the floriculture sector including "Integrated Development of Commercial Floriculture" which aims at improvement in production and productivity of traditional as well as cut flowers through availability of quality planting material, production of off season and quality flowers through protected cultivation, improvement in postharvest handling of flowers and training persons for a scientific floriculture. Many state governments have set up separate departments for promotion of floriculture in their respective states.

Research work on floriculture is being carried out at several research institutions under the Indian Council of Agricultural Research and Council of Scientific and Industrial Research, in the horticulture/floriculture departments of State Agricultural Universities and under the All India Coordinated Floriculture Improvement Project with a network of about twenty (20) centres. The key focus areas are crop improvement, standardization of agro-techniques including improved propagation methods, plant protection and postharvest management. In recent years, however, technologies for protected cultivation and tissue culture for mass propagation have also received attention. A large number of promising varieties of cut flowers have been developed. All these efforts indicate the government's commitment for improving the sector and creating a positive environment for entrepreneurship development in the field.

Demand and Supply

The demand for flowers is seasonal as it is in most countries. The demand for flowers has two components: a steady component and a seasonal component. The factors which influence the demand are to some extent different for traditional and modern flowers.

(i) Traditional Flowers

The steady demand for traditional flowers comes from the use of flowers for religious purposes, decoration of homes and for making garlands and wreaths. This demand is particularly strong in Kerala, Karnataka, Tamil Nadu, Odisha and West Bengal, as the use of flowers for above mentioned purposes is part of their local culture. The bulk of seasonal demand comes from festivals and marriages. The demand is generally for specific flowers.

(ii) Modern Flowers

The bulk of the steady demand for modern flowers comes from institutions like hotels, guest houses and marriage gardens. The demand is concentrated in urban areas. With increasing modernization and globalization the demand for modern flowers from the individual consumers is likely to grow enormously as the trend of "say it with flowers" is increasing and the occasions which call for flower giving will continue to present themselves. Although there is an increasing demand for modern flowers from individuals, institutions continue to be the dominant buyers in the market. The price of these flowers also depends on their demand and varies accordingly.

Greenhouse Technology for Flower Production

In present scenario of increasing demand for cut flowers protected cultivation in greenhouses is the best alternative for using land and other resources more efficiently. In protected environment suitable environmental conditions for optimum plant growth are provided which ultimately provide quality products. Greenhouse is made up of glass or plastic film, which allows the solar radiations to pass through but traps the thermal radiations emitted by plants inside and thereby provide favourable climatic conditions for plant growth. It is also used for controlling temperature, humidity and light intensity inside. On the basis of basic material used, building cost and technology used, greenhouses can be of three types:

(i) Low-cost Greenhouse

The low-cost greenhouse is made of polythene sheet of 700 gauge supported on bamboos with twines and nails. Its size depends on the purpose of its utilization and availability of space. The temperature within greenhouse increases by 6-100C more than outside.

(ii) Medium-cost Greenhouse

With a slightly higher cost greenhouse can be framed with GI pipe of 15 mm bore. This greenhouse has a covering of UV -stabilized polythene of 800 gauge. The exhaust fans are used for ventilation which are thermostatically controlled. Cooling pad is used for humidifying the air entering the chamber. The greenhouse frame and glazing material have a life span of about 20 years and 2 years respectively.

(iii) Hi-tech Greenhouse

In this type of greenhouse the temperature, humidity and light are automatically controlled according to specific plant needs. These are indicated through sensor or signal-receiver. Sensor measures the variables, compare the measurement to a standard value and finally recommend to run the corresponding device. Temperature control system consists of temperature sensor heating/cooling mechanism and thermostat operated fan. Similarly, relative humidity is sensed through optical tagging devices. Boiler operation, irrigation and misting systems are operated under pressure sensing system. This modern structure is highly expensive, requiring qualified operators, maintenance, care and precautions. However, these provide best conditions for export quality cut flowers and are presently used by large number of export units.

Employment Opportunities in Floriculture

Floriculture has emerged as an important agribusiness, providing employment opportunities and entrepreneurship in both urban and rural areas. National Horticulture Board helps one to establish a flower business. Agricultural and Processed Food Products Export Development Authority helps entrepreneurs with cold storage facilities and freight subsidies. It has been found that Commercial Floriculture has higher potential per unit area than most of the field crops and therefore a lucrative business. During the last decade there has been a thrust on export of cut flowers. The export surplus has found its way into the local market influencing people in cities to purchase and use flowers in their daily lives. Floriculture thus, offers a great opportunity to farmers in terms of income generation and empowerment. Small and marginal farmers may also use every inch of their land for raising the flower and foliage crops.

Floriculture also offers careers in production, marketing, export and research. One can find employment in the floriculture industry as a farm manager, plantation expert, supervisor or project coordinator. Besides, one can work as consultant or landscape architect with proper training. In addition, floriculture also provides career opportunities in service sector which include such jobs as floral designers, landscape designers, landscape architects and horticultural therapists. Research and teaching are some other avenues of employment in the field.

Post globalization. Floriculture has become an important commercial in agriculture. Floriculture activity has evolved a viable and profitable alternative, with a potential to generate remunerative self employment among small and marginal farmers, and earn the much needed foreign exchange in the developing countries such as India.

List of Flower Names and Meanings of Flowers

Here is a list of flower names along with their symbolic meanings. Find out how to use the language of flowers to express deep-felt emotions like love, longing, anticipation, jealousy, hope, and so on. As the saying goes, 'say it with flowers!' And indeed, people have been saying it with flowers since time immemorial, using the special meanings of flowers to express all kinds of emotions. Whether to woo somebody special on Valentine's Day, display affection on Father's Day or Mother's Day, or to express a 'get well soon' message for somebody ill, flowers have been used to say it all. In fact, the symbolic meaning of flowers has given birth to a language of its own, known as floriography.

Apart from each of the names of flowers being imbued with symbolism such as romance or sympathy or affection, all flowers express specific phrases as well. As a matter of fact, the Victorians were the ones who used flower symbolism most profusely to communicate subtly what they wanted to say, but could not speak aloud since decorum would not allow it. With flowers, and their symbolic meanings, they could verbalize communication without saying anything or evince a feeling. Thus, the Victorians would often make a bouquet that they could use as an expression of an unvoiced message to the person receiving it, using various kinds of symbolism like the number, the arrangement, the color, and of course the type of flower.

So, If you want to express your deep feelings to someone special in your life, in a thoughtful and unique way, check out the list of flower names given below, and find out what message you can devise using the special language of flowers. Or, if it is a gift of flowers that you have received, check out the flower names and their meanings to find out what it means. Is it love? Or is it ecstasy? Or perhaps it is a sign of hope, jealousy, anticipation, friendship, or even good-bye. Truly, you can say it all with flowers!

Name Meaning

Acacia: Hidden love, Beauty in withdrawal, Ambrosia Love requited, Amaryllis Pride: Aster: Symbolizing love, Delicacy, **Anemone:** A love that is diminishing, Vanishing hopes, **Almond Blossom** : Symbolizing hope, Delicacy, Sweetness, **Apple Blossom:** Good fortune, Harbinger of better things, A strong liking, **Arum:** Intense feeling of love, **Arbutus** : I love only thee, Azalea A symbol of womanhood in China, Passion, Fragility, Take care

Baby's Breath: Purity of heart, Innocence, Begonia: Be cautious, A fanciful mind, **Bachelor Button:** Celibacy. The blessedness of being single, **Balsam:** Fervent love **Balm** : Compassion, Empathy, **Bittersweet:** Truth, Loyalty, **Bird of Paradise:** Given by a woman to a man to symbolize faithfulness, **Bluebell** : Gratitude, Constancy, Humility, **Buttercup:** Riches, Childishness, **Black Bryont:** Be my support.

Caladium : Immense delight and joy Camellia Perfection, Gratitude Carnation (Red) Achingheart, Admiration Christmas Rose Allay my disquiet, Chrysanthemum Joy, Optimism, Perfection, Crocus Good cheer, Happiness, Do not treat me badly, Cyclamen Good-Bye, Resignation Clover (Four-leaf) Will you be mine? Clover (White) Think of me Clematis Ingenuity, Artifice, Cornflower Refinement, Delicacy.

Daffodil: The sun is bright when I am with you, Respect, Sunshine, Unrequited love, Regard Dahlia: Elegance and Dignity, Forever thine, Daisy Beauty, Innocence, I will never tell, Loyal love, Purity, Day-Lily (Yellow) Coquetry, Dandelion Love's oracle, Happiness, Faithfulness Date-Plum Resistance Delphinium Fun, Big-hearted, Dog Rose: Pleasure and Pain, **Dogwood** Am I indifferent to you? **Dragon Root** Ardor.

Edelweiss Noble purity, Courage, Daring, Endine: Frugality, Elder flower: Symbolizing Zeal, Eglantine Poetry, Everlasting (Immortal Flowers) Never ending memory, Unfading emembrance Euphorbia Persistence Eupatorium Delay Epigaea: Budding, Eucalyptus Protection

Forget-Me-Not: As its name suggests - Forget me not, Memories, True love Forsythia:, Expectation, Anticipation, Fuchsia Good taste, Fern Fascination, Filbert Reconciliation Flax Symbol of domesticity, Flora's Bell: Without pretentiousness, Flowering Reed Confide in heaven.

Gladiolus: Strength of character, Love at first sight, Generosity, Gardenia: Secret love, You are lovely, Galax : Encouragement, Geranium Folly, Stupidity, Gerbera: You are the sunshine of my life, Gloxinia: Love at first sight, Goldenrod: Be cautious, Goats-Rue Reason, Goosefoot: Insult Globe Amaranth: Unfading Love.

Heliotrope: Devotion, Hibiscus: Delicate beauty, Consumed by love, Holly Am I forgotten, Domestic happiness, Defense, Domestic Happiness, Hollyhock: Fruitfulness, Honey flower: - Secret love, Sweetness of disposition, Sweet, Affection, Honeysuckle: Bond of love, **Honeysuckle** (Coral) I Love You Hyacinth: Flower dedicated to Apollo, Rashness **Hyacinth (Yellow)** Jealousy.

Iris : Wisdom and Valor, Faith, Promise in love, Hope, **Iris (German)** Flame, **Iris (Yellow)** Passion, **Ivy** Affection, Friendship, Fidelity, Wedded love, **Ivy (Sprig of white tendrils)** : Affection, Anxious to please, **Indian Cress:** Resignation, **Ipomen Scarlet** Embrace, **Indian Cane** : Rendezvous, Ipomoea (Morning Glory) I attach myself to you, Affection

Jasmine (Indian) : Love, Attachment, **Jasmine (Yellow):** Elegance, Grace, Modesty, **Jasmine (Spanish)** Sensuality, **Japonica**: Symbol of love, Sincerity, Jerusalem Oak Your love is reciprocated, **Jonquil:** Sympathy, Desire, Affection returned, Love me, Affection Returned **Juniper** Chastity, Eternity.

King's Spear: Regret, Kennedia Intellectual beauty,

Laburnum: Blackness, **Lady's Slipper** Capricious beauty, Win me Lucerne Life: Larkspur: Lightness, An open heart, Levity, **Larkspur (White):** Happy-go-lucky, Joyful, **Larkspur (Purple)** Sweet disposition, **Lantana** Rigor, **Lily** Purity of heart, Majesty and Honor, **Lemon Blossom** I promise to be true, Fidelity in love, **Lilac** Pride, Beauty **Lily:** of the Valley Humility, Happiness, Tears of the Virgin Mary, Sweetness **Lotus**: Forgetful of the past, Estranged love **Laurel** Glory **Lavender**: Devotion, Love

Marjoram: Happiness, Joy Madder: Calumny, Magnolia: Perseverance, Nobility, Marigold: Sorrow, Jealousy, Caress, Pretty love, Affection, Sacred, Affection, **Maidenhair** : Discretion, Manchineel Betrayal, Mimosa - Secret Love, Mugwort: Happiness, Mulberry: Prudence, Mallow Sweetness, Delicate beauty, Mandrake Scarcity, Myrtle Hebrew emblem of Marriage, Love, Maple Reserve.

Nasturtium: Victory in Battle, Conquest, **Narcissus (White)** Selfishness, **Narcissus**: Stay as sweet as you are, Formality, Nightshade Truth, Nettle Cruelty, Nosegay Gallantry.

Orchid (Cattleya): Mature charm, **Orchid :** Symbol of many children in China, Refinement, Beauty, Love Orange Blossom: Eternal Love, Innocence, Marriage and Fruitfulness Oak: Hospitality Osmunda: Revere, Ophrys Spider Dexterity, **Ophrys Fly**: Mistake Olive Peace Oleander Caution, **Orange-Tree:** Generosity, **Orange (Milkweed): Deceit.**

Phlox : Harmony, A good partnership, **Passion-Flower**: Belief, **Peppermint:** Warmth of feeling, Cordiality **Periwinkle :** Sweet remembrance Poppy (General): Imagination, Oblivion, Eternal sleep, **Primrose (Evening)**: Inconstancy, **Petunia:** Your presence soothes me, Anger, Resentment, Anger, Pansy Merriment, Thoughtful reflection Peach Blossom I am your captive.

Quaking Grass: A symbol of agitation., Queen Anne's Lace Fantasy.

Rose (Red) : Passion, Love, **Rose (White):** Purity, Virginity, **Rose (Yellow)** Infidelity, Jealousy **Rose (Coral)** Desire, Rose (Pink): Grace and Sweetness, Secret love, Perfect happiness, Secret Love, Rose (Dark pink) Thankfulness, Rose (Pale pink) Joy, Grace, Rosebud (Red) Pure and Lovely, Rosemary Remembrance, Rhododendron I am dangerous, Danger, Beware.

Sunflower (Tall) Pride, False riches, Sunflower (Dwarf) : Adoration, Snowdrop: Consolation, Hope **Sage**: Great respect, Wisdom, Female fidelity **Snapdragon**: Strength, Gracious lady, **Satin-Flower:** Sincerity Spider Flower Elope with me, **Sweet-William** Gallantry, Grant me one smile, **Sweet Pea** Thank you for a lovely time, Blissful pleasure, Departure, Good-by, **Star of Bethlehem Reconciliation,** Atonement.

Tulip (General): Fame, Perfect lover, Flower Emblem of Holland **Tulip (Yellow):** Hopeless love, There's sunshine in your smile **Tulip (Variegated):** Beautiful eyes, Tulip (Red): Declaration of love, Believe me, **Thornapple:** I dreamed of thee, Tuberose Dangerous pleasures **Teasel** Misanthropy, **Trumpet Flower** Separation.

Valerian: Accommodating disposition, Verbena: Sensibility, Pray for me **Violet:** Modesty **Violet (White):** Let´s take a chance **Violet (Blue):** Love, Faithfulness, Watchfulness Virgin's-Bower Artifice, **Venus Flytrap:** Caught at Last **Viscaria** Will you dance with me? Veronica Fidelity, Vervain Enchantment.

Water Lily: Purity of heart, **Water Lily (Yellow):** Growing indifference, **Water Lily (White):** Eloquence, **Wallflower Lasting beauty,** Fidelity, Faithful in adversity, Wisteria Welcome, **Windflower :** Sincerity, Love, Abandonment, Witch Hazel: A spell Wild Rose-Tree Poetical person Wood-Sorrel Joy Whortleberry Treachery.

Xeranthemum: Cheerful in adversity

Yarrow: Healing, Health, Yew Sadness.

Zinnia: Thoughts of absent friends **Zinnia (Yellow) :** Daily remembrance **Zinnia (Magenta):** Lasting Affection, Zinnia (White): Goodness, Zinnia (Scarlet): Constancy **Zephyranth** Fond Caresses.

Ice Age Flower Blossoms again

Plant Resurrected from Fruit Collected by Squirrels 30,000 Yrs Ago

Ageless Beauty

It was an Ice Age squirrel's treasure chamber, a burrow containing fruit and seeds that had been stuck in the Siberian permafrost for over 30,000 years. From the fruit tissues, a team of Russian scientists managed to resurrect an entire plant in a pioneering experiment that paves the way for revival of other species.The Silene stenophylla is the oldest plant ever to be regenerated, the researchers said, and it is fertile, producing white flowers and viable seeds.

The experiment proves that permafrost serves as a natural depository for ancient life forms, said the Russian researchers, who published their findings in Tuesday's issue of 'Proceedings of the National Academy of Sciences' of the United States. "We consider it essential to continue permafrost studies in search of an ancient genetic pool, that of pre-existing life, which hypothetically has long since vanished from the earth's surface," the scientists said in the article.

Canadian researchers had earlier regenerated some significantly younger plants from seeds found in burrows.

Svetlana Yashina who led the regeneration effort, said the re-The oldest plant regenerated from tissue of fossil fruit found in a squirrel burrow in Siberia vived plant looked very similar to its modern version, which still grows in the same area. "It's a very viable plant, and it adapts really well. The team recovered the fruit after investigating dozens of fossil burrows hidden in ice deposits on the right bank of the lower Kolyma River in northeastern Siberia, the sediments dating back 30,000-32,000 years. The sediments were firmly cemented together and often totally filled with ice, making any water infiltration impossible — creating a natural freezing chamber fully isolated from the surface.

Sheer Poetree

A 1,000-year-old Champaka tree, we passed through the most beautiful evergreen forest, 6,000 feet above sea level. Forest so thick, green and impenetrable, it felt like a different world. Countless butterflies kept us company, fluttering alongside the safari jeep, as if taking us in procession. Called Dodda Sampige Mara or Big Champaka tree, it is worshipped by the Soliga tribe who inhabit these hills and in whose sacred grove it grows. The valley reverberated with singing cicadas. Tall trees, some more than a century old, vied with each other to reach the sky. The Soligas worship the tree as Mahadeshwar and celebrate the annual festival Jatra every April when they perform the fire dance around it. Their settlements are all over and I saw them cooking and going about domestic chores. They collect forest produce like gooseberries, honey, and lichens that are used as a spice. During coffee harvesting season, they work on the bushes to earn extra money. Some own small pieces of land where they grow their own coffee.

The tree was huge, filled with green and abundant foliage. Having witnessed a 1000 years, it had a mystical aura and was simply overwhelming to behold. The base was gnarled and enormous. Thousands of birds sheltered here, millions of champaka flowers blossomed in its branches, and it had witnessed endless cycles of life. It stood there, proud and wise with age, enormously tall, next to Bhargavi, a tributary of the Kaveri. Parasurama is said to have washed his axe in this river after slaying the Kshatriya rulers who had strayed from the path of dharma. Soligas believe the tree was planted by sage Agastya 3,000 years ago and that Bhargavi is actually Renuka, Parasuram's mother and the wife of sage Jamadagni. The legends add to its aura.

The resort manager described the flowers, which appear in April — yellowish-orange and fragrant. Its botanical name, Michaelia Champaca, and other statistics did not matter. It stood there, magnificent and imperial. I imagined it covered with hundreds of blooms and heady with scent. The beautiful and scented Champaka is often grown in temple precincts and is considered sacred to Krishna. It forms one of the five flower-darts of Manmadha, the Indian Cupid. Poets have celebrated its beauty over the ages. Kalidasa addressed a beautiful woman thus, "O Beauteous One! Maha-Brahma has formed thy eyes with lilies, thy face with lotus… thy limbs with the petals of the champaka. How is it that thy heart alone is cast in stone?"

At the foot of the tree were several small lingams, most smeared with holy ash. Some tridents with lemons pierced on them and bunches of brass bells also stood about. I went around the tree, peeping into the huge hollows that looked like homes for elves. Dodde Sampige Mara has a younger sister nearby called Chik Sampige Mara (small Champak tree). The path to this sibling was narrow and unused. We had to stop a couple of times to clear the way but it was worth it. The younger sister is another splendid specimen, maybe 800 years old, with an impressive trunk and still flowering.

Yes, I will wait for spring and come back, passing through the lovely, deeply wooded forest, and feast my eyes on the grand dame when she is bejewelled with countless flowers, and I will carry back the fragrance and the joy.

2

Development of Floriculture in India

Floriculture in India

The production of flowers is an age-old occupation. This does not find a place in the literature on horticultural crops. Until last decade, the growing and selling of flowers was confined to a few families. They grew a variety of flowers on the same land which were close to the their residence, as they could not survive a long journey. The situation in the last decade has however, changed. Now, different farmers are growing different flowers both for domestic market and export purposes. The flowers were, until 1960s, confined to domestic markets. These flowers are now moving long distances due to the availability of airfreight and hi-tech cooling systems.

The economic reforms and liberalization policies introduced from 1991 and modified EXIM policies of 1995-96 and 1999-2002 have given fillip to this sector. After liberalization, the Government of India identified this activity as a sunrise industry and accorded it 100 per cent export-oriented status. Later, many writers have termed this industry as "Rosy Business sector", a Global Concern, Blossoming Industry, Thrust Area, Money Spinning, Lucrative export-oriented sector *etc.* Growing demand and much higher return per unit of land than any other agricultural activity has prodded farmers to take to this sector.

The growing demand for this product has also increased on account of rapid urbanization, increase in individual purchasing power among middle-income groups, increase in the number of IT Units, Hotels, Tourists, Temples, increase in

GDP, Per capita Incomes, change in life-styles/social values of the people, greater awareness among the people to improve the deteriorating environment and economic up liftment of the people's conditions.

Government of India has identified floriculture as a sunrise industry and accorded it 100 per cent export oriented status. Owing to steady increase in demand of flower floriculture has become one of the important Commercial trades in Agriculture. Hence commercial floriculture has emerged as hi-tech activity-taking place under controlled climatic conditions inside greenhouse. Floriculture in India, is being viewed as a high growth Industry. Commercial floriculture is becoming important from the export angle. The liberalization of industrial and trade policies paved the way for development of export-oriented production of cut flowers. The new seed policy had already made it feasible to import planting material of international varieties. It has been found that commercial floriculture has higher potential per unit area than most of the field crops and is therefore a lucrative business. Indian floriculture industry has been shifting from traditional flowers to cut flowers for export purposes. The liberalized economy has given an impetus to the Indian entrepreneurs for establishing export oriented floriculture units under controlled climatic conditions.

Agricultural and Processed Food Products Export Development Authority (APEDA), is responsible for export promotion and development of floriculture in India.

Floriculture Development in India

About two decades back or so, the floriculture was just a pastime of rich people and hobby of flower lovers, but now it has opened a new vista in agri-business i,e, commercial floriculture. With the increase in buying capacity of people, the flower lovers have now started buying them from the markets to beautify their home as well as to adore some one they love simply because they don't have time and enough space to grow flowers particularly in urban areas and in metropolitan cities. Flowers, it seems, is the most wanted item in any social occasions for conveying one's status and aesthetic sense. Flower is now so indispensable that one may cancel his her birthday celebration or Yama may postpone the death of a dying person in case flowers are not available at that time. No nuptial is performed and honeymoon of a young couple is not consummated till garden fresh rose and rajanigandha or tuberose with lingering and stupefying aroma are made available. Warm welcome cannot be offered to VIPs in the public functions without bouquet – flowers are so indispensable! All these, no doubt, have set flower business on a top gear. One may wonder, the global market on flower is at present, carrying a business worth 2000 crores US dollar (1992) par annum. India is also having a business worth ₹ 280 crores in her domestic market (1992-93).

Indian Council of Agricultural Research (ICAR) conducted a survey of assessment on the possibilities of cut flowers trade in India during 1960-62. An important conclusion was that an internal sale as ₹ 9.26 Crores worth flower weighing 10,460 tones grown in an area of 4000 hector. Flowers like Rose, Gladiolus,

Tuberose, Chrysanthemum, Aster, Carnation, Orchids, Marigold are most popular in cut flower market all over the World.

Varieties

Floriculture products mainly consist of cut flowers, pot plants, cut foilage, seeds bulbs, tubers, rooted cuttings and dried flowers or leaves. The important floricultural crops in the international cut flower trade are rose, carnation, chrysanthemum, gargera, gladiolus, gypsophila, liastris, nerine, orchids, archilea, anthuriu, tulip, and lilies. Floriculture crops like gerberas, carnation, *etc.* are grown in greenhouses. The open field crops are chrysanthemum, roses, gaillardia, lily marygold, aster, tuberose *etc.*

The major loose flower producing states showing in **Table 2.1**. It indicate that in order to loose flower production, Tamil Nadu tops with 25 per cent in year 2007-08, followed by Karnataka 19 per cent, Andhra Pradesh 14 per cent, Punjab 9 per cent and Maharashtra occupies fifth place with 8 per cent. It is important to see that Punjab occupies forth place that was 0.5 per cent in 2005-06. As regard to growth in total, it indicates that net 33 per cent growth over the year 2005-06.

Table 2.1: Major Cut Flower States

Sl.No.	States/UTS	Production (Lakh Nos.)		
		2005-06	2006-07	2007-08
1	West Bengal	9347(32)	12966(35)	19680(45)
2	Maharashtra	3410(12)	4774(13)	5728(13.1)
3	Karnataka	5239(18)	5660(16)	5550(12.7)
4	Gujarat	4392(15)	5063(14)	5063(12)
5	Uttar Pradesh	3668(13)	3746(10)	3752(9)
6	Uttarakhand	575(2)	1229(3)	1455(3)
7	Delhi	1038(3)	1038(2)	1038(2)
8	Others	1533(5)	2639(5)	1151(6)
	Total	**29203(100)**	**37156(27)**	**43417(18)**

Figures in the bracket indicate percentage to respective total.

* All figures are rounded.

Source: Indian Horticulture database (2008), Ministry of Agriculture Government of India Gurgaon.

Table 2.1 presents major cut flower states. A closer look at the table reveals that the production trend of cut flowers shows increasing trends. It showed 48per cent growth in the year 2007-08 over the year 2005-06 of cut flower production. However, state wise production indicate that West Bengal tops with 45 per cent, followed by Maharashtra got second place with 13 (13.19) per cent, Karnataka third with 13 per cent. No doubt, West Bengal obtained top rank; predominantly it is noticeable growth of cut flower production in West Bengal and Maharashtra, Karnataka and Delhi showed increasing with constant.

Major Floriculture Crops in India

Rose is the principal cut flower grown all over the country, even though in terms of total area, it may not appear so. The larger percentage of the area in many states is used for growing scented rose, mainly to be sold as loose flowers. These are used for offerings at places of worship, for the extraction of essential oils and also used in garlands. For cut flower use, the old rose varieties, such as Queen Elizabeth, Super Star, Montezuma, Papa Meilland, Christian Dior, Eiffel Tower, Kiss of Fire, Golden Giant, Garde Henkel, and First Prize are still popular. In recent times, with production for export gaining ground in the country, the latest varieties like First Red, Grand Gala, Konfitti, Ravel, Tineke, Sacha, Prophyta, Pareo, Noblesse, Virsilia, and Vivaldi are also being grown commercially.

Gladiolus is the next most important cut flower crop in the country. Earlier it was considered a crop for temperate regions and its growing was restricted to the hilly areas, particularly in the north eastern region, which still continues to supply the planting material to most parts of the country. However, with improved agronomic techniques and better management, the northern plains of Delhi, Haryana, Punjab, Uttar Pradesh, as well as Maharashtra and Karnataka have emerged as the major areas for production of Gladiolus. Tuberose, a very popular cut flower crop in India is grown mainly in the eastern part of the country *i.e.* West Bengal, and also in northern plains and parts of southern India. Both single and double flower varieties are equally popular. Tuberose flowers are also sold loose in some areas for preparing garlands and wreaths.

The other main cut flowers include Asters, Gerbera, Carnation, Anthodium, Lilium, and Orchid. Production of Orchids is restricted mainly in the northeastern hill regions, besides parts of the southern states of Kerala and Karnataka. The main species grown are Dendrobiums, Vanda, Paphiopedilums, Oncidiums, Phalaenopsis and Cymbidiums.

Among the traditional crops grown for loose flowers, the largest area is under Marigold, grown all over the country. In most parts of the country only local varieties are grown for generations. African Marigolds occupy more area as compared to the small flowered French types. Jasmine flowers in view of its fragrance are also very popular as loose flowers, and find a very important place both in the perfume industry and flower market. The major areas under this crop are in Tamil Nadu and Karnataka in South, and West Bengal in East.

The varieties are mainly improved clones of *Jasminum grandiflorum, J. auriculatum* and *J. sambac.* Chrysanthemum is recognized as a potent flower crop in India. It is used as a cut flower for interior decoration and as loose flowers for making garlands. The Chrysanthemum, particularly the white varieties are much in demand as loose flowers during the autumn period of October- December when other flowers like Jasmine, Tuberose are not available for use in garlands and other similar purposes. Bougainvillea is an important and popular flowering plant grown widely throughout the country. The dwarf and bushy ones are mainly grown as pot plants. Among other traditional flowers grown in large areas are Crossandra in southern states of Tamil Nadu, Karnataka and Andhra Pradesh, and Aster in Maharashtra.

Marigold, Aster, Roses, Tuberose, Gladiolus, Jasmine, Crossendra are grown generally in the open fields, while Gerbera, Carnation, Roses, Anthurium, Lilium, and Orchids are cultivated under greenhouse conditions.

Major flowering plants largely produced and sold in pots for indoor use are Poinsettias, Orchids, florist Chrysanthemums, and finished florist Azaleas. Foliage plants such as ferns, crotons, bamboos, and palms are also produced and sold in pots and hanging baskets for indoor and patio use, including larger specimens for office, hotel, and restaurant interiors.

Consumption and Domestic Trade

While exports remain the prime motivator for Indian flower cultivators, the demand in the domestic market is also enormous and is on the rise. Modernisation and growing western cultural influences has resulted in consumers, especially the young population in the country buying flowers on occasions like Valentine's Day, Friendship Day, Mother's Day, Father's Day and so on. There is also a huge spurt in demand for flowers during religious festivities. Flower retailing is also undergoing a sea change in the country. Besides the small vendors selling flowers from buckets on roadsides, there has been an emergence of many up market shops and flower boutiques in major metros, in the recent years that witness demand round the year. The large supermarket/hypermarket retail chains that are mushrooming across the country are expected to further stimulate this growth.

The major markets for flowers are situated in the states, which produce significant quantities of flowers. Kerala is one state that has a fairly large market without any significant production of flowers. Some states, particularly those in the Southern India, have more than one large market in the state as the area under flower production is fairly widely distributed. The major markets in terms of number of traders involved and the quantity traded are in the peninsular region and east India.

The major markets in peninsular India are Chennai, Coimbatore and Madurai in Tamil Nadu; Bangalore, Mysore and Dharwad in Karnataka; Hyderabad and Vijayawada in Andhra Pradesh; Trivandrum and Cochin in Kerala; and Mumbai and Pune in Maharashtra. The city of Mumbai itself has three large markets. Kolkata is the biggest market in the eastern India. In addition to the market in the city of Kolkata, there are several fairly large regional markets in West Bengal. In the north, Lucknow/Kannauj and Delhi are the major markets for flowers besides locations in Rajasthan.

Employment in Floricultural Crops Cultivation

Floriculture is a capital-intensive industry with long gestation period. According to a report on floriculture by the Parliamentary Standing Committee on Commerce, Government of India, investment worth over ₹500 crores has been made in the floriculture industry in the country during the last decade. There are wide variations in the cost of establishing the export-oriented floriculture units in the country. However, according to the report, the cost of flower production under controlled condition is estimated at around ₹1.75 crores to ₹2.25 crores per hectare.

Floricultural crops are highly labour intensive and have the capacity to generate more direct and indirect employment both in rural areas as well as in urban areas. Estimates across different states in India indicate that the employment generation of flower crops cultivation was higher than other horticulture crops, food crops and commercial crops. According to estimates, the employment generation of floricultural crops cultivation was 913 man-days per hectare in Crossandra and 1,210 man-days in Jasmine. A Study by UAS Bangalore, in Chitradurga district showed that the employment generation of one hectare of Crossandra was 1,461 man-days per year. Of this, 65 accounted for female workforce. It was estimated that conventional (traditional) floriculture provide a decent standard of living for nearly 10,000 farm households and employment to 80,000 farm labourers and 2.5 lakh small retailers and flower vendors in the state of Karnataka. In contrast to the traditional floriculture, the modern floriculture generated more employment. The range of employment per hectare in this activity was 7,121 man-days to 7,468 man-days/hectare including technical labour, whereas the food crops, generated 860 man-days per hectare per annum as against 143 man-days for cereal crops, Paddy 175 man-days, and Sugarcane 285 days, to 305 man- days, Groundnut 105 to 225 per hectare of land, Paddy 175 man-days, and Sugarcane 285 days, to 305 man- days, Groundnut 105 to 225 per hectare of land.

The high-tech floriculture employs more labour. However, the cost of generating a humanLabour Day is ₹ 1886 in high-tech compared to ₹87 in conventional floriculture and ₹217 in commercial floriculture. This means that cost incurred by high-tech floriculture to generate one human day is capable of generating employment to nine labourers in the cultivation of field roses and 22 labourers in conventional floriculture. There are no estimates on total labour required in modern floriculture. But some case studies indicate that the total labour in 10 sample units in Bangalore were 981 which worked out to 98 persons per unit, the proportion of permanent labourers was 43.73 per cent. But the study of the proportion of permanent labour at 8 per cent and they were given industrial type of benefits. Prakash's study also shows that nearly 72 per cent of the labourers working in hi-tech floricultural units were in the age group of less than 25 years, 50 per cent of the total workers belonged to dalits and 24 per cent of them were the previous owners of the land which was in the hands of the corporate floriculturists.

Floricultural crops are highly labour intensive and have the capacity to generate more direct and indirect employment both in rural areas as well as in the urban areas. Estimates across different states in India indicate that the employment generation of flower crops cultivation is higher than other horticulture crops, food crops and commercial crops. A study conducted by University of Agricultural Sciences (UAS) Bangalore, in Chitradurga district of Karnataka showed that the employment generation of one hectare of Crossandra was 1,461 man-days per year. Of this, 65 per cent accounted for female workforce. It was also estimated that conventional (traditional) floriculture provided a decent standard of living for nearly 10,000 farm households and employment to 80,000 farm labourers and 2.5 lakh small retailers and flower vendors in the state of Karnataka in 2002. In contrast to the traditional floriculture, the modern floriculture generated more employment. The range of employment per hectare was 7,121 man-days to 7,468 man-days/hectare including technical labour, whereas the food crops, generated employment of 860 man-days per hectare per annum.

Hi-tech floriculture employs more labour. There are no definite estimates on total labour required in hi-tech/modern floriculture, which largely depends on the capacity of the units. However, the UAS study estimates the proportion of permanent labour at 8 per cent in the hi-tech floriculture units in Bangalore in 2002, and that they were given industrial benefits. In tandem with employment generation, the flower crops have the inherent advantage of providing higher productivity per unit of land resulting in higher income. Studies reveal that income obtained from floriculture is 4, 5, 10, and 20 times higher than sugarcane, fruits and vegetables, paddy, and ragi, respectively.

The cut flower industry makes an important contribution to economy. These are briefly summarized here. Cut flower industry provides important contributions to fertiliser and agricultural chemical industry since it is an intensive agricultural production activity by having short-term production, requiring intensive fertiliser and plant protection techniques. Packaging is one of the most important elements in postharvest processes for cut flower products. During supplying of products to market and for transport, lots of packaging material originated from cellulose and plastic is used, by this way cut flower industry also provides an important contribution to this industry.

The Domestic Floriculture Industry

The domestic floriculture industry has been witnessing an unprecedented growth during the past years and has also been getting increased acceptability in world markets, currently estimated at US$ 50 billion. The floriculture industry has been growing at an annual rate of 17 per cent, which has also seen a number of corporate houses entering the fray during the past three to five years. Higher standards of living and the growing desire to live in an environment-friendly atmosphere have led to a boom in the domestic market as well. The most important flower traded in the international market is still Rose. However, Dendranthema, Dianthus, Chrysanthemum, Carnation, Gerbera, Dahlia, Poinsettia, Orchids, Lily, Tulip are the flowers emerging as close competitors to Rose.

India is also the second largest consumer base and has unlimited opportunities for growth in flower retailing. India's flower trade is attracting a large demand from an estimated 300 million middle class people. Flower consumption in the cities and major towns is reportedly growing at 40 per cent per annum. Flower retail shops and boutiques have mushroomed all over the cities and towns. The demand will get further impetus with the growth of modern retailing concepts. Alongside, India has become world's attractive tourism destination, creating a new boom in flower consumption in Hospitality industry. Cover Story Flower designing is a centuries old art, craft and skill in India by way of displaying flower arrangements in temples of worship, marriages and many religious and social functions. A surging per capita income and progressive lifestyles have led to a phenomenal increase in florist and floral designing market in this country. With the ever-growing gift-lovers' interest in flowers, the floral industry is growing at a very good pace.

While exports remain the prime motivator for cultivators, local demand is also growing by leaps and bounds, especially in cities. Modernization and growing western cultural influences has resulted in consumers–especially the young–buying flowers on occasions like Valentine's Day, Friendship day, Mother's day, Father's day and so on. Of course, there is a huge spurt in demand for flowers during religious festivities. Flower retailing is also undergoing a change. Many of the new shopping malls and large format retailers have exclusive flower shops that witness demand round- the-year.

As far as domestic floriculture is concerned, it is constrained by lack of awareness about its potential, lack of quality planting material, weak infrastructural support, lack of post- harvest facilities, lack of good markets, exploitation by middlemen, weak database, and absence of information on income generation and employment generation from different flower cultivation and export barriers. It is also viewed that a majority of the flower growers belong to small and marginal farmers' category, facing many problems. No comprehensive study has been undertaken to cover all these aspects in the state.

Floriculture has tremendous potential in India. The different types of climatic conditions provide for the possibility of growing almost all the major cut flowers. Species of the world, whether of tropical, sub-tropical or temperate climate origin. However, flowers in India are produced in open field conditions, mostly during the mild winter months without use of any advanced technology. As a result, the quality and quantity available for marketing are heterogeneous and vary according to the prevailing weather conditions.

India has relatively better opportunities for development of the floriculture sector for the following reasons:

☆ Diverse agro-climatic conditions and geographical locations suited for growing various types of flowers

☆ Skilled manpower to absorb the technology and implement the same at a relatively low cost

☆ Soil and water supply at most locations

☆ Good radiation/sunlight leading to healthier plant growth and better quality flowers

☆ Light rains and salubrious climate during winter, the prime export season, leading to sustained high yields

☆ Good period of sunlight even during the heavy rains leading to continued plant growth and proper yield

☆ India is located centrally for catering to European and Far Eastern markets, as well as being close to the South East Asian and Middle East Asian markets that have high consumption requirement of flowers India has a conducive environment for floriculture exports due to :

- [ ] increasing labour cost that is putting pressure on the cost of cultivation of major flower producing countries like The Netherlands, Japan, Taiwan and Israel

- [ ] environment degradation and cost of land which impede further expansion of cut- flowers production in EU countries

- [ ] increasing demand in the nearer markets of West Asia and South East Asia, where the rising standards of living are pushing-up demand for floriculture products

- [ ] dependence of Europe, which is one of the biggest markets on flowers imports during winter months better prices expected to be offered by the other potential markets *viz.* West Asia, South East Asia, Japan, Hong Kong. Singapore and Korea throughout the year than European market, provided they are supplied with good quality flowers on continuous basis.

What do Floriculture Products Look Like when One Uses Them?

Flowers play an essential role in people's celebrations and every day lives. Weddings, graduations, funerals, Mother's Day, St. Valentine's Day, Easter and Christmas are all peak periods of demand for flowers and plants. Cut flowers are combined into elaborate arrangements and bouquets, or several stems are packaged together for impulse cash-and-carry purchases. Flowering and foliage plants are combined together in baskets or planters, or sold individually with pot covers and sleeves to accent their beauty. Cut flowers, potted plants and bedding plants are available at florists, supermarkets, corner grocery stores, mass-market outlets and garden centers. More people are buying flowers at their supermarket as part of their weekly grocery shopping. Another shift in marketing is the move towards more direct farm marketing. Several growers have retail outlets on the farm where you can buy products such as long stem roses, potted orchids and bedding plants.

CHALLENGES

Growers face many challenges including:

Declining Margins

While prices have remained steady over the past several years, most input costs have risen steadily. To remain profitable, growers have had to become more efficient in production and management. Environment – Environmental issues are a major concern for growers. Growers have responded by re-using irrigation water, reducing pesticide and fertilizer use and reducing greenhouse runoff.

Pest Control

Concerns over pesticide use by the public and producers alike, along with pesticide resistance and the loss of approved pesticides, have prompted growers to adopt alternative pest control methods. Integrated pest management (IPM) is playing a larger role in greenhouse pest control. Many growers are now using biological or bio-rational control methods to supplement or replace existing pesticides.

Employment

Labour is an important element in production. Bedding plant and cut flower growers face labour costs of up to one third of gross sales. Although increased mechanization is a necessary element of global competition, the industry continues to be a major agricultural employer.

Urban-rural Conflicts

Urban-rural conflicts are a fact of life for most agriculture in the Province. some municipalities look upon floriculture as more of a factory production industry rather than agriculture. Most municipalities have zoning regulations concerning the maximum site coverage for greenhouses.

Capital Costs

Modern, state-of-the-art greenhouse operations can cost up to $200 per square metre. This represents a barrier to entry for many potential growers. Field-grown cut flowers and bedding plant production have much lower capital costs, so they are often entry level crops.

Seasonal Demand

The demand for fresh floriculture products is seasonal and the product is very perishable. Large numbers of people want to buy flowers for special occasions or holidays like St. Valentine's Day, Easter, Mother's Day and Christmas. Growers must time their production to meet these periods of high demand. Some growers have 30 per cent of their annual sales in a three week period in spring.

Who's Involved in Producing Floriculture Products?

☆ Growers

☆ Greenhouse and field employees

☆ Wholesalers

☆ Florists

☆ Garden centres

☆ Supermarkets

☆ Corner stores

☆ Mass-market outlets

☆ Retail clerks

Interesting Fact about Floriculture

Some of our important floriculture crops originate as weeds in other parts of the world. For example, gerberas (Transvaal Daisies) in South Africa and eustoma (Prairie Gentian) in Texas. Some countries grow dandelions commercially as a salad crop. Floriculture is a world-wide industry: the flowers you buy today could have been picked in South America, Europe or Israel two days ago. To compete with imports, local growers must be able to provide a fresh, high quality product for less money.

Flower Trade Across the World

World trade on floriculture produces like cut flowers, ornamental plants, flowering plants, flower seeds and plantlets gaining tremendous momentum. Many countries, particularly the developed ones, are importing flowers to meet their internal demand. It will be worthwhile to mention that the annual import figures of some of the largest importers on flowers – USA (232 crores US dollar) Japan (192 crores US $), Germany (180 crores US $) France (77 Crores Us Dollar), Italy (55.6 Crores US Dollar), Holland (50 Crores US Dollar). The other importers like Switzerland, Sweden, Denmark, Belgium, Middle-east countries *etc.* also import a sizable amount of cut flowers. In recent past, Israel has come up as the biggest grower of flowers, using modern agro-techniques like glass-house culture, drip irrigation, liquid pesticides and fertilisers application along with drip irrigation channels, Tissue Culture. It may be mentioned that the roses of Israel adjudged to be the best in the World. via-a-vis such a huge market potential of floriculture produce, India's contribution is not at all encouraging as its flower export amount to 30 lakh us dollar only, hence India has to do a lot to exploit this agro-business.

Government Incentives and Schemes

The union government has recognized floriculture as a thrust area for export and announced several concessions/incentives for its development in the country. Main features of government incentives and schemes as below. Zero import duty in seeds, bulbs, cut on import duties for machinery, flower, and tissue culture seed on OGL: fifty per cent domestic sales allowed for EOU's, simplification of plan quarantine procedure and airfreight subsidiary.

The Ministry of Commerce, Government of India has identified four states, *viz.* Kerala for orchids, Maharashtra for carnation and roses, Karnataka for chrysanthemums, and Andhra Pradesh for roses to organized flower production. The National Horticulture Board (NHB) have two major schemes: soft loan up to a maximum limit of ₹ 1crore advance for setting up infrastructural facilities like cold storage, pre-cooling units, packing and grading sheds, refrigerated transport, etc and loan facilities are available for the integrated projects involving production and disposal of floricultural products.

In order to increase production and enhance the quality of cut flowers, a central sector scheme with an outlay of ₹1 crore during Eighth Five Year Plan and ₹ 2.17crore for 1993-94 remained operative during the year. The concept of the scheme is to introduce improved varieties of flowers of commercial importance, to intensify production of planting material, to introduce modern system of postharvest handling and to impart training to farmers and field staff. Nine model centres with tissue culture facilities were set-up in public sector in various states of India.

The Government implanting ₹10 crore scheme for commercial floriculture during 8ᵗʰ Five year plan involving the settings up of model centers, both in the public and private sectors, opening up 38 tissue culture, and expansion of areas under commercial floriculture by 2000. It proposed to establish model floriculture centre in the public sector, one each at Mohali, Calcutta, Lucknow, Bangalore,

Pune, Srinagar, Trivandrum, Gangtok and Chennai. Eight such centers are also being set-up in the private sector, one each at all the above places except Srinagar. Settings up of wholesale market cum auction centre's for trading in flowers were planned at Delhi, Bangalore, and Mumbai, in association with one of the agencies of the concerned State Governments. The Government of India will meet 50 per cent of the cost and concerned State Governments will meet the remained 50 per cent. APEDA have several schemes to support export activity. Aim of this scheme has- a) development of infrastructure and services. b) Development of postharvest infrastructure c) packaging development and export promotion and market development.

During Ninth Plan, various centrally sponsored schemes were implemented to overcome constraints and improve productivity of the crops. These related to the integrated development of fruit, vegetable and commercial floriculture and separate scheme for horticulture development through plasticulture intervention implemented during the Ninth Plan. It was aim that promotion of protected cultivation through greenhouse technology. Technology mission for integrated Development of Horticulture also introduced towards the end of 1999-00. A central sector scheme for development for harvest management and commercial horticulture is also in operation through NHB.

Measures to Promote Floriculture

The Indian floriculture is riddled with such problems as exhorbitant airfreight costs, high interest rates and insensitive government policy. Also, the research turned out is not of much help for commercialization. The floriculture sector is getting help from several institutions. The National Horticulture Board (NHB) has introduced several new schemes to strengthen the infrastructure for floriculture. Financial aid provided by the NHB is used for setting up floriculture units.

The Agriculture and Processed Foods Export Development Authority (APEDA) has initiated promotional programmes to boost floriculture exports. The FAO sponsored project o greenhouse technology for small scale farmers has been implemented at Bangalore, Pune and confidence of the farmers.

APEDA plants to develop common infrastructure facilities for floriculture clusters in Belgium and Kodagu districts of Karnataka. The intention is to help small growers. Each common infrastructure facility would have cold store and pre cooling facilities, besides grading and packing halls. Floriculture has got a boost due to the setting up of an agri-export zone (AEZ) in Bangalore. But, the unorganized sector, comprising largely small and individual growers, has not got any support. The setting up common infrastructure facilities would provide the much needed boost.

Of course, APEDA had provided common infrastructure facilities in different parts of the country, notably at airports and seaports. These facilities came up to provide support for exports of perishable commodities. Commercially, floriculture can provide ample opportunities for our poor farmers. But, floriculture urgently requires government help. The flower growers should be organized into societies which should be linked to a network of retail stores in the big cities, hotels and markets.

National Horticulture Mission

The centrally sponsored scheme of National Horticulture Mission (NHM), which launched during 2005-06, it has continued for development of the horticulture sector in India, with an outlay of ₹ 1000.00 crores. Eighteen states and two UT are covered under the NHM during the year. The thrust of the NHM is on an area based regionally differentiated cluster approach for the development of horticulture crops, having comparative advantage. In all 259 districts have been taken up under the cluster approach and this includes 32 new districts added during 2006-07. Activities carried out through the state horticulture missions of various states and UT?s. Further, some components of the NHM are being implemented through various national level agencies such as the National Committee on Plasticulture Applications in Horticulture (NCPAH), which co- ordinates and monitors activities relating to precision farming and hi tech horticulture.

For the development of horticulture, duly ensuring an end-to-end approach having and forward linkages covering research, production, postharvest management, processing and marketing. The focus area of the mission is as under.

☆ Capacity building for production and supply of adequate quality planting material

☆ Increased coverage of crops under improved/yielding cultivars

☆ Enhanced production and productivity of horticulture crops

☆ Strengthening of infrastructure facilities such as soil and leaf analysis labs, survey and surveillance of pest and deceases, greenhouse, micro irrigation, plant health clinics, vermicompost, etc

☆ Building adequate infrastructure for in farm and postharvest handling

☆ Strengthening infrastructure facilities for marketing and export

☆ Enhanced production of high value and low volume horticulture products for exports

☆ Enhanced production of high value processed products

☆ Building a strong base to enhance efficiency in adoption of technologies

A technology driven cluster approach with focused attention on competitive horticulture crops, is the underlying approach of the NHM.

Agriculture Foreign Trade Policy 2004-09

The agriculture ministry of Government of India initiatives for agriculture in this policy includes:

Vishesh Krishi Upaj Yojana: This scheme introduced to boost the export of fruits, vegetables, flowers, minor forest produce and their value added products. EPCG-Duty free import of capital goods under the export promotion capital goods scheme. Capital goods imported under EPCG for agriculture permitted to be installed anywhere in the Agri-export zones. ASIDE-Assistance to states for infrastructure development of exports funds to be also utilized for the development of Agri-export zones.

Export/Import Liberalization

Import of seed, bulbs, tubers and planting material has liberalized. Export plant portions, derivatives and extracts have liberalized with a view to promote exports of medicinal and herbal products.

Remove the Cess

The produce cess laws (Abolition) Act 2006 notified, in order to remove the cess on export of agricultural products and to encourage the export of agricultural products.

Indian Floriculture Mission 2010

Recently the second International Flora Export 2006 organized by Indian Flowers and Ornamental Plants Association in Delhi inaugurated by the President of India Dr. A.P.J. Kalam. He address regarding Mission Floriculture 2010 as below.

☆ An annual target of the one billion Dollars(₹-5000cr) of floriculture export by the year 2010

☆ Quality and standard of floriculture produce with enhanced sell life must be maintained with active participation from research institutions and Bureau of Indian Standard

☆ Infrastructure development including roads, uninterrupted electricity, water, cold storage, facilities of airport and training of customs officials for faster and careful clearances for floriculture units

☆ Development of better and high yielding varieties of flowers and ornamental plants for domestic market

☆ Pest management through indigenous research and knowledge

☆ Establishment of Special Floriculture Zones(SFZ) in various part of the country to provide economical as well as technical support to farmers

☆ Floriculture business must be under insured from production to market.

The Role of Supporting Agencies

It is inappropriate to discuss floriculture development without knowing the role of supporting agencies such as NHB, APEDA and NABARD. National Horticulture Board has given support through various programmes to development horticulture including fruits, flowers and vegetables. APEDA nurtured the floriculture industry by various strategic measures. NABARD gave refinance facilities to banks and financial institutions *i.e.* State Co-operative Banks, Commercial Banks, and State Agricultural Development Finance.

National Horticulture Board and Floriculture Development

The NHB has considered two-pronged strategy; one of them is promotional activities to give boost to the process of employment generation, increase in income of small and marginal farmers and backward communities in the horticulture development process. Then second one is, catalytic activities for commercialization of horticulture through production, postharvest management and processing with enhanced productivity, processing and marketing related programmes.

According to this strategy government has allocate funds, till 91-92, the allocation of funds to N.H.B. was quite less. In fact, from 1985-86 to 1991-92 the fund allocation to N.H.B. was only about ₹15 crore. However, a major thrust was given to N.H.B. programmes in the Eighth Plan (1992-93 to 1996-97) and allocation was raised to 200 crore. The scheme was launch in the Eighth Plan with the following features:

☆ Soft loan schemes for exploring commercial horticulture, postharvest management and marketing.

☆ Schemes to provide support services like market information and transfer of technology

☆ Schemes to create awareness and updating the state of horticulture through innovative ideas and professional concepts

The schemes during Ninth Plan have been made broader based to comprise every aspect of horticulture promotion in the country. Now the importance of the scheme would be result oriented and have been substituted with back ended capital-intensive subsidy. The N.H.B. has decentralized and simplified the procedures and subsidy linked with performance. Structure of the schemes consist both type of components.

A) Production Related Components

☆ High quality commercial horticulture crops

☆ Indigenous crops, herbs, aromatic plants, seed and nursery

☆ Biotechnology, tissue culture, bio- pesticides, organic foods

☆ Establishment of horticulture Health Clinic/laboratory (for agri/horti unemployed graduates)

☆ Consultancy services, bee keeping

B) Postharvest Management/Primary Processing Related components

☆ Grading/washing/sorting/drying/packing centres.

☆ Pre cooling unit/cool stores

☆ Refer van/containers

☆ Special transport vehicle

☆ Retail outlets, auction platforms, market yard, rope ways

☆ Ripening/curing chamber

☆ Radiation unit/dehydration unit/vapour heat treatment unit

☆ Primary processing of products fermentalisation, extraction, distillation, juice vending pulping, dressing, cutting, chopping, etc

☆ Hort. Ancillary Industry *e.g.* tools, equipment, plastics, packaging, *etc.*

☆ Crates, cartons, aseptic packaging and nets(50 per cent subsidy) pattern of assistance

N.H.B. has given assistance with following pattern-Back ended capital investment subsidy not exceeding 20 per cent of the project cost with a maximum limit of ₹ 29 lakh per project (₹, 30 lakh for North-East tribal area).

Eligible Organizations

NHB has given benefit to following organizations through above scheme:

The eligible promotes under the above scheme shall include NGO's, Association of Growers, Individuals, Partnerships/Proprietary Firm, Companies, Corporations, Cooperatives, Agricultural Marketing Committees, Marketing Boards/Committee's, Municipal Corporations, Agro Industries Corporation and other concerned Research and Development Organizations.

Agricultural and Processed Food Products Export Development Authority (APEDA) and Floriculture Development

The Agricultural and Processed Food Products Export Development Authority (APEDA) established by the Government of India under the Ministry of Commerce on February 13, 1986 under the Agricultural and Processed Food Products Export Authority Act, 1985. The main responsibility of APEDA is the export promotion of processed horticultural products.

Objectives: The main objectives of establishing APEDA are:

☆ To maximize foreign exchange earnings through increased agro-export for providing higher incomes to the farmers through higher unit value realization.

☆ To create employment opportunities in rural areas by encouraging value added exports of farm products, and

☆ To implement schemes for providing financial assistance to improve postharvest facilities to boost their export.

☆ The Role of APEDA in the development of floriculture has been important which prove the facts. APEDA has initiated the following strategic measures, which have greatly enhanced the export performance of the floriculture industry. There are set cold storage and cargo handling centre for perishable cargo at Chhatrapati Shivaji International (C.S.I.) airport Mumbai that APEDA has contributed by way grant-in-aid. This would enable growers to realize better prices because of the large potential and growing domestic market during the season when exports are not made

The Government of India had already approved setting of a flower auction centre at Bangalore and sanctioned its contribution of ₹ 3.5 crores through APEDA. For improving the export performance, planting material subsidy proposed for Tenth Plan with an outlay of ₹ 20 crores. This would facilitate the unit to replant the current varieties in demand attractive higher realization. However, research and development efforts to be made in order to develop own varieties keeping in view consumer preferences.

APEDA has set-up a marketing centre at Alasmeer, in Netherland to promote Indian produce, which become operation from November 2001. APEDA will consider providing assistance for the infrastructure and day-to-day management will be responsibility of the growers so that centre will be self-sufficient in meeting all operational costs. APEDA had conducted the study on viability of floriculture industry. The study identified several constraints *i.e.* high marketing cost, absence of critical infrastructure for export facilitation, high cost of finance *etc.*, Suggestion taking into consideration with several measures by the government to floriculture industry to these constraints.

The National Bank for Agriculture and Rural Development (NABARD) and Floriculture Development

The National Bank for Agriculture and Rural Development set- up on 1 July, 1982 under an act of Parliament by merging the Agriculture Credit Department and Rural Planning and Credit Cell of Reserve Bank of India and the entire undertaking of Agriculture Refinance and Development Corporation (ARDC). NABARD is an apex development bank of the country for supporting and promoting agriculture and rural development. The Government of India and Reserve Bank of India contribute the share capital of NABARD. NABARD operates through its head office at Mumbai, 28 regional offices situated in state capitals, a sub office at Port Blair and 320-district offices. A credit function of NABARD to floriculture is important as regard to development roles. Its mission to promote sustainable and equitable agriculture and rural prosperity through effective credit support. As a refinance institution, NABARD has been playing important role to develop floriculture industry.

Credit Function of NABARD

☆ Preparation of potential linked credit plans annually for all districts for identification of potential for exploitation through bank credit

☆ Monitoring the flow of ground level rural credit

☆ Issuing policy and operational guidelines to rural financing institution

☆ Providing credit facilities to institutions

Financing to Floriculture

Finance is one of the most critical inputs particularly for floriculture development, which is capital-intensive activity. NABARD has made a significant contribution in this direction through its various functions. The developments of floriculture have needed more finance for various activities for its cultivation. The development of land, drip irrigation, machinery and equipment, greenhouse cold storage, planting material, postharvest handling and others. This whole factors required finance. NABARD has studied all these aspect and made finance structure for floriculture project. NABARD has financing to various banks by way of providing refinance to floriculture activities in different parts of the country. NABARD has been playing important role in promoting and supporting hi-tech floriculture projects in different states, predominantly Maharashtra, Karnataka, Tamil Nadu and Andhra Pradesh. NABARD has provided huge finance to hi-tech project.

Indian Floriculture Industry: Present Status and Scope

After liberalization, the Govt. of India identified floriculture as a sunrise industry and accorded it 100 per cent export oriented status. Owing to steady increase in demand of flower floriculture has become one of the important Commercial trades in Agriculture. Hence commercial floriculture has emerged as hi-tech activity-taking place under controlled climatic conditions inside greenhouse. Floriculture products mainly consist of cut flowers, pot plants, cut foilage, seeds bulbs, tubers, rooted cuttings and dried flowers or leaves. The important floricultural crops in the international cut flower trade are rose, carnation, chrysanthemum, gargera, gladious, gypsophila, liastris, nerine, orchids, archilea, anthuriu, tulip, lilies.

Present status and growing trade is still in infancy. Floriculture in India, is being viewed as a high growth Industry. Commercial floriculture is becoming important from the export angle. The liberalization of industrial and trade policies paved the way for development of export oriented production of cut flowers. The new seed policy had already made it feasible to import planting material of international varieties.

The government of India offers tax benefits to new export oriented floriculture companies in the form of income-tax holidays and exemption from certain import duties. Agricultural and Processed Food Products Export Development Authority (APEDA), responsible for export promotion and development of floriculture in India, grants subsidies for establishing cold storage, precooling units, refrigerated vans and greenhouses, and air freight subsidy to exports. It has been found that commercial floriculture has higher potential per unit area than most of the field crops and is therefore a lucrative business.

Conductive Conditions

India is endowed with diverse agro-climatic conditions like good quality soils, suitable climate, abundant water supply, low labour cost, proximity to market in Japan, Russia, South-East Asia, Middle-East Countries. Subsidy on airfreight for export of cut flowers and tissue-cultured plants is allowed by the Government. Freight rates are ₹10 per kg for export to Europe and ₹ 6 per kg for export to West Asia, SouthEast Asia whichever is less.

Import duties have been reduced on cut flowers, flower seeds, tissue-cultured plants, *etc.* Setting up of walk in type cold storage has been allowed at the International airports for storage of export produce. Direct subsidy upto 50 per cent of the precooling and cold storage units is available, as well as subsidy for using improved packaging material is given by APEDA. Eleven-model floriculture centre units and two large centres, 20 tissue culture units have been established by Ministry of Agriculture. Refinance assistance is available from NABARD to a number of hi-tech units at reasonable interest rate.

Export Constraints

In spite of an abundant and varied production base, India's export of floricultural product is not encouraging. The low performance is attributed to many constraints like non-availability of air space in major airlines, since most of the airline operators prefer heavy consignments. The existing number of flights during the peak seasons is not sufficient for export purpose. Exporters for infra-structural problems like bad interior road, inadequate refrigerated transport and storage facilities. Lack of professional backup of delivery and supporting companies, which resort into high cost of technology for Indian entrepreneurs. Tedious Phyto-sanitary certification and an unorganized domestic market.

In order to overcome these problems, attention must be focussed on:

Reduction in Import Duty on Planting Material and Equipment

Airfreight should be reduced to a reasonable level. Sufficient cargo space may be provided in airlines. Establishment of model nurseries far supplying genuine planting material. Co-operative florist organizations should be established at regional level. Training centres for diploma course on the pattern of ITI for training the personnel in floriculture should be set up. Exporters should plan and monitor effective quality control measures right from production to postharvesting, storage, and transportation. An analysis of strengths, weaknesses, opportunities of the floricultural industry shows that India has immense potential for export of floricultural products.

Though the global floriculture industry is growing comparatively at a faster pace than in India, still a scope exists to bridge the demand and supply gap.

Traditional Flower Cultivation

The domestic consumption of loose flowers, especially of marigold, China aster, jasmine, crossandra, barleria, *etc.* has been increased tremendously. It is clear that area under traditional flowers has increased significantly (>90 per cent of total flower crops). This sector, inspite of its potential, is still an unorganized and often does not get proper importance. Research is required on developing high yielding varieties, year round production of chrysanthemum, China aster, marigold *etc.* and in promotion of crops like annual chrysanthemum, *desi* rose, *etc.*

Hi-tech/Protected Cultivation

The cut flowers, which are being exported from India, are from these hi-tech floricultural units. Protected cultivation, although is in limited area (5 per cent of total flower crop area), its contribution to total floricultural exports is significant. At present, there are about 110 export- oriented floricultural units (EOUs) in operation, covering an area of 500 ha. These units are growing mostly roses, but can be diversified into orchids, Anthurium, gladiolus and tuberose as the demand for tropical flowers is increasing worldwide. India has several advantages and great potential to increase the acreage under intensive production and ultimately to increase the floricultural exports provided the units should be opened in ideal locations with sound technological back-up.

Dry Flowers

Dry flowers constitute more than two-thirds of total floricultural exports. For making dry flowers and plant parts can be collected from wild sources or some flower crops like Dahlias, marigold, jute flowers, wood roses, wild lilies, helichrysum, lotus pods, *etc.* some flowers that are air-dried and used include Dahlias (*Dahlia hortensis*), poppy seed heads (*Papaver somniferum*), roses (Rosa), Delphinium, larkspur (*Consolida ambigua*), lavender (*Lavendula augustifolia*), African marigold (*Tagetes erecta*), strawflower (*Helichrysum bracteatum*), globe amaranth (*Gomphrena globosa*), lotus pod *etc.* dry flowers constitute nearly 15 per cent of the global floriculture business and form the major share in Indian floricultural exports as well. At present, the industry is not well organized and depends on plant material available in forests and no systematic growing of specialized flowers exists anywhere in the country. The demand for dry flowers is increasing at an impressive rate of 8-10 per cent and therefore there is a great scope for the Indian entrepreneurs.

Flower Seed Production

Seed production of seasonal flower crops is a lucrative business and practiced in considerable area in Punjab and Haryana. This offers higher returns from unit area. Of late, demand is increasing in domestic market also. Research work is required to develop high-yielding varieties including F1 hybrids, agro-techniques for producing uniform seed with higher certification standards.

Nursery Industry

Lack of quality planting material is the major hindrance for not realizing the full potential of floriculture in India. Plant material of various kinds (seedlings, budded plants, rooted cuttings, bulbs, tubers, corms, annual seed, *etc.*) is required for commercial flower production, pot plant production for adding to home garden and for landscaping (corporate landscaping, bioaesthetic planting *etc.*).

Pot Pourri

Pot pourri is mixture of dried, sweet-scented plant parts including flowers, leaves, seeds, stems and roots. The basis of a pot pourri is the aromatic oils found within the plant. A significant component of dry flower export comprises pot pourries. In the recent past, floriculture has been considered as a viable option of diversification in agriculture. But now within floriculture itself, there are in a number of options a flower or a floriculturist can take.

Natural Dyes

Marigold pigments are widely used in the poultry industry to enhance the colour of the meat and yolk of the eggs and also used in food and textile industry. So far, isolation of xanthophylls from marigold has been standardized. More crops can be identified and procedures can be standardized for full exploitation. Technology development in all the areas mentioned above not only improves the situation of respective sub-sector of floriculture, but these become important avenues for diversification of floriculture, sources of income generation and means of employment to the youth.

Positive Signal

The global downtrend has delivered a positive signal to Indian floriculture industry by underlying the industry's inherent strengths. The overall situation, in fact, offers new opportunities for the Indian floriculture sector. This is the right time for the government departments trying to promote floriculture and the Indian industry to seize the opportunity to prepare a long term plan to strengthen this sector and facilitate domestic suppliers to gain a stronger foothold in the developed markets. To attain this objective, the domestic industry should not only have access to technology and good quality planting material at reasonable prices but also proper marketing support. The need of the hour is good cooperation among all segments that would include big growers engaged in high tech production, small growers, input suppliers, cooperatives and government departments. Policy makers would do well to organise the domestic flower market and provide at least minimum infrastructure facilities like sheltered market yards to enable the growers to sell their produce, unspoiled by the harshness of summer sun or pouring rain. Basic facilities to handle perishables like flowers are needed in all major cities and towns, along with adequate support from the state marketing departments. Needless to say, such facilities are now woefully lacking.

"Surprisingly the Indian growers have responded positively to floriculture plantations this year which is very unlike in this recession worldwide, says Kishore Rajhans, Vice-President of KF Bioplants. We have lot of booking for Gerbera and other plants for greenhouse plantation. "We have overwhelming response mainly from Uttarakhand, Himachal Pradesh, Sikkim and Assam from North East. We also have fantastic response from Tamil Nadu and Karnataka," says Rajhans. He feels the company, which is a leading biotech plants supplier, has been quite fortunate in not having faced any setback of the current meltdown till today. However, we have kept our fingers crossed, said Rajhans. Although the company's principals are playing it safe, as they have been hit directly by the economic meltdown worldwide, the response in India not being negative is quite encouraging, says Rajhans. According to him, the biggest strength of India is its manpower which can sustain any kind of adverse conditions, while being tolerant. "We can also control the cost of production to a certain extent, to become the most competitive player in the market." This is unlike other countries where the standard of living is quite high (of course not in Asian and African countries) and they are the ones who crumble first in this kind of adversity.

India has Largest Floriculture (Flowers) Industry in the World - Presentation

Floriculture or flower farming as it is popularly called is a discipline of Horticulture, and is the study of growing and marketing flowers and foliage plants. Floriculture includes cultivation of flowering and ornamental plants for sales or for use as raw materials in cosmetic and perfume industry and the pharmaceutical sector. The persons associated with this field are called floriculturists.

Officially Floriculture began in the late 1800's in England where flowers were grown in large estates, and now has spread to most other countries as well. The floral industry today has grown to much larger proportions and offers a wide scope for growth and profits. The countries involved in the import of flowers are Netherlands, Germany, France, Italy and Japan while those involved in export are Columbia, Israel, Spain and Kenya.

In India, Floriculture industry comprises flower trade, production of nursery plants and potted plants, seed and bulb production, micro propagation and extraction of essential oils. Though the annual domestic demand for the flowers is growing at a rate of over 25 per cent and international demand at around ₹ 90,000 crore India's share in international market of flowers is negligible. India has a blooming future as far as floriculture is concerned. Enormous genetic diversity, varied agro climatic conditions, versatile human resources *etc.* offer India a unique scope for judicious employment of existing resources and exploration of avenues yet untouched.

Karnataka is the leader in floriculture, accounting for 75 per cent of India's total flower production. The state has the highest area under modern cut flowers, and 40 flower growing and exporting units. The expert committee set up by Govt. of India for promotion of export oriented floriculture units has identified Bangalore, Pune, New Delhi and Hyderabad as the major areas suitable for such activity especially for cut flowers. APEDA (Agricultural and Processed Food Products Export Development Authority) is the registering authority for such units.

The employment opportunities in this field are as varied as the nature of work itself. One can join the field of floriculture as farm/estate managers, plantation experts and supervisors, project coordinators *etc.* Research and teaching are some other avenues of employment in the field. Marketing of Floriculture products for different ventures is emerging as a potential segment of this field. Besides one can work as consultant, landscape architect etc with proper training. One can also work as entrepreneur and offer employment to others. In addition to these careers which involve research and actual growing of crops, floriculture also provides service career opportunities which include such jobs as floral designers, groundskeepers, landscape designers, architects and horticultural therapists. Such jobs require practitioners to deal directly with clients.

Professional qualification combined with an inclination towards gardening and such other activities produces efficient floriculturists and landscaping professionals. The skills and knowledge required are imparted under the professional courses of floriculture and landscaping

Floriculture Industry to Cross ₹ 8000 Crore by 2015

Growing at a compounded annual growth rate of about 30 per cent, India's floriculture industry is likely to cross ₹ 8,000 crore mark by 2015, apex industry body ASSOCHAM. Currently, the floriculture industry in India is poised at about ₹ 3,700 crore with a share of a meagre 0.61 per cent in the global floriculture industry which is likely to reach 0.89 per cent by 2015, according to a study titled, 'Indian Floriculture Industry: The Way Ahead' released by The Associated Chamber of Commerce and Industry of India (ASSOCHAM).

Besides, the global floriculture industry is likely to cross ₹ 9 lakh crore mark by 2015 from the current level of about ₹ 6 lakh crore and is growing at a compound annual growth rate of 15 per cent, said a statement issued by the chamber. With a share of about 65 per cent, market share of rose alone in the flower industry in India accounts for over ₹ 2,400 crore of the overall floriculture industry and 75 per cent of the global floriculture industry, said the study.

Rising demand from tier II and III cities apart from urban centres is likely to spur demand for roses this Valentine's Day as price of export quality cut rose is likely to quadruple from its current average ruling price of about ₹ 15 to ₹ 20 per stem. The chamber interacted with about 250 rose merchants including the cultivators, exporters, wholesale flower dealers and florists in Bangalore, Chennai, Delhi, Mumbai and Pune to gauge the scenario vis-a-vis business of rose flower during the Valentine's week considering India is also world's biggest rose grower.

Demand for roses has spiralled upwards by over 25 per cent in domestic and by about 30 per cent in international markets as the V-Day draws closer, said over 55 per cent of all the respondents. Fall in the value of rupee against major currencies is the prime reason behind this upsurge in demand for roses in international markets of Australia, Germany, Greece, Italy, New Zealand, the Netherlands, the United States, the United Kingdom and other countries of Europe and the Middle East.

Besides, majority of flower growers also said they are hoping for about 30 per cent rise in terms of revenue during February alone with a turnover of about ₹ 10 crore. About 40 crore cut roses are grown across India every year and Karnataka alone accounts for about 75 per cent followed by Maharashtra, Tamil Nadu, Bihar, West Bengal, Uttar Pradesh, Haryana, Punjab, Jammu and Kashmir, Andhra Pradesh and Madhya Pradesh, said ASSOCHAM.

3

Area and Production of Flowers in India

In India flowers are mostly grown in open. Though Floriculture International, 1997 indicated that India's area under floriculture is around 34,000 hectares, the National Horticulture Board (NHB) indicates much higher area under flower alone (**Table 3.1**) with a faster increasing trend. In fact, according to the NHB statistics, during 1990s there was a phenomenal increase in area under flowers from 53 thousand hectares in 1993-94 to 89 thousand hectares in 1999-2000. The increasing trend of area persisted in the new millennium as well and in 2001-2002, the area under flowers was as high as 106 thousand hectares. Accordingly, the production of loose and cut flowers also shows a significant increasing trend.

The commercial cultivation of flowers is predominant in Karnataka, Tamil Nadu, Andhra Pradesh, West Bengal, Maharashtra, Delhi and Haryana. More than two-third of the area is under production of traditional flowers like marigold, jasmine, roses, chrysanthemum, aster, tuberose, *etc.* which are mainly for domestic consumption. The rest of the flowers, such as, rose, gladiolus, carnation, and orchids used in banquets and arrangements.

By 2013-14, area touched 255 thousand hect. From among all the States Tamil Nadu with 55 thousand hect. (21.6 per cent) topped the list. West Bengal, Karnataka, Andhra Pradesh and Maharashtra are the other important states growing flowers. Tamil Nadu has also been at the top in loose flowers also at 343.7 thousand tones (19.5 per cent). It is followed by Karnataka, Andhra Pradesh and Maharashtra. As regards cut flowers, West Bengal is the leader with 145.2 thousand tones – over 26 per cent. It is followed by Karnataka, Odisha, U.P and Maharashtra. Tamil Nadu is a poor 12.9 thousand tones. All this shows that from among the major flower growing States. It is different States which specialize in loose or cut flowers (Tables 3.1 and 3.2).

Table 3.1: All India Area and Production of Flower

Year	Area in 000'ha	Production	
		Loose (n '000 mt.)	Cut in million no.)
1993-94	53	233	555
1994-95	60	261	519
1995-96	82	334	537
1996-97	71	366	615
1997-98	74	366	622
1998-99	74	419	643
1999-00	89	509	681
2000-01	98	556	804
2001-02	106	535	2565
2002-03	70	735	2060
2003-04	101	580	1793
2004-05	118	659	2071
2005-06	130	656	2921
2006-07	144	880	3716
2007-08	166	868	4365
2008-09	167	987	4794
2009-10	183	1021	6667
2010-11	191	1031	6904
2011-12	254	1652	7507
2012-13	233	1729	7673
2013-14	255	1754	543

Source: National Horticulture Board, Indian horticulture Database-2015.

Table 3.2: State-wise Area and Production of Flowers in India During 2002-2005

State	Area (in '000 ha.)			Production (Loose in 000 MT and Cut in lakh Nos)					
	2002-03	2003-04	2004-05	2002-03 Loose	2002-03 Cut	2003-04 Loose	2003-04 Cut	2004-05 Loose	2004-05 Cut
Tamil Nadu	17676	20274	23233	135221	-	161655	-	187342	-
Karnataka	19097	18182	18458	151953	-	143286	5591	145890	4134
Andhra Pradesh	13310	12902	13909	72205	87	49130	44	57875	71
Haryana	3600	4296	4810	32500	1200	58333	461	55583	508
Maharashtra	--	8422	8660	-	-	48538	-	51705	-
West Bengal	13870	17328	17925	33749	7020	43575	8767	44674	8963
Gujarat	4917	4917	6956	30187	-	30187	-	41811	1969
Delhi	4490	4490	4490	25007	-	25007	-	25007	-
Odisha	0	282	314	-	-	78	11	17252	12
Uttar Pradesh	6325	6325	7968	9753	2650	9753	2650	11905	3527
Punjab	600	600	615	3000	-	3000	-	3075	-
Chattisgarh	11	11	1508	60	-	60	-	2829	-
Rajasthan	1505	1949	312	986	-	2161	-	2604	-
Himachal Pradesh	245	311	407	999	283	1504	380	2243	182
Bihar	96	103	103	1757	11	1757	11	1757	11
Madhya Pradesh	-	-	1829	-	-	-	-	1097	-
Jammu and Kashmir	69	98	226	38	7	207	3	922	110
Manipur	93	535	535	30	-	701	-	701	-
Uttarakhand	409	96	525	254458	-	545	-	558	-
Daman and Diu	-	2	2	-	-	7	-	7	-

State	Area (in '000 ha.)			Production (Loose in 000 MT and Cut in lakh Nos)					
	2002-03	2003-04	2004-05	2002-03 Loose	2002-03 Cut	2003-04 Loose	2003-04 Cut	2004-05 Loose	2004-05 Cut
Mizoram	0	56	56	-	-	0	1	0	1
Assam	0	-	-	1000	0	-	-	-	-
Jharkhand	91	-	-	2000	347	-	-	-	-
Pondicherry	-	-	-	-	-	-	-	-	-
Sikkim	6	6	80	-	9	-	9	-	28
TOTAL	86409	101185	115921	754903	11614	579484	17926	654837	19515

Source: National Horticulture Board, Indian horticulture Database-2005.

State-wise Area and Production of Flowers in India During 2005-06 to 2007-08

State	Area (in '000 ha.)			Production (Loose in 000 MT and Cut in lakh Nos)					
	2005-06	2006-07	2007-08	2005-06 Loose	2005-06 Cut	2006-07 Loose	2006-07 Cut	2007-08 Loose	2007-08 Cut
ANDAMAN and NICOBAR	0.01	0.01	0.03	2.54	-	2.54	-	4.70	-
ANDHRA PRADESH	17.51	21.66	23.52	88.81	67.07	116.24	65.87	126.27	67.82
BIHAR	0.20	0.20	0.20	2.30	10.60	2.30	10.60	2.30	10.60
CHHATTISGARH	1.55	2.03	2.36	-	-	7.84	-	6.91	-
DAMAN and DIU	0.00	0.00	0.00	0.01	-	0.01	-	0.01	-
DELHI	5.50	5.50	5.50	5.67	1038.20	5.70	1038.00	5.70	1038.00
GUJARAT	7.12	8.42	9.74	42.18	4392.00	49.50	5063.00	49.50	5063.00
HARYANA	5.40	5.65	6.11	26.30	622.70	52.15	1404.04	61.76	10.53
HIMACHAL PRADESH	0.41	0.58	0.58	3.01	434.35	3.63	530.75	3.40	565.55
JAMMU and KASHMIR	0.33	0.33	0.33	1.34	217.90	1.34	218.00	1.34	218.00

State	Area (in '000 ha.)			Production (Loose in 000 MT and Cut in lakh Nos)					
	2005-06	2006-07	2007-08	2005-06		2006-07		2007-08	
				Loose	Cut	Loose	Cut	Loose	Cut
JHARKHAND	0.00	0.21	0.11	0.00	0.00	0.33	273.00	0.23	73.00
KARNATAKA	21.10	23.02	22.34	156.20	5239.00	192.07	5660.00	169.12	5550.00
MADHYA PRADESH	3.67	2.50	2.55	2.00	-	1.40	-	1.53	-
MAHARASHTRA	9.44	14.76	16.74	56.08	3410.00	88.90	4774.00	69.45	5728.00
MIZORAM	0.04	0.04	0.05	0.00	17.96	0.00	30.91	0.00	36.59
NAGALAND	0.00	0.00	0.02	0.00	0.00	0.00	0.00	0.00	16.50
ODISHA	0.59	0.59	2.40	1.79	129.64	1.93	0.00	7.00	129.60
PONDICHERRY	0.48	0.33	0.33	2.67	-	2.67	129.60	2.67	-
PUNJAB	0.80	0.95	1.00	4.10	-	74.00	-	77.90	-
RAJASTHAN	3.01	2.73	3.34	2.26	-	3.26	-	4.61	-
SIKKIM	0.10	0.10	0.01	-	33.10	0.09	16.78	0.09	22.82
TAMILNADU	24.75	26.73	26.74	202.00	-	218.06	-	214.38	-
UTTAR PRADESH	8.25	8.39	8.41	12.18	3668.00	12.34	3746.00	12.36	3752.00
UTTARAKHAND	0.56	0.71	0.90	0.36	575.00	0.46	1229.74	0.70	1455.45
WEST BENGAL	17.89	18.56	27.42	42.29	9347.90	43.68	12966.00	48.45	19680.00
TOTAL	128.68	144.01	160.72	654.08	29203.42	880.43	37156.29	870.37	43417.46

State-wise Area and Production of Flowers in India During 2008-09 to 2010-11

State	Area (in '000 ha.)			Production (Loose in 000 MT and Cut in lakh Nos)					
				2008-09		2009-10		2010-11	
	2008-09	2009-10	2010-11	Loose	Cut	Loose	Cut	Loose	Cut
ANDAMAN and NICOBAR	0.035	0.0	0.0	0.335		0.3		4.7	
ANDHRA PRADESH	19.5	21.4	21.8	125.0	3	130.3	6202.0	133.7	6202.0
ARUNACHAL PRADESH		1.2	1.2				2860.0		2860.0
ASSAM									
BIHAR	0.2	0.2	0.2	2.3	11	2.3	11.0	2.3	11.0
CHANDIGARH									
CHHATTISGARH	2.4	4.1	6.9	6.9		13.5	0.0	27.1	
D and N HAVELI									
DAMAN and DIU	0.0	0.0	0.0	0.007		0.0		0.0	
DELHI	5.5	5.5	5.5	5.7	1038	5.7	1038.0	5.7	1038.0
GOA									
GUJARAT	9.7	12.5	12.5	49.5	5063	49.5	5063.0	49.5	5063.0
HARYANA	5.5	6.2	6.2	53.9	929	60.3	1084.0	60.3	1084.0
HIMACHAL PRADESH	0.6	0.7	0.7	3.4	566	0.6	605.0	0.6	605.0
JAMMU and KASHMIR	0.065	0.1	0.1	0.011	20	0.2	66.3	0.2	66.3
JHARKHAND	1.600	1.6	1.6	22.0	1711	22.0	1711.0	22.0	1711.0
KARNATAKA	26.0	27.0	27.0	203.9	5867	203.9	5860.0	203.9	5860.0
KERALA									
LAKSHADWEEP									
MADHYA PRADESH	3.0	6.6	7.7	1.8		5.0		6.0	
MAHARASHTRA	16.400	17.5	17.5	89.4	5728	91.1	7914.0	91.1	7914.0
MANIPUR	0.0	0.0	0.0	0.000		0.0		0.0	

State	Area (in '000 ha.)			Production (Loose in 000 MT and Cut in lakh Nos)					
				2008-09		2009-10		2010-11	
	2008-09	2009-10	2010-11	Loose	Cut	Loose	Cut	Loose	Cut
MEGHALAYA									
MIZORAM	0.2	0.0	0.1	0.000	168	0.0	142.0	0.0	162.0
NAGALAND	0.0	0.0	0.0	0.000	17	0.0	17.0	0.0	17.0
ODISHA	5.65	7.1	7.4	23.4		25.3	5356.0	3.7	5911.0
PONDICHERRY	0.29	0.3	0.3	2.368		2.4		2.4	
PUNJAB	1.70	1.7	1.7	82.0		82.0		82.0	
RAJASTHAN	3.35	3.3	5.4	4.9		4.9		9.6	230.0
SIKKIM	0.15	0.2	0.2		66		200.0		
TAMILNADU	29.14	32.0	32.0	233.7		247.3		247.3	
TRIPURA									
UTTAR PRADESH	13.53	10.4	10.4	24.3	3467	17.6	2958.0	17.6	2958.0
UTTARAKHAND	0.85	1.3	1.3	0.6	2056	1.0	3414.0	2.3	3416.0
WEST BENGAL	21.07	21.9	23.1	52.008	21232	55.2	22170.0	59.2	23919.0
TOTAL	166.52	182.9	190.9	987.4	47942	1020.6	66671.4	1031.3	69027.4

Source: National Horticulture Board, Indian horticulture Database-2011.

State-Wise Area And Production of Flowers in India During 2011-12 to 2013-14

State	Area (in '000 ha.)			Production (Loose In '000 Mt And Cut In Lakh NOS)-Cut 2013-14 In '000 Mt					
	2011-12	2012-13	2013-14	2011-12		2012-13		2013-14	
				Loose	Cut	Loose	Cut	Loose	Cut
West Bengal	23.9	24	24.9	63.91	25042.1	65.14	25429.1	66.50	145.2
Karnataka	29.2	30	30.6	211.5	10388.0	207.5	9441.8	211.5	71.5
Odisha	7.5	8	7.4	26.1	6020.0	26.2	6040.0	37.4	57.4
Uttar Pradesh	14.5	16	16.6	27.05	4194.0	31.49	4908.0	32.16	54.1
Maharashtra	18.9	22	23.0	104.0	7914.0	119.0	7914.0	122.7	44.0
Telangana			6.9					40.7	35.1
Assam	0.0	2	3.0	0.0	0.0	11.7	3750.0	20.0	32.7
Andhra Pradesh	64.2	35	20.4	389.0	7099.4	224.4	6909.0	136.3	30.0
Tamilnadu	32.3	29	55.0	332.81	0.0	312.97	1168.0	343.65	12.9
Himachal Pradesh	0.9	1	0.8	35.3	1948.1	37.7	1760.3	28.1	12.4
Haryana	6.3	6	6.5	64.2	1269.5	64.7	1270.6	65.5	11.3
Jharkhand	1.6	2	1.6	22.0	1711.0	22.0	1711.0	22.0	9.5
Uttarakhand	1.5	2	1.4	1.81	3567.6	1.82	3633.0	2.02	8.5
Delhi	5.5	6	5.5	5.7	1038.0	5.7	1038.0	5.7	5.8
Bihar	0.9	1	0.8	8.7	1285.0	10.2	324.0	7.6	2.7
Meghalaya			0.1					0.0	2.4
Sikkim	0.2	0	0.2	25.95	209.1	26.50	214.1	16.00	1.9
Arunachal Pradesh	1.2	0	0.0	0.0	2860.0	0.0	297.0	0.0	1.9
Jammu and Kashmir	0.2	1	0.8	1.1	155.9	0.4	222.1	0.4	1.8
Mizoram	0.1	0	0.2	0.0	349.0	166.8	605.2	171.6	1.2
Nagaland	0.0	0	0.0	0.0	15.4	0.0	96.7	0.0	0.4
Goa		0	0.0					0.0	0.1

	Area (in '000 ha.)			Production (Loose In '000 Mt And Cut In Lakh NOS)-Cut 2013-14 In '000 Mt						
	2011-12	2012-13	2013-14	2011-12		2012-13		2013-14		
				Loose	Cut	Loose	Cut	Loose	Cut	
Manipur		0.0	0	0.8	0.0	0.0	0.0	0.0	0.3	0.0
Madhya Pradesh		15.6	17	17.1	150.7	0.0	193.0	0.0	200.4	0.0
Gujarat		16.0	17	17.3	135.5	0.0	149.3	0.0	163.6	0.0
Chhattisgarh		8.4	10	10.1	32.9	0.0	37.8	0.0	45.7	0.0
Punjab		2.1	2	1.4	10.1	0.1	10.5	0.0	10.5	0.0
Rajasthan		2.5	3	2.5	2.7	0.0	3.7	0.0	2.7	0.0
Puducherry		0.1	0	0.1	0.4	0.0	0.4	0.0	1.2	0.0
Andaman and Nicobar		0.0	0	0.1	0.3	0.0	0.4	0.0	0.3	0.0
Daman and Diu		0.0	0	0.0	0.0	0.0	0.0	0.0	0.0	0.0
Total		**253.6**	**232.6**	**255.0**	**1650.87**	**75066.0**	**1728.4**	**76731.9**	**1754**	**543**

Source: National Horticulture Board, Indian horticulture Database-2015.

Leading cut flower producing states (2013-14)

HIMACHAL PRADESH 2%
OTHERS 9%
TAMILNADU 2%
WEST BENGAL 27%
ANDHRA PRADESH 6%
ASSAM 6%
TELANGANA 6%
KARNATAKA 13%
MAHARASHTRA 8%
UTTAR PRADESH 10%
ODISHA 11%

India is a leading grower of roses. Karnataka continues to be the leader, accounting for more than 50 per cent of the natural rose production. Bangalore has around 35 floriculture units producing roses.

Karnataka used to have around 35 units earlier. By June 2003, their number came down to around 15. It is said that only those companies which have gone

Leading loose flower producing states (2013-14)

OTHERS 13%
TAMILNADU 19%
CHHATTISGARH 3%
HARYANA 4%
WEST BENGAL 4%
KARNATAKA 12%
MAHARASHTRA 7%
ANDHRA PRADESH 8%
MADHYA PRADESH 11%
GUJARAT 9%
MIZORAM 10%

through the bad phase and learnt cost saving methods and efficient marketing have survived. Karnataka produces 131,000 tonnes every year. In fact, Bangalore contributes 60 per cent of India's floriculture trade, with 900 tonnes of just-cut roses exported every year. Bangalore has about 140 hectares under cultivation for roses only. This segment is the most predominant cut-flower in the market, export or domestic. Pune, the second largest production centre in the country, cultivates only 50-60 hectares for roses.

Floriculture is at present confined to Karnataka, Tamil Nadu, Andhra Pradesh, West Bengal, Maharashtra, Rajasthan, Uttar Pradesh, Delhi and Haryana. However, in Maharashtra, Madhya Pradesh and Uttar Pradesh, there is much higher level of floricultural activities than what is indicated by the area **(Table 3.2)**. Of the total area, two-third is devoted to production of traditional flowers. Almost all of the area reported is under open-field flower production. Greenhouse cultivation of flowers in India is increasingly practiced for off-season production and also for improved quality. The annual turn over of the industry, at the national level, is around ₹ 10 billion. However, India's share in the global market at $ 40 billion is negligible. At present, the internal demand for cut-flower and ornamental plants is mainly met through production at coastal areas of Kerala, mainly Cochin and Thiruvananthapuram. There is a huge potential for growing these plants on commercial scale along the coastal region in Maharashtra, Andhra Pradesh, Tamil Nadu, and Karnataka, besides the Western Ghats.

Karuturi Net Works Company

Amudhagondapally in Hosur in Krishnagiri district (50km) of Bangalore is now a floriculture address in the state, from where 10 lakh roses flew to UK, Germany, Australia, Singapore and West Asia (February, 2007). The turnaround credit goes to Tanflora Infrastructure Park, which made flowers good business sense. The agropark projected an annual production capacity of 70 million roses in its 50 hectares. Grows a potential remains for another 30 million flowers. Small entrepreneurs of Thally, Berigai and Kelamangalam villages, which have the ideal climate and land for growing roses, can use this opportunity. In fact, our estimate is that we have 3,000 hectares lying vacant which can be used.

> *Karuturi Networks; a little known Bangalore company, is close to acquiring the Netherlands-based Sher, the world's largest producer and supplier of roses, to emerge as the global leader in the field. Sher's greenhouses in the Netherlands, Kenya and Ethiopia produce 600 million roses annually.*

While the Tanflora Park project was sanctioned in 1999, it took nearly seven years before the flowers began to bloom in its greenhouses. The joint venture project promoted by the state-run Tamil Nadu Industrial Development Corporation Ltd. (TIDCO) and the Bangalore-based MN Ahmed and Associates, has state-of-the-art centralized infrastructure facilities for postharvest. As of now, 25 growers have two hectares each on a cooperative farming pattern. The infrastructure is common to all-ranging from centralized water sourcing and supply, eco-friendly cooling facilities, free refrigeration equipment with dock shelters and bouquet making

Tanflora Park at Amudhagundapalli, a village in Tamil Nadu's hosur subdivision, is the biggest agri-export zone (AEZ) for floriculture. Ahead of Valentine's Day, roses have been exported from Amudhagundapalli to Europe, West Asia, Southeast Asia and Australia.

1 million

The number of roses (30 exotic varieties) sent to Europe, West Asia, Southeast Asia and Australia this week. (February 2007)

☆ *$678 million India's Flower exports in 2005-06*

☆ *$1 billion Projected flower exports by 2010*

☆ *$800 million India's total annual flower production.*

machines, logistics arrangement, marketing and refrigerated transportation. The future, say growers, is targeting buyers as far as Frankfurt, London, Sydney, Tokyo and Dubai. Overseas offices are also on the cards.

Once all the 25 units spread over 50 hectares, with an annual production capacity of 67.7 million flowers, are fully functional, the park's turnover will touch ₹50 crore. December to March is a good time for flower producers in India as roses grow only in temperatures above 12 degree centigrade. So, European countries with their chilly winters depend on flowers produced outside. Set up at about ₹22 crore, Tanflora aims to provide infrastructure for the floriculture industry (cut roses) by centralizing postharvest facilities. The park's contribution eventually will be about one fifth of the country's total production. About 25 greenhouses covering 50 per cent of the total area have already been constructed and the 10 lakh roses are being produced from about 15 greenhouses. While trial shipments began early last year, the 10 lakh roses leaving from February 3 up to February 10. is the largest consignment so far.

> *December to March is a good time for us because European countries with their chilly winters depend on flowers from outside.*

A day after Valentine's Day, there's news of a deal that'll make the global Indian takeover come out smelling of roses. Next Vallentine's day, when lovers and spouses gift roses across the world, chances are most of the flowers will have an Indian imprint. Karuturi networks, a little known Bangalore company, is close to acquiring the Netherlands-based Sher, the world's largest producer and supplier of roses, for about $50 million (₹220 crore) to emerge as the global leader in roses. Sher's greenhouses in the Netherlands, Kenya and Ethiopia produce 600 million roses annually.

MD of Karuturi Networks, confirmed that his company was about to wrap up to deal. The acquisition will be funded through internal accruals and the proceeds of a $25 million foreign currency convertible bond issue. UTI Bank and the London based Silverdale Services are involved in the deal. It all began on a Valentine's Day 12 years ago when Karuturi, a mechanical engineer, was sniffing around Bangalore for roses for his wife. Finding none, he decided to step out of the cable business he was running and plant high-value stem roses for a living. He set up Karuturi Networks, which now processes 12 million roses annually. Karuturi Networks has 60 hectares of greenhouses in India and Ethiopia for rose cultivation. Apart from the European company's facilities, Karuturi will get a strong brand in Sher.

Aravalli Biodiversity Park

When you stand on Delhi's Aravalli mountains, the oldest mountains in the world, you are standing on a piece of geological history. Now, couple the undulating red terrain with the rare orchids from Kumaon and Western Ghats, and you have Delhi's first orchidarium. But this is just one component of an exciting education-cum-tourism destination at the Aravalli Biodiversity Park, between Vasnt Kunj and Vasant Vihar. Developed by the Centre for Management of Degraded Ecosystems (CEMDE), Delhi University, an orchidarium, a butterfly park, a fernery, a medicinal garden and a wetland have just been completed. The Aravalli Biodiversity Park has been built on 692 acres skirting Jumbo Point and attempts to recreate small ecosystems as well as preserving the natural vegetation of the Aravallis and the Delhi Ridge.

The area had highly degenerated soil, thanks to its original usage – mining Badarpur sand, mica and china clay. Work is now in progress to create a small tropical rainforest as well, in time for the 2010 Commonwealth Games. We have used the depressions created by the extensive mining to create small ecosystems. Orchids, for instance, reuire 80 per cent relative humidity to grow. For this, we have made use of an Israeli technique; mist irrigation, which propels very fine jets of water through motors on plants. This creates and maintains a very humid climate.

The orchids grown at the park include vanilla, wanda and the butterfly orchids. Next, the park is trying to

> **Floral Facts**
>
> ☆ The orchids at Delhi's florist shops do not come from North India but are usually grown in Bangalore
>
> ☆ With over 1,100 species growing in India, the orchid family is the largest flowering family in the country: found naturally in inaccessible areas in the Himalayas, the Western Ghats and the Andamans
>
> ☆ Most popular species worldwide are the Vanilla orchids (pure Vanilla is extracted from it) and the Blue Wanda.

obtain orchid species endemic to the Andaman and Nicobar Islands. The CEMDE is also actively replanting Aravalli's native vegetation at the site. The area was infested with the exotic Vilayati Kikar, which has become a major problem in Delhi. We are replanting the park with native vegetation. We have grown medicinal plants, which are endemic to the Aravallis but the becoming extinct in the wild, like the Jangali Mali, which is used to make brain tonic, and the Shikakai, which is grown naturally only on Mount Abu. At the fernery, we have grown 70 species, which first appeared on Earth 400 million years ago and provided food for dinosaurs. The species grown here will give an idea of Indian varieties of fern, among other varieties.

Shades of Holland in India

About 50 different varieties of colourful tulip flowers are in full boom at Siraj bagh, known as Asia's largest tulip garden, that is being projected as an added attraction for both the tourists to the valley and the people of Kashmir. Already, the department has imported over 3.50 lakh tulip bulbs from Holland and plans are a foot to import 10 lakh bulbs by the next season 2008 for expansion purposes. With the tulips first to bloom in spring, other varieties of flowers are being introduced across the valley. The Asia's largest tulip garden will remain in full bloom. This is spread over 570 kanals, with tulips grown in bulk for the first time in 2007 spread over 35 kanals. Other varieties of flowers are being developed in the remaining portion, while the tulips would be grown mostly over the garden in the coming years. The Floriculture Department also plans to develop private growers for tulip culture so as to make more and more areas under its cultivation and preservation of the bulbs, which otherwise have to be imported from Holland.

Protected Cultivation

Several export oriented enterprises have come up in the country since mid 1990s, and approximately 200 hectares area has been brought under protected cultivation, *i.e.* plastic greenhouses, for growing cut flowers. These units are mainly export oriented, with nearly 60 to 80 per cent of their production being destined for export. Mention may be made in this context that most important cut flower under protected cultivation is rose, and large headed Hybrid-T roses are cultivated in polyhouses. Initially imported technology was used for cultivation, but farmers have now modified techniques more suitable to Indian conditions. All together 44 cultivars are grown in various greenhouses.

Low Tech Producers

Indian flower producers are generally very small belonging to the category of 1-4 hectare units. According to a rough estimate, there are hundreds and thousands of such growers, cultivating on a small piece of land and growing different varieties of traditional flowers (loose flowers) and occasionally stem flowers. The produce is grown outdoors, in the open. They normally do not possess sophisticated equipment and their families have perhaps been working in the same way for several generations.

Advantage India

Oriented Status

India has, in fact, all the favourable conditions for becoming the hub of world floriculture. However, despite having the most impressive varied agro-climatic profile, vast land resources, and the availability of abundant labour and agricultural scientists. India's contribution towards the global floriculture market is negligible. The annual demand for flowers domestically is growing at a rate of over 25 per cent and the international market at almost ₹90000 crore, but the supply is less than half of the demand. The recent past has been a shift from traditional flowers to cut flowers for export purposes. The liberalized economy has somewhat helped the Indian entrepreneurs, for establishing export oriented floriculture units under controlled climatic conditions. Interestingly, in India the most promising area in the industry is the dry flower segment. Dried flower and plants have been exported for the last 30 years, and today, India is one of the leading countries in the field. In Rajasthan, more than 7,000 tonnes of rose are produced, and about 75 per cent of this are exported to West Asian countries in the form of dry petals. Apart from flowers, parts of plants like leaves, stems and pods are also used in the dry flower industry.

Recently large-scale commercial companies have started investing in the sector. Many are also considering joint venture agreement with foreign companies, with the aim of producing and supplying high quality flowers both for local and international markets throughout the year.

India enjoys some advantage in respect of floriculture. First of all, there is cost advantage. In Holland, which is the main competitor for India, floriculture is done under controlled conditions like glasshouses to obtain products of the highest standard. However, in India, the cost of running glasshouses is cheaper than in Holland. For instance, in a place like Bangalore, the climate does not require lighting, cooling or heating system. Also, India has a large number of agricultural universities which produce enough qualified engineers. Besides potted plants, Indian roses, carnations, orchids and chrysanthemums are important cult flowers. Of late, illium and other bulbous plants are becoming important for cut-flower trade.

Popular Types

India has many exclusive varieties of ornamental flowers, which makes it an ideal destination for the floriculture industry. Indian flowers are exported to world over in the form of seeds or capsules, apart from dried flowers. Rose, marigold, Chrysanthemum and Jasmine are some of the mostly sold varieties, and about 10,000 hectares of the total cultivated area is devoted to modern flowers like rose, carnation, orchid, *etc.*

Requirements

It all looks colourful enough, but the blooms are not easy to appear. It is a lot of hard work, including the scientific management of soil preparation, administration of nutrients and controlling climate. Collecting and distributing the flowers without damaging the beauty, grace and fragrance is challenging indeed. Careful packaging for international markets in a way that best ensures a long shelf-life is most important.

Growth Potential

Floriculture has an annual growth potential of 25 to 30 per cent. Of late, large scale commercial companies have started joint ventures with foreign companies to invest in floriculture sector. Some major corporate leaders, namely TATAs, Birlas, Thapars, MRF Group and R.P. Goenka Group, have taken a lead [2]. Many non-traditional states have also been interested in floriculture, for example, floriculture has gained momentum in Haryana. The state achieved a record production of 1,200 lakh cut flowers and 32,500 tonnes of flowers valued at ₹30 crore during 2002-03 financial year. The State Director of Horticulture reported that the area under cultivation of commercial flowers had risen to 3,600 hectares during the said year and the State Horticulture Department proposed to increase further this to 3,800 hectares.

Cultivation of commercial flowers is picking up rapidly in Haryana state and the demand for flowers and cut flowers have increased immensely. The soil and climate of the state is quite suitable for growing flowers like gladiolus, roses, marigold, tuberoses, chrysanthemum, garbera, carnation and lillium. Farmers of the State, especially those based in Sonipat, Gurgaon, Faridabad and Karnal have taken up cultivation of flowers in a big way as there is a good market in Delhi. Farmers of Yamunanagar and Panchkula have also switched over to cultivation of flowers because of the lucrative earning potential. In fact, presently farmers are earning up to ₹1.25 lakh per acre from flowers.

The Horticulture Department has established a model floriculture centre at Karnal to give impetus to the cultivation of commercial flowers with the state-of-the-art technology. Further, a hi-tech poly greenhouse has been established at Uchani in which high quality lillium flowers and other varieties are being grown. Regular training and seminars on floriculture are being organized at this centre, and technical know-how is also being provided by the exporters to the farmers.

Cut Flower Production in India

India has a long tradition of floriculture. References to flowers and gardens are found in ancient Sanskrit classics like the Rig Veda (C 3000-2000 BC), Ramayana (C 1200-1300 BC), Mahabharata (prior to 4th Century BC), Shudraka (100 BC), Ashvagodha (C 100 AD), Kalidasa (C 400 AD) and Sarangdhara (C 1200 AD). The social and economic aspects of flower growing were, however, recognized much later. The offering and exchange of flowers on all social occasions, in places of worship and their use for adornment of hair by women and for home decoration have become an integral part of human living. With changing life styles and increased urban affluence, floriculture has assumed a definite commercial status in recent times and during the past 2-3 decades particularly. Appreciation of the potential of commercial floriculture has resulted in the blossoming of this field into a viable agri-business option. Availability of natural resources like diverse agro-climatic conditions permit production of a wide range of temperate and tropical flowers, almost all through the year in some part of the country or other. Improved communication facilities have increased their availability in every part of the country. The commercial activity of production and marketing of floriculture products is also a source of gainful and quality employment to scores of people.

Present Situation of Cut Flower Production

Inspite of the long and close association with floriculture, the records of commercial activity in the field are very few. The information on the area under floriculture and the production generated is highly inadequate. As commercial floriculture is an activity which has assumed importance only in recent times, there are not many large farms engaged in organised floriculture. In most part of the country flower growing is carried out on small holdings, mainly as a part of the regular agriculture systems.

Production Areas

The estimated area under flower growing in the country is about 255,000 hectares. The major flower growing states are Karnataka, Tamil Nadu and Andhra Pradesh in the South, West Bengal in the East, Maharashtra in the West and Rajasthan, Delhi and Haryana in the North. It must, however, be mentioned that it is extremely difficult to compute the statistics of area in view of the very small sizes of holdings, which very often go unreported. This perhaps would be the reason for unrealistically small areas reported for floriculturally active states like Maharashtra, Uttar Pradesh and Madhya Pradesh.

More than two thirds of this large area is devoted for production of traditional flowers, which are marketed loose *e.g.* marigold, jasmine, chrysanthemum, aster, crossandra, tuberose *etc.* The area under cut flower crops (with stems) used for bouquets, arrangements *etc.* has grown in recent years, with growing affluence and people's interest in using flowers as gifts. The major flowers in this category are rose, gladiolus, tuberose, carnation, orchids and more recently liliums, gerbera, chrysanthemum, gypsophila *etc.*

The production of flowers is estimated to be nearly 1.75 million tonnes of loose flowers and over 500 million cut flowers with stem. In the case of production also, the estimates could be at variance from the actual figures as some of the flowers like rose, chrysanthemum, and tuberose are used both as loose flowers and with stem.

It may be mentioned that almost all of the area reported here is under open field cultivation of flowers. Protected cultivation of flowers has been taken up only in recent years for production of cut flowers for exports. The estimated area in production is about 200 hectares, which is likely to increase to over 500 hectares by the year 2020.

Recognising the potential for low cost production for export, in view of cheap land, labour and other resources, several export oriented units are being set up in the country. These projects, located in clusters around Pune (Maharashtra) in the West, Bangalore (Karnataka) and Hyderabad (Andhra Pradesh) in the South, and Delhi in the North, are coming up in technical collaboration with expertise mainly from Holland and Israel. More than 90 per cent of these units are for rose production, on an average size of 3-hectare farm, while some projects for orchid, anthurium, gladiolus and carnation are also being set up. Nearly one third of over 200 proposed projects, have already commenced production and export.

Major Cut Flower Crops

Rose is the principal cut flower grown all over the country, even though in terms of total area, it may not be so. The larger percentage of the area in many states is used for growing scented rose, usually local varieties akin to the Gruss en Tepelitz, the old favourite to be sold as loose flowers. These are used for offerings at places of worship, for the extraction of essential oils and also used in garlands. For cut flower use, the old rose varieties like Queen Elizabeth, Super Star, Montezuma, Papa Meilland, Christian Dior, Eiffel Tower, Kiss of Fire, Golden Giant, Garde Henkel, First Prize *etc.* are still popular. In recent times, with production for export gaining ground in the country, the latest varieties like First Red, Grand Gala, Konfitti, Ravel, Tineke, Sacha, Prophyta, Pareo, Noblesse. Virsilia, Vivaldi *etc.* are also being grown commercially.

Gladiolus is the next most important cut flower crop in the country. Earlier it was considered a crop for temperate regions and its growing was restricted to the hilly areas, particularly in the north eastern region, which still continues to supply the planting material to most parts of the country. However, with improved agronomic techniques and better management, the northern plains of Delhi, Haryana, Punjab, Uttar Pradesh, as well as Maharashtra and Karnataka have emerged as the major areas for production of gladiolus.

Tuberose, a very popular cut flower crop in India is grown mainly in the eastern part of the country *i.e.* West Bengal, and also in northern plains and parts of south. Both single and double flower varieties are equally popular. Tuberose flowers are also sold loose in some areas for preparing garlands and wreaths.

The other main cut flower item is orchid. Its production is restricted mainly in the north-eastern hill regions, besides parts of the southern states of Kerala and Karnataka. The main species grown are Dendrobiums, Vanda, Paphiopedilums, Oncidiums, Phalaenopsis and Cymbidiums.

Among the traditional crops grown for loose flowers, the largest area is under marigold, grown all over the country. In most parts of the country only local varieties are grown for generations. African marigolds occupy more area as compared to the small flowered French types. Jasmine flowers in view of its scent are also very popular as loose flowers and for use in garlands and Veni (ornament for decoration of hair by women). The major areas under this crop are in Tamil Nadu, Karnataka in South and West Bengal in East. The varieties are mainly improved clones of *Jasminum grandiflorum, J. auriculatum and J. sambac.* The chrysanthemum, particularly the white varieties are much in demand as loose flowers during the autumn period of October-December when other flowers like jasmine, tuberose are not available for use in garlands *etc.* Among other traditional flowers grown in large areas are crossandra in southern states of Tamil Nadu, Karnataka and Andhra Pradesh and aster in Maharashtra.

Research Support

Research work on floriculture is being carried out at several research institutions under the Indian Council of Agricultural Research and Council of Scientific and Industrial Research, in the horticulture/floriculture departments of State Agricultural Universities and under the All India Coordinated Floriculture Improvement Project with a network of about twenty (20) centres. The crops which have received larger attention include rose, gladiolus, chrysanthemum, orchid, jasmine, tuberose, aster, marigold *etc.* The thrust till recently had been on crop improvement, standardization of agro-techniques including improved propagation methods, plant protection and postharvest management. In view of the fact that most of the cut flower production is being done under open field conditions, the research efforts generally relate to open cultivation. In recent years, however, technologies for protected cultivation and tissue culture for mass propagation have also received attention. A large number of varieties suitable for cut flower use, as well as garden display have been developed. Production technology, particularly the agronomic requirements and control methods for important diseases and insect pests have also been developed. Contribution by the private sector in research activities in floriculture is negligible.

Planting Material

The requirement of planting material to cater to the large area under flower crops, is largely met from domestic production. Since efforts to set up large commercial farms generally suffered due to lack of quality planting material in sufficient quantities, this aspect has received greater attention in recent years in the breeding centres, which are producing sufficient quantity of planting material. Most of the nurseries propagating planting material are in the private sector. In the absence of any mechanism to register nurseries, it is very difficult to ascertain their exact number, but at a very conservative estimate there are more than 100,000 nurseries, spread out all over the country, producing seeds and other planting materials for flower growers. The states with larger numbers of nurseries include Maharashtra, West Bengal, Karnataka and Tamil Nadu. Most of the nurseries are small, with little or no improved facilities like mist propagation unit, greenhouses/ net houses *etc.* For meeting the demand of flower seeds, several large seed companies have production units in Punjab, Himachal Pradesh and Jammu and Kashmir in the North, Karnataka in the South and West Bengal in the East. A few of the leading multinational seed companies have tied up with local seed companies or producers for custom production of seeds of their varieties. In the case of bulbous plants, most of the planting material is produced in the north eastern hilly regions of West Bengal (Kalimpong) and Sikkim, though for some crops, it is also produced in hilly regions of northern India. The introduction of a revised seed policy by the government of India in 1989 has enabled unrestricted introduction of many new and superior varieties into the country, increasing the variety in the floral basket.

Tissue culture has, in recent years, been recognized as an important tool in agriculture development. With its diverse climatic zones and qualified manpower, India is well placed to exploit the benefit of tissue culture based applications to floriculture crops. Most popular application of tissue culture has been micropropagation using in vitro technique for mass multiplication of planting material. Tissue culture plants of ornamentals have found ready acceptance by the commercial growers and their production increased significantly from 130 million plants in 1985-86 to 680 million in 1994-95. At present 30 commercial tissue culture units with annual capacities of 0.5 to 15 million plants each are in operation, resulting in total capacity of about 110 million plants. While most of it is exported, a small percentage of cut flower crops like carnation and gerbera are finding good market within the country.

Marketing

Marketing of cut flowers in India is very unorganised at present. In most metropolitan cities, with large market potential, flowers are brought to wholesale markets, which mostly operate in open yards. A few large flower merchants generally buy most of the produce and distribute them to local retail outlets after significant mark up. The retail florist shops also usually operate in the open on-road sides, with different flowers arranged in large buckets. In the metros, however, there are some good florist show rooms, where flowers are kept in controlled temperature conditions, with considerable attention to value added service. The government is now investing in setting up of auction platforms, as well as organized florist shops with better storage facilities to prolong shelf-life.

The packaging and transportation of flowers from the production centres to the wholesale markets at present is very unscientific. The flowers, depending on the kind, are packed in old gunny bags, bamboo baskets, simple cartons or just wrapped in old newspapers and transported to markets by road, rail or by air. The mode of transportation depends on the distance to the markets and the volume. Mostly, flowers are harvested in the evening time and transported to nearby cities by overnight trains or buses. In recent years, the government has provided some assistance for buying refrigerated carriage vans. A large number of export oriented units have built up excellent facilities of pre-cooling chambers, cold stores and reefer vans and their produce coming for domestic market sales are thus of very good quality and have longer vase life and command higher price. The government programmes for floriculture development include creating common facilities of cool chain in large production areas to be shared on cooperative basis. Formation of growers' cooperatives/associations are being encouraged.

In view of the unorganized set up, it is difficult to estimate the size of flower trade, both in terms of volume and value. A study conducted in 1989 estimated the trade to be worth ₹ 2050 million. It is in the period of the last five years or so that this business has really boomed in India, which is reflected in the number of new florist outlets in all cities and increase in the public's purchase of flowers as gifts. This would put the current trade at several times the earlier estimate. A recent study of Delhi market alone put the value of flowers traded on wholesale as ₹ 500 million.

The loose flowers (traditional crops like marigold, jasmine *etc.*) are usually traded by weight. The average price of different flowers in major markets varies considerably depending on the period of availability (Table 3.3).

The net returns to the growers depend on the packaging and transportation costs. The cut flowers with stem have a limited overall market in terms of volume. The share of cut flowers has almost doubled from 30 to 60 per cent in the last decade.

Table 3.3: Average Market Price for Major Flower Crops

Flowers	Unit	Price (US$1 = ₹40) ₹/kg or doz or each stem
Marigold	kg.	3-60
Jasmine	kg.	15-150
Crossandra	kg.	20-120
Chrysanthemum	kg.	5-25
Tuberose	kg.	5-30
Rose	kg.	6-60
Gladiolus	doz.	20-75
Carnation	doz.	30-75
Gerbera	doz.	36-75
Orchids	each stem	10-45
Liliums	each stem	10-45
Anthuriums	each stem	15-45

The value of cut flower export from India has increased twenty five fold during the last five years (Table 3.4). With more export oriented units coming into operation, exports are likely to grow further in the coming years. The major share of the export trade is for roses, in addition to orchids, gladiolus *etc.* The major markets are Europe (Holland, Germany and U.K.) and Japan. The exports of roses to Japan, have really picked up in the three years from ₹ 360 million in 1993-94 to ₹ 6090 million in 1995-96. As per the estimates for 1996-97, India has been the largest supplier of roses to Japan (volume wise).

India

Some recent data for cut-flower production in India are tabulated below (Table 3.4). They will look unfamiliar to European growers with the importance of marigold and the sale of cut-flowers 'loose', both connected to the floral decorations that play a large part in Hindu festivals such as Diwali. Marigold and jasmine production currently occupy about 40ha each, and are sold 'loose'. Rose, chrysanthemum, gladiolus and tuberose are also important, with about 70ha in all and generally on the increase, and only chrysanthemum being sold loose in any quantity. Anthurium, carnation, gerbera and tulip are grown in small quantities and are sold conventionally. The production of 'other' cut-flowers is extensive and makes up some 40 per cent of the total area, being sold both conventionally and 'loose', mostly the former.

No trends can be described based on only two years' data, but there is a suggestion of a slightly expanding production area with somewhat lower yields.

**Table 3.4: Production Areas and Quantities of
Cut and 'Loose' Flowers in India, 2011/12-2012/13**

	Total Area (1,000ha)		Cut-flowers (millions)		Loose Flowers (1,000t)	
	2011/12	2012/13	2011/12	2012/13	2011/12	2012/13
Total of which:	234	254	7,673	7,507	1,729	1,651
Marigold	43	44	0	0	360	382
Jasmine	10	42	0	0	51	207
Rose	28	28	1,990	2,740	76	66
Chrysanthemum	18	19	3	148	176	195
Gladiolus	9	12	707	1,061	12	11
Tuberose	8	12	156	1,401	28	39
Anthurium	0	0	32	12	0	0
Carnation	0	0	15	37	0	0
Gerbera	0	0	25	32	0	0
Tulip	0	0	5	4	0	0
'Other'	116	97	4,741	2,072	1,027	751

Potential for Cut Flower Production Development

The availability of natural resources like favourable and diverse climatic conditions permit production and availability of a large variety of flower crops round the year. Cheap labour leads to reduction in production costs, increasing access of the consumer to good quality flowers at affordable prices, besides increasing our competitiveness in the export markets. Being a new concept in the agri-business, it took some time for scientific commercial flower production to take roots, but with the appreciation of its potential as an economically viable diversification option, its growth is slowly stabilising. The government also has, during the last few years, recognized floriculture as an important segment for developmental initiatives. Model Floriculture Centres being set up in 11 major production zones, to serve as focal units for development in the region, have a mandate of making available quality planting material, new/improved production technologies and also to provide training in production and postharvest management. There are also special government programmes for area expansion in floriculture with state assistance. The National Horticulture Board, a major developmental agency for horticulture, also makes available finances as soft loan for setting up integrated projects for production and marketing. As mentioned earlier, the government is investing in improving the infrastructure for marketing in the domestic sector.

Production of cut flowers for exports is also a thrust area for support. The Agricultural and Processed Food Products Export Development Authority (APEDA), the nodal organization for promotion of agri-exports including flowers, has introduced several schemes for promoting floriculture exports from the country. These relate to development of infrastructure, packaging, market development, air freight subsidy *etc.* The 100 per cent Export Oriented Units are also given benefits like duty free imports of capital goods.

All these efforts indicate the government's commitment for improving the sector and creating a positive environment for entrepreneurship development in the field.

Constraints in Cut Flower Production Development

Being a new concept, the requirements of scientific and commercial floriculture is not properly understood in the country. The developmental initiatives of the government have to keep in mind the low knowledge base, small land holdings, unorganized marketing and poor infrastructural support.

While long experience of flower growing in the open field conditions enable sufficient flower production for domestic markets, the quality of the produce, in view of its exposure to various kinds of biotic and abiotic stresses, is not suitable for the ever growing export market. The production technology for flowers under protected environment of greenhouses needs to be standardized. There is hardly any postharvest management of flowers for the domestic market. Availability of surplus flowers from exports for sale in the domestic market, has increased the appreciation of quality produce and the demand for good quality flowers is increasing. With the introduction of new varieties of crops in the country, facilities for generating their planting material for large scale production need strengthening. Special attention needs to be paid to strengthen the marketing infrastructure like organised marketing yards, auction platforms, controlled condition storage chambers *etc.*

Greater research efforts are also needed for integrated pest management, development of location specific package of practices for traditional flowers, value addition to traditional flowers *etc.* The initial cost and availability of finance is a critical matter in the development of large commercial projects requiring heavy investments. More options for developmental finance, such as the soft loan scheme of the National Horticulture Board need to be identified. In the initial years of commercial floriculture development, the governmental support in terms of subsidies *etc.* needs special attention.

The potential for growth of export market is always linked to the strength of domestic market - its capacity to absorb surplus and over production, and quality consciousness of consumers. Though we have a large domestic market, the marketing system and facilities need to be modernized.

The production for exports at present has suffered due to a few constraints. While our growers have been successful in producing world class quality at low cost, high air freight rates, low cargo capacity available, imposition of import duties, inadequate export infrastructure *etc.* have reduced their competitiveness.

There is also a shortage of trained manpower to handle commercial floriculture activity. The demands of the growing export oriented industry would require adequate attention to be paid for human resource development, particularly at the supervisory level.

Table 3.5: Cut Flower Exports from India

Country	1993-94	1994-95	1995-96
Japan	322.50	8255.58	35932.56
Netherlands	1004.61	9102.49	24799.90
U.S.A.	1175.38	2495.21	17652.50
Germany	957.61	2538.63	9256.00
U.K.	1420.93	1113.78	3345.86
U.A.E.	2120.40	3388.19	2459.29
Italy	210.28	164.96	2200.49
Hongkong	730.02	903.15	1504.02
Singapore	78.45	437.37	1190.54
Nepal	11.86	36.09	292.64
Kuwait	24.22	5.00	274.13
Saudi Arabia	413.98	169.52	272.77
Switzerland	136.96	258.20	242.83
Hungary	-	286.69	181.82
Thailand	-	86.49	177.82
Australia	-	-	132.89
Russia	368.46	20.96	119.85
Others	988.68	719.39	9293.82
Total	9964.34	29981.60	109329.73

The Cut Flower Trade

Worldwide

The cut flower trade is a multibillion dollar world industry. Flowers are grown in many countries and, due to their perishability, must be rapidly transported by air often to far-flung destinations. The Netherlands largely dominates this industry; more than 60 per cent of the international trade in cut flowers is conducted from there, much of it at the flower auctions. Although the Netherlands is itself a major producer of cut flowers, some of the flowers it exports are imported to it from other countries and pass through the auctions before heading to their export destinations. In 2003 Kenya was by far the biggest supplier to the Netherlands auctions (Table 3.6).

Kenya and Zimbabwe are the leading flower exporters in Africa. South Africa, Uganda, Tanzania and Zambia are also major producers, while some other African countries export a much smaller volume.

Two options are open to African growers: to directly market their flowers to consumer countries, or export them through the Netherlands auctions. The more-developed cut flower operations (such as Kenya's) have the means to directly market some of their product, most of which is exported to Europe and the U.S. In less-developed African countries, however, such as Tanzania, where most of the growers are small-scale, the Netherlands auctions take 90 per cent of the flower production.

Table 3.6: Turnover Realized in Imported Cut Flowers

Country	Turnover (in euros) 2003
Kenya	160,553,000
Israel	119,503,000
Zimbabwe	50,380,000
Ecuador	26,379,000
Uganda	13,797,000
Other countries	89,498,000
Total	460,110,000

Source: VBN (Dutch Flower Auctions Association)

Africa

Just a few flower types dominate African flower exports. Roses make up 70 per cent to 95 per cent of the exports in some African countries, particularly the less-developed ones. Other top flowers exported are *Dendranthema* (chrysanthemums), *Dianthus* (carnations), and *Limonium* (statice). A broad range of other flowers (not just those native to Africa) are grown in smaller quantities.

United States imports

The U.S., along with Germany, the Netherlands, the United Kingdom, Switzerland, Italy, France, and Japan are the major world consumer markets for cut flowers. Asia, particularly China, is expected to become the largest consumer market in the world within the next few decades. In the U.S. alone, more than $13 billion retail worth of flowers are sold annually. Although the U.S. itself produces flowers, over half of the flowers sold are imported (Table 3.7).

Table 3.7: Top Eight Countries Exporting to the U.S. in 2004

Country	Percent of Total Imports
Mexico	24
Netherlands	10
Thailand	8
Colombia	6
Ecuador	5
Costa Rica	4
Australia	2
Israel	2
Total of top 8	61

Source: USDA/APHIS.

Cut flower imports directly from Africa account for about 0.5 per cent of the total. Another way that African flowers come to the U.S., however, is through the

Netherlands auctions. Other countries, such as Mexico, also import and reexport flowers to the U.S.

Miami International airport handles over 85 per cent of the country's cut flower imports. Most of these shipments come from South America, and consist primarily of roses and relatively few other major flower types. A larger variety of flowers, including flowers from Africa, arrive at New York's JFK airport, Chicago O'Hare and other airports. Customs and Border Protection (CBP) agricultural specialists inspect all shipments of imported cut flowers for pests and diseases at U.S. air and border ports. Pests and diseases carried by cut flowers are a concern because if they were to become established here in the U.S., they could seriously harm American agriculture and the environment, particularly the domestic floriculture and nursery industries.

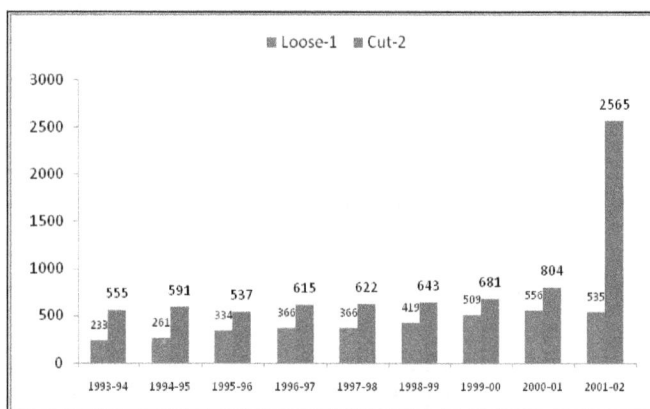

Production Loose '000 mt. Cut in Million Number.

Conclusions

India has a long floriculture history and flower growing is an age old enterprise. What it has lacked is its commercialization. The growing demands of flowers in the domestic as well as the export market will require a concerted effort on the part of the government as well as the private entrepreneurs to develop floriculture on scientific lines. Paying attention to the input needs, better resource management and making various policies entrepreneur friendly would lead to a balanced growth of the industry.

Floriculture Development in States

Floriculture could be considered as the most colourful sector of horticulture, which includes flowers, foliage, potted plants, ornamentals and greens. With urbanization and increase in disposable income level, the demand for floriculture products has increased significantly. As a result, there has been an increasing demand for cut flowers like rose, gladiolus, carnation, orchid, gerbera, lilium *etc*. There is a equally good demand for the traditional flowers like jasmine, marigold, chrysanthemum, tuberose *etc*. This has led to the transformation of floriculture sector from household activity to a commercial venture.

The total world area under floriculture is 6,20,000 hectares in which Asia -Pacific occupies 4,53,000 ha (nearly 73 per cent). India occupies 51 per cent of area under floriculture in Asia – Pacific region. (AIPH/Union Fleurs International statistics flowers and plants 2013). World floriculture market is growing significantly at the rate of 10-15 per cent per annum, estimated to be worth over $17 billion (Floriculture Today). In this, fresh cut flowers and foliage contributes 49.1 per cent (US$ 83.1 billion) and live plants, bulbs and cuttings contribute to 50.9 per cent (US$ 8.60 billion). Developed countries in Europe, America and Asia account for more than 90 per cent of the total world trade in floriculture products. Netherlands continues to dominate the world floriculture industry, accounting for 60 per cent (US$ 4.73 billion) of world floriculture exports in 2013. (COMTRADE, United Nations).

Major exporting countries of floricultural products are Netherlands, Germany, Italy, Belgium, Denmark, USA. Major importing countries in floriculture are Germany, France, Netherlands, USA, United Kingdom, Italy, Belgium, Switzerland, Austria, Japan. In India area under cultivation of flower crops was 2,33,000 ha with a production of 76,732 lakh of cut flowers and 1.72 million tones of loose flowers in 2012-13. The country has exported 27,121 MT of floriculture products to the world to the worth of ₹ 423.44 crore during the year 2012-13. Major Export destinations (2012-13) were United States, Germany, United Kingdom, Netherland, and United Arab Emirates.

In India, Maharashtra, Karnataka, Andhra Pradesh, Haryana, Tamil Nadu, Rajasthan, West Bengal have emerged as major floriculture centers. West Bengal (33.1 per cent), Karnataka (12.3 per cent) and Maharashtra (10.3 per cent) are the leading cut flower production states. Tamil Nadu (18 per cent), Andhra Pradesh (12.98 per cent) and Karnataka (12 per cent) are major loose flower production states. (NHB data base – 2013). Jammu and Kashmir could play a vital role in this field Floricultural exports from India comprises of fresh cut flowers to Europe, Japan, Australia, Middle East and USA, loose flowers to the Gulf, cut foliage to Europe, dry flowers to USA, Europe, Japan, Australia, Far East and Russia and potted plants limited to very few countries. Dry flower and ornamentals have great export and potentiality as nearly 70 per cent of total export of floricultural commodities from India consists of dried products.

The floriculture exports from India dropped marginally in value terms, due to local consumption of flowers growing at a phenomenal speed, high import tariff, low availability of dedicated perishable carriers, high freight rates. The industry is also faced by several challenges at the production level mostly related to availability of basic inputs, including seeds and planting materials, quality irrigation and skilled manpower, ageing plantations *etc.* At the marketing stage, major challenges faced by the Indian flower exporters are related to low level of product diversification and differentiation, lack of integration and innovation, low quality flowers and challenges associated with quality and environmental issues. Inadequate cold chain management is not only affecting the future floriculture trade in the country, but also having a negative impact on the present produce and on its marketability.

Strategies for Improvement

☆ The first and foremost requirement is to develop a Integrated Cold Chain for flowers right from the point-of-origin (growers) to the point-of-consumption (customers)

☆ Developing the non-traditional production areas may help in meeting the growing needs of both domestic and international market

☆ Develop and propagate varieties indigenously to ensure regular supply of quality inputs such as planting material

☆ Establish training centers and organize programmes for development of skilled manpower Establishing a network of support systems with the involvement of government, private sector and public institutions

☆ Government may enhance its efforts in negotiating preferential tariff regimes with countries to reduce import tariff

☆ Periodic re-plantation

☆ Shifting to Integrated Supply Chain Model *i.e.* integrating small and medium scale growers into large-scale producer supply chains, may help in attaining economies of scale in the industry

☆ Increasing the frequency of International flights and chartered flights besides, creating additional cargo space, specific to floriculture

☆ A lot has been done but we have miles to go before we can boast of achievements in the floriculture sector. As of now, all we can affirm is that, slowly but steadily India is moving towards a color Revolution.

Flowers are high value commodities used in various ways in domestic and social activities and in industries such as essential oils, dry flowers, natural dye extraction *etc.* Cultivation of flowers provides opportunity to farmers to earn a better livelihood and harvest more profit per unit area. Floriculture has received considerable interest in India in recent years from the policymakers, researchers, agricultural and horticultural planners. The enhancement in per capita income and urbanization have led to increased demand for flowers. A provisional estimate of National Horticulture Board puts the area under flower crops at 2,42,000 hactare with a production of 1847 MT of loose flowers and 79432 lakh cut flowers during 2013-14.

List of Indian State Flowers

India, officially the **Republic of India** is a country in South Asia. It is made up of 29 states and 7 union territories. All Indian states have their own government and Union territories come under the jurisdiction of the Central Government. As most of the other countries India too has a national emblem-The lion capital. Apart from India's national emblem, each of its States and Union Territories have their own state seals and symbols which include state animals, birds, trees, flowers *etc.* A list of state flowers of India is given below.

State	Common Name	Scientific Name
Andhra Pradesh	Lotus	*Nelumbo nucifera*
Arunachal Pradesh	Lady's slipper orchid	*Cypripedioideae*
Assam	Kopou phul	*Rhynchostylis gigantea*
Bihar	Genda	*Calendula officinalis*
Chhattisgarh	Not declared yet	*Not declared yet*
Goa	Not declared yet	*Not declared yet*
Gujarat	African Marigold	*Tagetes erecta*
Haryana	Lotus	*Nelumbo nucifera*
Himachal Pradesh	Common rhododendron	*Rhododendron ponticum*
Jammu and Kashmir	Common rhododendron	*Rhododendron ponticum*
Jharkhand	Palash	*Butea monosperma*
Karnataka	Lotus	*Nelumbo nucifera*
Kerala	Golden shower tree	*Cassia fistula*
Madhya Pradesh	Lilium candidum	*Lilium candidum*
Maharashtra	Jarul	*Lagerstroemia speciosa*
Manipur	Siroi lily	*Lilium mackliniae*
Meghalaya	Lady's slipper	*Cypripedioideae*

State	Common Name	Scientific Name
Mizoram	Red Vanda	*Renanthera imschootiana*
Nagaland	Rhododendron	*Rhododendron arboreum*
Odisha	Ashoka	*Saraca asoca*
Punjab	Lilium candidum	*Lilium candidum*
Rajasthan	Rohira	*Tecomella undulata*
Sikkim	Noble orchid	*Cymbidium goeringii*
Tamil Nadu	Glory lily	*Gloriosa superba*
Telangana	Ranawara	*Senna auriculata*
Tripura	Nag Kesar	*Mesua ferrea*
Uttar Pradesh	Palash	*Butea monosperma*
Uttarakhand	Brahma Kamal	*Saussurea obvallata*
West Bengal	Night-flowering Jasmine	*Nyctanthes arbortristis*

Himachal Pradesh

Himachal Pradesh is located in North Western part of India between latitude 300 22' 40° N to 330 12' 20° N and longitude 750 45' 55? E to 790 04' 20° N. In Himachal Pradesh, the per capita cultivated land is only 0.12 hectares while per capita irrigated land is a meager 0.02 hectares. This situation necessitates a cropping pattern that would ensure highest income per unit area/labour/investment. Commercial floriculture perfectly caters to this necessity. The agro-climatic conditions prevailing in the State of Himachal Pradesh offer excellent opportunities for the development of floriculture both to serve the internal off-season market and also for exports, an avenue yet to be tapped. A large variety of floriculture products, *viz.*, cut flowers, bulbs, seeds, live plants, *etc.* can be produced as economic cash crops. Although flowers from different agro climatic zones of the State can be made available all through the year for domestic market, export quality flower produce can be ensured only by cultivation under controlled environment conditions of greenhouses.

Commercial Floriculture in Himachal Pradesh

Agro Climatic Zones for Floriculture

Zone Description	Elevation Range (Meters msl)	Rainfall (cms)	Suitable Flower Crops
Low Hill and Valley Areas near the plains	350 – 900	60 - 100	Gladiolus, Carnation Lilium, Marigold, Chrysanthemum, Rose
Mid Hills (Sub Temperate)	900 – 1500	90 – 100	Carnation, Gladiolus, Lilium, Marigold, Chrysanthemum, Alstroemeria, Rose
High Hills and Valleys in the interiors (Temperate)	1500 – 2750	90 - 100	Gladiolus, Carnation Lilium, Marigold, Chrysanthemum
Cold and Dry Zone (Dry Temperate)	2750 – 3650	24 - 40	Seed/Corm/Bulb production

Advantages of Floriculture

☆ The Agro climatic conditions prevailing in the State of Himachal Pradesh offer excellent opportunities for the development of floriculture both to serve the internal off-season market and also exports.

☆ A large variety of floriculture products, *viz.*, cut flowers, bulbs, seeds, live plants, *etc.* can be produced.

☆ The natural agro climatic conditions offer ideal production environment for flowers and the planting material *i.e.*, expensive heating and cooling systems in the greenhouses are not required

☆ Power required for running the greenhouses is charged at domestic rates in the State.

☆ Flowers from different agro climatic zones of the State can be made available from open field cultivation all through the year for domestic market, however, export quality flower produce can be ensured only by cultivation under controlled environmental conditions of greenhouses.

Services Provided by the Department of Horticulture

Infrastructural Support

The Department of Horticulture has established seven Floriculture Nurseries in various Districts, *viz.*, Navbahar and Chhrabra in Shimla District, Mahog Bag and Parwanoo in Solan District, Bajaura in Kullu District and Dharamshala and Bhatoon in Kangra District.

Model Floriculture Centre

The "Model Floriculture Centre" has been established at Mahog Bag (Chail), District Solan and a Tissue Culture Laboratory is being set up for the propagation of planting material of commercially important floriculture crops. The present infrastructure at the "Model Floriculture Centre" consists of 1706.5 sq. m of Greenhouse area, one Handling Unit for postharvest handling of flowers and 3 Nos. of Cool Chambers for forcing and storage of planting material. The building, which shall house the Tissue Culture Laboratory, Training Hall and other infrastructure of the Centre, has been constructed at an estimated cost of ₹ 94.22 lakhs and taken over by the Department of Horticulture in July 2004.

Postharvest Infrastructure

Collection, Grading and Packing House and cool chamber facilities have been established by the District Rural Development Agency for postharvest management of floriculture produce in the districts of Bilaspur, Mandi and Kangra.

Research and Development

The following organizations provide the necessary R and D support in the field of floriculture:

1. Dr. Y.S. Parmar University of Horticulture and Forestry, Solan. This University has a separate Department of Floriculture and Landscaping as its head quarters at Nauni. The location specific research work is being carried out at the regional Research stations of the university located in various Agro climatic Zones of the State.

2. Institute of Himalayan Bio-resource Technology, Palampur, District Kangra

3. ICAR Research Station at Katrain District Kullu H.P.

4. National Bureau of Plant Genetic Resources, Phagli, Shimla, H.P.

Research and Development Support

☆ Dr. Y.S. Parmar University of Horticulture and Forestry, Solan. This University has a separate Department of Floriculture and Landscaping at its head quarters at Nauni. The location specific research work is being carried out at the Regional Research Stations of the University located in various Agro climatic Zones of the State.

☆ Institute of Himalayan Bio-resource Technology, Palampur, District Kangra.

☆ ICAR Research Station at Katrain District Kullu H.P.

☆ National Bureau of Plant Genetic Resources, Phagli, Shimla, H.P.

Stepts Initiated to Promot Horticulture

☆ Creating more public awareness regarding use of floriculture produce through media and other agencies as well as more exposure of floriculture products during consumer exhibitions.

☆ Retailing of flower produce through super markets in addition to Florist shops to encourage flower consumption especially in metropolitan cities.

☆ Organizing postharvest infrastructure for marketing needs at the domestic terminal markets, particularly the Delhi market.

☆ Promotion of interaction between growers and scientific Institutions for effective lab to land technology transfer.

In Himachal Pradesh, farmers are taking to floriculture in a big way to supplement their income

growing flowers. However, till recently the farmers were reluctant to go for commercial floriculture because of the absence of an assured market for flowers. But the ever-increasing demand for flowers in Delhi, Chandigarh, Amritsar and other major cities of the northern region and the growing market within the state has provided the much needed boost of floriculture in the state.

Already nearly 325 hectares of the land has come under floriculture and the government plans to increase it 10 times by 2008. So far, about 1500 progressive farmers have taken to floriculture. They have been marketing their produce through

48 flower growers cooperative societies. The State Cooperative Department will set up an apex cooperative flower and production and marketing federation. A model floriculture centre is also being established at Mahon Bag in Chail to provide various facilities to the floriculturists.

As many as seven floriculture nurseries have been established at Navbahar and Chharabra in Shimla, Mahon Bag and Parwanoo in Solan, Bajaura in Kulu and Dharamsala and Bhatoon in Kangra with the objective of imparting training to commercial flower growers. The Department of Horticulture in the state has been providing financial incentives to the growers, including subsidy to individual growers under the area expansion programme. In addition, model schemes for the cultivation of gladiolus, carnation and lilium have been approved by Nabard to provide funds. Besides, subsidy to the tune of 40 per cent of cost, subject to a maximum of ₹ 40,000 per farmer for a maximum area of 500 square metres is being given for the establishment of a greenhouse. Plastic crates are also being provided at 50 per cent subsidy for handling flowers on the farm.

Churah Valley Cooperative Society

Churah in Himachal Pradesh's Chamba district is growing flowers that give the Netherlands and Taiwan a run for their money. The 350 odd small farmers here are a major supplier of cut flowers to the north Indian market, specifically Delhi and Chandigarh. Notwithstanding the daunting distance of 750km and 500 km respectively, the highly perishable consignments carnations, liliums, tulips, gladioli, liatris and orchids – reach their destinations in perfect shape to be sold from the same counters that deal with flowers imported from the Netherlands, China, Thailand and Kenya.

The venture is proving to be both commercially viable and socially sustainable. The annual turnover of the cut flowers business has already touched ₹ 2.50 crores and its still growing. The success is further drawing in farmers from neighbouring villages, indicating that the flower revolutions is unlikely to be limited to Churah. "The scope is immense. All that's needed is the will to work hard, a little bit of technical know-how and a spirit of togetherness among the small farmers, Sharma himself a progressive farmer attributes the success to the economic revolutions wrought by the Churah Valley Cooperative Society, which began with 13 poor farmers in 1995. Today it has 350 members and is looking at a headcount of 1,000 soon.

The society plays a pivotal role in the flower enterprise, by helping prepare the land, setting up polyhouses (collapseible greenhouses), packing, grading and marketing flowers.

The greatest natural advantage for the venture, of course, happiness to be the local climate. Flowers grow in abundance both on the open slopes as in the 200 –odd poly houses that dot the mountains, prompting individual annual profits between ₹35,000 and ₹51,000, depending on the variety and market response. A central processing centre is the nucleus for the grading and packing processes. The farmers' cooperative takes the responsibility for transporting the flowers via public transport everyday to Delhi and Chandigarh.

To polish its competitive edge against big landholders and flower producers in other states, the society recently imported 2,000 orchid bulbs from Taiwan. The bulbs will be propagated and multiplied under local conditions for distribution among the

> *To polish their competitive edge, Churah farmers recently imported 2,000 orchid bulbs all the way from Thailand.*

farmers. Earlier, the Society had imported tulip and lilium bulbs from Holland. The society also mobilizes finances and assistance from central government agencies, banks and state government schemes and links up with universities and local institutions for expert advice and necessary back-up.

Interestingly, through the society is taking the lead now in the flower power movement, the seeds were sown by a young IAS officer hailing from this region. "When I was posted in Kangra and later in Mandi as the district collector, I began to wonder why small farmers couldn't come together for economic growth," says Jagdish Chandra Sharma, now with the Himachal State Electricity Board in Shimla. In 1995, the idea took off at Churah. Now the experiment can be replicated in the entire state".

Damask Rose

The scent of a sweet-smelling Himalayan rose is luring a Parisian couture house. House of Chanel, the French fashion brand name synonymous with feminine sartorial elegance and signature perfumes like No.5, has been sniffing out aromatic fragrances in Himachal Pradesh. A year ago, a two-member team from Chanel visited the government-run Institute of Himalayan Bioresource Technology in Palampur on the slopes of Dhauladhar range to check out the perfuming potentials of its rose stock. They (Chanel) are testing these products to develop perfumes and we are trying to match their specific requirements.

During August, 2006 the institute sent a 10 kg batch of rose concrete to a Mumbai-based company to forward it to Chanel's Paris office. Price:₹60,000 – ₹75,000 per kg. The rose concrete, a highly expensive wax-like substance, has been extracted from a rose variety called Damask rose, or Rosa Damascena. In the secretive world of perfume business, it is used in cosmetics by select few makers. The stock of Damask rose was originally brought from Bulgaria in the late 1970s, but the institute started conducting trials on rose concrete a few years ago. Besides, it has developed two varieties of damask – Himroz and Jwala – that can grow in high altitude as well as in the plains. The Damask rose grown in the hills is better in quality and has higher notes of fragrance. The fashion house is keen to procure it.

Himachal Pradesh, endowed with different agro-climatic conditions ranging from sub-tropical to dry temperate zones, has premium potential for growing cut flowers, which are off-season wrt plains as well as bulbs and seed multiplication. Area under floricultural crops in the state has arisen from only 25 ha in 1993-94 to 914 ha during 2012-13. Farmers are earning 99.98 crores annually from the cultivation of flower crops. Ornamental crops grown according to maximum area are chrysanthemum, marigold, gladiolus, carnation, rose, potted plants, annuals, lilium, gerbera, daffodils, flower seeds and alstroemeria.

Important Commercial Flower Crops

- ☆ Alstroemeria
- ☆ Aster
- ☆ Antirrhinum
- ☆ Bird of Paradise
- ☆ Carnation
- ☆ Chrysanthemum
- ☆ Daffodil
- ☆ Freesia
- ☆ Gladiolus
- ☆ Godetia
- ☆ Gerbera
- ☆ Iris
- ☆ Lilium
- ☆ Marigold
- ☆ Ornithogalum
- ☆ Orchids
- ☆ Rose
- ☆ Tuberose

Mandate

- ☆ Impart education at undergraduate and post graduate levels
- ☆ Introduction, evaluation and development of new cultivars of flower crops
- ☆ Standardize technology for production and management of flower crops, pot plants and landscape plants
- ☆ Standardization of seed production and propagation technology of ornamental crops
- ☆ Standardization of dehydration technology for important cultivated and wild ornamentals
- ☆ Explore possibilities for commercialization of wild ornamentals of Himachal Pradesh
- ☆ Standardization of postharvest technology for ornamental crops
- ☆ Transfer of technology for commercialization of ornamentals

Area of Specialization

The department offers specialization in the following areas at the post graduate level

☆ Flower regulation

☆ Production Technology

☆ Germplasm conservation, biodiversity and improvement

☆ Biotechnology

☆ Propagation

☆ Protected cultivation

☆ Flower regulation

☆ Postharvest technology

☆ Dehydration technology cum value addition

☆ Bio-aesthetic planning

Uttarakhand

The Valley of a Million Blooms

The Valley of Flowers known for its cascading post-monsoon blossoms has finally been recognized as world heritage site. Unesco along with the International Union for Conservation of Nature (IUCN) has given the Valley of flowers located in Chamoli district the WHS (World Heritage Site) status. The new heritage status is expected to put the valley on the world tourist map and help Uttarakhand attract tourists from around the globe. The Valley, a glacial corridor located at an altitude between 3,250 and 6,750 meters above sea level is also the country's smallest national park. It gets carpeted with wild flowers during monsoons, bursting into bloom between mid-July and mid-August, with almost 300 species of wild flowers making it a riot of colours. The valley was discovered in the 1930s by an Englishman but it was a woman Indian Forest Service (IFS) officer from the North-east, Jyotsna Sitling, who rescued it. First, she worked tirelessly with the locals to clear the 87-tonne garbage from the buffer zone of the Nanda Devi Biosphere near the valley. This led to regeneration of some flowers which were considered extinct.

The Valley's most abundantly seen flower is Himalayan blue poppy. One can also feast one's eyes on rare primulas and orchids which bloom during June, Impatiens, Potentillas and Campanulas paint the Valley in different hues of pink, red and purple during July and August. Most of these flowers have medicinal value too. There is an abundance of Asmanda as well. This national park was established in 1982 with the aim of protecting the catchment area of Pushpavati river.

The park, a huge expanse, starts from Ghangharia, passes the snow-clad peak of Ratban Parvat and then turns towards Kunt Khal and Nar Parvat. Traditionally, the Valley of Flowers has been used by shepherds for grazing their sheep.

The beautiful valley, situated in the upper reaches of Bhyunder Ganga in Chamoli district, is spread over 87.5 square kilometers. It has also found mention in ancient Indian history and literature. Legend has it that the valley also had the precious Sanjeevani booti. From October onwards, the valley gets snowbound for about five months. In fact, because of its near inaccessibility, the valley disappeared from the tourist map, until 1931, when Franksmith, a British mountaineer, having lost his way while returning from a successful expedition to Mt. Kamat, stumbled upon the Valley. Out of six natural World Heritage Sites in India, Uttarakhand now has two-the second being Nanda Devi Biosphere Reserve of which the Valley of Flowers is a part. Situated in the upper Himalayan ridges at a height of 3,200 to 6,675m over 87.5 sqkm in Chamoli, it has 521 varieties of flowering plants. Of these, six are not found anywhere else.

Uttarakhand's Unique Climate makes it Holland's new Hothouse

Acting on a report by Yes Bank and the Punjab, Haryana and Delhi Chamber of Commerce and Industry (PHDCCI), which was submitted to Chief Minister ND Tiwari in July 2005, Uttarakhand is working to exploit its unique weather conditions to re-export flower bulbs and, in the process, provide and economic leg-up to its farmers. Located in Chaffe near Bhimtal, in the Nainital district of the Kumaon, the floriculture BPO project is being set up by the State Industrial Development Corporation (SIDCUL). Initially, the plant material will be imported from Holland which partnered PHDCCI to highlight key growth areas that could be harnessed for sustainable socio-economic development of Uttarakhand.

In the first leg, 15 varieties of flowers, like gladioli, tulips, daffodils, hyacinths, freesias, geraniums will be taken up Seeds cuttings of gladioli, tulips, daffodils, Hyacinths, freesias, geraniums will be flown in from Holland and transplanted in Uttarakhand. After they reach full maturity, they go back to Europe for sale plant material will be distributed among farmers in Nainital, Almora, Bageshwar and Pitthoragarh after training them. They will grow the flowers for export.

The farmers will be under contract to the private partner of the enterprise, who will be provided all the necessary infrastructure including machinery and land on lease – by the state government for a period of 10 years. The bidding process for running operations and maintenance is underway; the project is expected to get off the ground by the year-end. State Industrial Development Cooperation (SIDCUL) also plans to set up a plant tissue culture laboratory and bulb storage capacity at Chaffe. As per anticipated estimates, the plant tissue laboratory and bulb storage capacity will handle 50 lakh bulbs and 500,000 cut flowers with gradual increase in the next five years.

The state government already provides 75 per cent transportation subsidy to hill farmers; officials hope that farmers keen on export-oriented floriculture will also avail of the benefit to transport their produce up to the nearest railhead at Katgodam. The emphasis is on growing high quality bulbous material which has a demand in the international market. The flower export market is conservatively estimated will at between ₹400 crore and ₹500 crore, most of which is monopoly of floriculturists in south western India, especially Pune and Bangalore.

West Bengal

To develop floriculture, the government of West Bengal is stated to be going in for public-private partnerships to improve infrastructure and extend cultivation. Large exports from the state, particularly, tube roses, have boosted overall production, which is increasing by 20 per cent since a couple of years. Flowers are mainly cultivated in the South Bengal districts of Borth 24 Parganas, Howrah and East Miduapore and in North Bengal in the hills of Darjeeling.

The Budding Flowerbed of North-East India

The north-eastern states of the country are ill-famed for the sound of shootings of smoking guns. But here, the shootings of buds, of flowers, shrubs, orchids and other local greens are being highlighted. At present being sold in local markets across the region, the vendors of these colourful blooms hope to make a mark in the floral export trade in the near future.

On a recent stocktaking trip organized by Agricultural and Processed Food Products Export Development Authority (APEDA), an autonomous body under the Ministry of Commerce, all one has seen around in the States of Nagaland and Meghalaya are beds and beds of anthuriums, lilliums, roses, dry flowers and even strawberries and passion fruits. Though most of the farms that one has visited are on the farmers' backyards, on not too big a scale, but there too exist now a few five-star farms with shades imported from Israel. APEDA's Navnesh Sharma, responsible for its floriculture wing in New Delhi, says the organization had facilitated one grower from Dimapur to visit the famous annual flower fair held in Amsterdam in 1994. On seeing the world of opportunity before her, the grower, Akruzo Putsure returned home only to devote her huge backyard to grow anthuriums. She also encouraged 10 more women in her neighbourhood to take up floriculture, who have now their own little poly houses.

In 2005 too, APEDA planned to send a few flower growers from the region to get a first-hand experience of world trends in the floral trade. Since fresh flowers need a lot of care and attention besides smooth infrastructure facilities, what can take off from the region right away is a good base for dry flowers. India, is now one of the top exporters of dry flowers. With a moderate climate, the locals traditionally being inclined to farming and a huge mass of manual labourers willing to get into a worthwhile business that can well-sustain them. The experts are of the opinion that it could very well be the hot bed of Indian floriculture.

Spread over 10 hectares of land along National Highway 37 as one travels out of Guwahati, a Daffodil Nursery is the most visited not to mention the most visible – nursery in the entire state of Assam. So long, production and supply of saplings and seeds of fruits, flowers and vegetables were either the forte of a handful of nurseries set up way back in the 1950s or in the grip of some MNCs who operated through their won net work of wholesalers and retailers. But the launch of the Technology Mission for Integrated Development of Horticulture a few years ago is finally impacting things on the ground. Earlier, farmers were just glued into traditional paddy and rabi crops. The concept of nursery was also not well-known. But with the Technology Mission coming in, things are gradually changing. Horticulture Department has supported at least 50 nurseries, half of them new, in producing various varieties of mother plants so that saplings and seeds are available locally for the people. For instance, this Daffodil Nursery which earlier produced Saplings and cuttings of flowers, today has diversified into 300 odd varieties of mother plants of various horticulture and flower species.

Assam Laxmi Nursery, owned by Tankuram Bora of Sonapur, about 25 km from Guwahati

Set up in 1996 as a homestead nursery, it now spreads over 15 bighas of land and has already attracted a ₹4 lakh subsidy under the Horticulture Technology Mission. Owners of nurseries with an area of two hectares and more are being given a subsidy of up to ₹8 lakh, while those with less than two hectares are entitled for subsidy up to ₹3 lakh. Apart from monetary support through subsidies, the Horticulture Technology Mission has also provided these nurseries access to good sources of seeds.

Maharashtra

In Maharashtra, of more than 370 acres under 'greenhouses'- precise statistics are lacking with officials conceding their record-keeping is "poor"-the corporates' share is some 160 acres. Corporates stepped in with intent only to export. Of the 77 who registered to grow exotic flowers in the country in the 1990s, as many as 49 were in Maharashtra. Of them, only 14 are working. Now that a Floriculture Park is coming up near Pune with common facilities like pre-coolers, some bigger farmers may 'corporatise' their approach to directly seek to export.

> **BOX 1**
>
> ☆ *Focus on domestic markets helps small farmers to remain in business*
>
> ☆ *Corporates are hamstrung by high overheads*
>
> ☆ *Floriculture Park coming up in Pune*
>
> ☆ *Sizes of 'greenhouses' have become larger*

Small holdings, which yield little gains from other crops, are shifting to floriculture, especially in Sangli, Pune, Satara and Kolhapur. The acreage may seem misleadingly small but each square metre in greenhouse annually yields 200 stems, peak season price per cut flower being ₹3. It sinks to 20 paise in lean season. With farmers coming to grips with shifts in preference, the sizes of the greenhouses have become larger from a standard 560 sq.m. to 1,000 sq.m. Locally they are called 'polyhouses', because polyester sheets are used for canopy.

Corporate forays into floriculture were hurt by high overheads on the ₹2-crore to ₹3-crore investment per unit and longer gestation periods, though not all have given in. Recently, Mody Exotica came in with a huge project to grow anthurium-a new variety here – but small farmers' focus on domestic markets at low prices helped them profit better and remain in business. They even reach buyers in State Transport buses to help grow the business at 11 to 12 per cent annually. Informal surveys have revealed that a self-employed farmer-floriculturist earns twice of those who employ others to do the work. He gets loans from cooperatives apart from up to ₹2.25 lakh as subsidy for a 560 Sq.m 'polyhouse'. If his project breaks even in 24 to 30 months, the corporates need five years or more to reach that stage.

Off the highway to Indore, down a dusty track, is small room. Set amidst field full of vegetables and flowers, the one room is frequented by ministers, government officials, old-fashioned farmers, pesticide makers, traders and journalists. They all come to see how 42 year – old Megha Borse has single-handedly created a flourishing flower business for herself. Just into its second year, Sheeman Flora is in full bloom, making Borse the "flower power woman" of Nashik. Her gerberas and marigolds attract visitors from across the region. Some come to see how a woman with an M.Sc. in organic chemistry manages a farm, others to know if they can do the same. As she gets ready to open her third greenhouse and diversify into coloured capsicum next year, the government wants her to help with the Horticulture Mission as it gears up to penetrate farmlands.

This is more than just farming or growing flowers, she calls it an industry because she spends ₹10 lakh for every 1008 sq. m of plot that she develops. These flowers are growing in a controlled climatic environment, so the natural seasons don't really affect them. Initially she was looking at the food processing industry. But after much research she settled on flowers. Interestingly, both Pune and Nashik were kind of pioneers in growing flowers in the late '80s. While in Pune 400 –odd flora farms sprung up in the next 10 odd years, Nashik never really took to its flowers. When she started out in 2004, there were only 10 for a farms. Since she opened, 10 to 15 open every year.

Extremely difficult but also very necessary, It has become essential for farmers to harvest a variety of crops and not just stick to traditional ideas. There are huge returns in flowers. But initially all a farmer is able to see is the fact that for ₹ 10 lakh, he can grow grapes in almost 10 acres of land instead of flowers in 1008 sq.m. There is a huge gap between policies planned in New Delhi and the ground level. So many things are being planned, new schemes launched but it is not penetrating to the grassroots. It will take a long time for that, which is very disheartening.

Maharashtra Jarul Flower

Jarul, also called as "Taamhan" in Marathi gets its name from a Swedish naturalist. The meaning of the flower is spectacular or showy. Jarul belongs to the family of Lythrum, also called as Lythracea. The flower holds a great significance as it finds use in preparing medications and other health care products.

Characteristics

Grill is a beautiful purple flower that smells really nice. The size of the tree varies as some of them grow to a small height while some of them may grow up to 20 meters. Its leaves are smooth, long and oval in shape. On an average one Jarul flower has about 6-8 petals which collectively form a complete flower. It looks really lovely and comes in white shade too. **Ecology and Habitat** Jarul finds its roots from the South East Asian countries including India. In India, it just grows as a flower but in the subtropical regions, this plant holds huge importance and grown as an ornamental plant.

Uses

Jarul finds use in a number of applications and is good for health too. Here are some of its uses.

The most important use of this flower is done in preparing medicines that can be used to control blood pressure, diabetes, bowel and other health problems. To lower cholesterol level also, this flower finds great use and shows quick results.

In Japan and Philippines, the leaves of this flower are used for tea preparation. It gives a different but enriching taste to the tea. According to the latest research it is found that Jarul can be used for lowering the sugar level. Owing to the number of benefits that are connected with this flower, Philippines have started promoting Jarul on a global level so that it can be cultivated in large numbers.

Flower Market in Delhi

Flowers are for all seasons but during springtime their presence is more visible than ever. The Capital's flower market is one of the biggest in the country. The Delhi Government has also woken up to its tax generating potential. A majority of the trade in Delhi is in cut flowers. According to a rough estimate, around ₹50 lakh worth of flowers are sold in Delhi on a daily basis. This does not include the big chunk that is exported or sold to different states. Flowers are supplied to UP, Haryana, Rajasthan, Himachal Pradesh and Punjab. In the 1970s, the flower business in the Capital was in its infancy. Earlier, only Rajnigandha (tube roses), local roses and gladioli were available in the city. The boom came in the last 10 years with foreign traders getting involved in the farming of exotic varieties in Bangalore and Pune. The government has also played an encouraging role by providing 40 per cent subsidy to flower growers. Delhi's flower supply mainly comes from Bangalore - roses, carnations, birds of paradise, chrysanthemums and enthusium. Pune is known for its gerberas and Coorg in Karnataka for enthusiums. The hills of Uttarakhand and Himachal Pradesh are known for their fine variety of lilies, carnations and gladioli. Lily which at present is being sold for ₹ 20 is one of the most expensive flowers with its price touching the three-figure mark. Tulips are from Himachal Pradesh and not imported from Holland as is widely believed. The Holland variety has a very short shelf life due to Delhi's harsh climate. Most of the supply of gladioli comes from Dehradun and Lucknow. All the flowers from this region reach Delhi in buses. And orchids don't come from Sikkim. A majority of orchids and lilies, are from Thailand where they are grown commercially and are cheaper than the ones from Sikkim. Tuberoses (Rajnigandha) comes from Meerut and the Muzzafarnagar belt and the local roses come from farms in Chattarpur (Delhi) and Gurgaon region. Marigold comes from the river banks of Yamuna and Rajasthan. The business in the flower trade is conducted at three levels. At the top are the big dealers (around 20 in Delhi) who source flowers from all over India and abroad. The next in line are mashkhors or the middlemen who take the flowers from the big dealers (around 70 of them in Delhi) and sell them to retailers. The Flower Market Committee, Mehrauli, which is the central committee of flower traders in

Delhi, has two markets (mandis) operating under it the connaught Place market and the Fatehpuri market. Three major dealers operate from the Fatehpuri mandi in Old Delhi. The Connaught Place flower mandi comes to life at 4.30 am. on Baba Kharak Singh Marg. It has been around for the last 10 years.

Delhi Government Plans

The Delhi Government plans to set up an exclusive flower market near the DTC Bus Depot, Srinivaspuri. But the plan is not making much headway. An office bearer of the Flower Market Committee, Mehrauli, says, "The committee needs to be reconstituted before any move". Traders feel that the location is not very well connected.

Tamil Nadu

Floriculture is the art and knowledge of growing flowers to perfection. It deals with the cultivation of flowers and ornamental crops from the time of planting to the time of harvesting. It also includes production of planting materials through seeds, cuttings, budding, grafting and marketing of flowers and flower produces. It includes cultivation of flowering and ornamental plants for sales or for use as raw materials in cosmetics, Perfume industry and also Pharmaceutical sector.

India is bestowed with several agro-climatic zones conducive for production of sensitive and delicate floriculture products. This era has seen a dynamic shift from sustenance production to commercial production. As per National Horticulture Database 2010 published by National Horticulture Board, during 2009-10, the area under floriculture production in India was 0.183 milliion hectares with a production of 1.021 million loose flowers and 666.7 million cut flowers. Floriculture is now commercially cultivated in several states with Tamil Nadu (25 per cent) Karnataka (20 per cent), Andhra Pradesh (14 per cent) having gone ahead of other producing states like Maharashtra, Punjab, Haryana, West Bengal, Gujarat, Odisha, Jharkhand, Uttar Pradesh and Chattisgarh.

India's total export of floriculture was ₹286.45 crores in 2010-11. The major importing countries were USA, Pakistan, Netherlands, Germany, Italy, Belgium and United Kingdom. There are more than 300 export-oriented units in India. More than 50 per cent of the floriculture units are based in Karnataka, Andhra Pradesh and Tamil Nadu. With the technical collaborations from foreign companies, the Indian floriculture industry is poised to increase its share in world trade.

In Tamil Nadu, out of the total area of cultivation under Horticulture and Plantation crops of 922005 ha the flowers occupy 25610 ha. Dindigul,Krishnagiri, Dharmapuri,Salem, Vellore, Madurai, Tiruvannamalai, Tirunelveli and Erode are the major flowers growing districts in our State.

The major flowers grown are Jasmine, Mullai, Rose, Crossandra, Chrysanthimum, Marigold, Tube Rose, Arali, Jathimalli *etc*. The Area, Production and Productivity for the major flower crops grown in Tamil Nadu are:

Sl.No.	Flowers	Area (in Ha)	Production (in Tonnes)	Productivity (Tonnes/ Ha)
1	Rose	1949	14130	7.25
2	Jasmine	10623	92951	8.75
3	Mullai	2769	23537	8.50
4	Jadhi malli	841	7569	9.00
5	Crossandra	1317	2634	2.00
6	Chrysanthimum	2240	20160	9.00
7	Marigold	1502	22530	15.00
8	Arali	1195	9261	7.75
9	Tube rose	1529	15290	10.00
10	Others	3174	34343	10.82
	TOTAL	25610	227115	8.87

Major Districts Growing Flower Crops

	JASMINE		MULLAI		ROSE		TOATL FLOWERS	
	Area	Prodn	Area	Prodn	Area	Prodn	Area	Prodn
Coimbatore	158	1383	207	1760	22	160	781	7151.62
Dharmapuri	400	3500	127	1080	121	877	2133	17583.46
Dindigul	703	6151	168	1428	399	2893	3499	31822.13
Erode	1061	9284	67	570	13	94	1437	12941.22
Krishnagiri	754	6598	307	2610	436	3161	2552	21833.9
Madurai	1220	10675	51	434	92	667	1658	14952.36
Salem	576	5040	16	136	184	1334	2097	19398.15
Thiruchirapalli	503	4401	48	408	40	290	927	8906.51
Thirunelveli	1267	11086	26	221	32	232	1596	14510.56
Thiruvallur	800	7000	55	468	146	1059	1134	9657.62
Tiruvannamalai	391	3421	248	2108	69	500	1620	15294.85
Vellore	492	4305	865	7353	79	573	1878	15947.57
STATE TOTAL	10623	92951	2769	23537	1949	14130	25610	227114.93

Varieties of Flower Crops

1. Barleria

CO 1 (1984)

It is a clonal selection from the local type. It bears attractive pink flowers, producing on an average 2.11 kg of flowers per plant in a year. It is produces flower early and flowers become available in about 210 days after planting.

2. Chrysanthemum

CO 1 (1985)

It is a selection made form a bulk population introduced from Hozur of Dharmapuri district. Flowers are medium sized (2.5g) and attractive (canary yellow) The flowers have thick, sturdy stalks, which are an added advantage for easy tying in the making of garland and other decoratives. It flowers early by about 15-20 days and the blooming period also lasts longer when compared to the other local cultivars. Average yield on main crop is 16.7 t/ha.

MDU 1 (1985)

It is a selection from the germplasm type. It is an early type, coming to first flowering in 104 days as against 120 days in the local type. The flowers are large and attractive sulphur yellow in colour with a diameter of 3.90 cm. It yields 30.59 tonnes per hectare per year in two crops (main and ratoon crop).

CO 2 (1989)

This is a clonal selection from among the germplasm type introduced from the National Botanical Research Institute, Lucknow. This selection recorded higher yields than CO.1 and MDU.1. It has several attributes like more number of flowering shoots per plant, more number of flowers per plant, invisibility of the disc in the flower (capitulum) which is considered a desirable feature in the trade circle and a novel new purple colour (Rhodamine purple–29) as compared to the conventional more familiar carmine yellow colour of CO.1 and other local varieties.

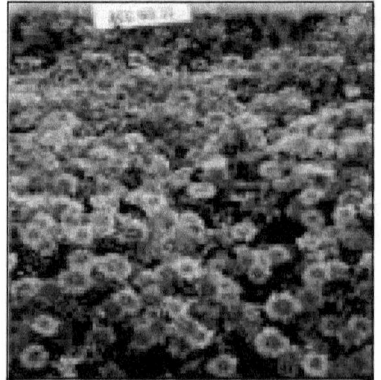

3. Gerbera

YCD 1 (1992)

Yecaud –1 Gerbera is a clonal selection from seedling from a mixed open pollinated seeds collected from germplasm of gerberas maintained at Horticultural Research Station, Yercaud. It is a dwarf, herbaceous perennial

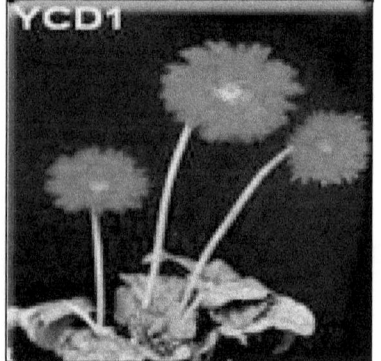

growing to a height of 39cm. The flowers are double in form with cherry red colour. Flowers are large (9.11cm diameter) with moderately prominent disc. Petals are dense, compact and arranged in concentric whorls. Flowers are borne on long (47-79 cm) and thick stalk. Free from the disorders like bent neck, petal necrosis (during vase life) and temporary wilting in field are absent. Plants flower earlier (within in 45 days after planting) and produce about 60 flowers per plant per year. Flowers have a retentivity of 8 days on the plant with a vase life of 7 days. The variety is suitable for use as cut flower, raising as borders in garden and for pot cultivation. It is suitable for growing in the hill ranges of Tamil Nadu situated at an altitude of 1000 – 2000m.

YCD 2 (1995)

It is a cut flower variety selected from among the germplasm collection at Horticultural Research Station, Yercaud. It blooms throughout the year with peak flowering during May – June. The flowers are attractive, rosy pink coloured, borne on long stalk without bend. The flowers have a vase life of 15 days in hills and 10 days in plains. The variety yields about 80 flowers/ clump in a year and suitable for cultivation in hilly regions of Tamil Nadu.

4. Gladiolus

KKL 1 (1993)

It is an improved selection evolved at the Horticultural Research Station, Kodaikanal from the population of the cv. American Beauty. The selection is adapted to the hilly regions of Tamil Nadu particularly to the Palani hills, the Nilgiris and the Shevroys. (1200 – 2200m above M.S.L). The crop can be raised during August – February under lower altitudes while in the high mountains the season is from February – August. The flower colour is an attractive red purple with white flushed throat. The plant grows to a height of 116cm. The mean spike length is 89.4cm with an average of 16.2 florets per spike. The average floret size is 13.5cm. The florets are open shaped. Each spike weighs on an average of 92.3g with a vase life of 12.1 days. The selection yields on an average of 21.1 spikes and 19.5-corms/sq.m. The spikes will be ready for the first harvest in 90 days after planting. The harvest of spikes will continue up to 120 days. The corms are harvested from 150 days after the leaves stared yellowing and drying. It yields 2,11,100 flower spikes and 1,95,000 corms/ha. It out yields its parent type (local) by 18.74 per cent (flower spike yield) and 34.68 per cent (corm yield).

5. Hibiscus

CO 1 (Thilagam) (1981)

It is an inter-generic hybrid between *Hibiscus rosasinensis* and*Malvaviscus arboreus*. It is a woody perennial shrub with an erect growth (2.3 meters) and attractive double flowers, which are having attractive carmine red petals without any throat colouration. There are 30-36 petals in a flower arranged in three whorls. It is highly floriferous yielding 3055 flowers per plant in a year. It is suitable for planting as a single specimen in lawn, foundation planting and for pot culture.

CO 2 (Punnagai) (1981)

It was evolved by selection from the open pollinated seedlings of 'Chandrika' variety. The plants are semi spreading, growing up to 1.5 m. The flowers are solitary, terminal or axilary, bigger in size with a diameter of 14cm. The flowers are attractive with apricot yellow colour having signal red throat. The petals have a crepe paper like texture with feeble veins radiating from the throat. The plant yields about 988 flowers per year. It is suitable for planting as single specimen in lawn and for pot culture.

CO 3 (1984)

It is a clonal hybrid between Bright Yellow and Red Gold cultivars. It produces apricot yellow flowers having 'signal red' throat. The flower colour gradually changes to Chinese yellow with Turkey red throat in the evenings. It is floriferous and produces on an average 1309 flowers per plant every year compared to 374 and 503 flowers by their female and male parents respectively. It takes about 105 days for flowering from planting.

6. Jathimalli

CO 1 (1980)

It is a secondary clonal selection from germplasm collection. The average flower yield

is 10,144kg per hectare in a year. The flower buds are pink tinged with long corolla tube. It is suitable for oil extraction with a concrete recovery of 0.29 per cent. The concrete yield is 29.42 kg per hectare.

CO 2 (1991)

It is an induced mutant (I.M.3) developed by treating the vegetative cuttings of CO.1 Pitchi with gamma rays @ 1.5 kR. This mutant is characterised by bold pink buds. The flower bud is 4.14 cm in length as against 4.00 cm in CO.1. The 100 buds weight is 10 g in CO.2 as against 9.4 g in CO.1. This variety is amenable for earlier and quicker tying of buds in garland making. It yields on an average 11.68 tonnes of flower buds per hectare, which is 19.5 per cent higher than CO.1 Pitchi.

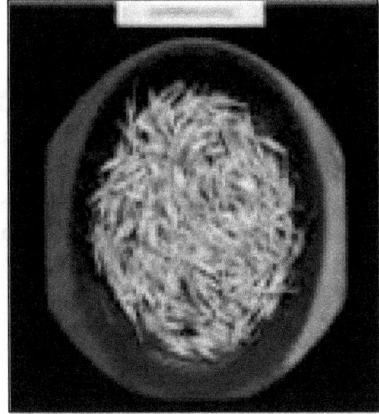

7. Marigold

MDU 1 (1986)

It is a selection from a germplasm type. The plants are medium tall with moderate branching habit. The plant produces on an average 97 flowers weighing 561.40 g/plant, with an estimated yield of 41.54 t/ha. The flowers are large with a stalk length of 8.39 cm. The light orange colour petals are compactly arranged and each flower has 210 petals. The flowers fetch premium price in the market.

8. Mullai

PARIMULLAI (1972)

It is a clonal selection from a germplasm clone. The plants exhibit resistance to gall mite. The yield is 7800 kg of flower buds per hectare per year. The buds are white with moderate corolla tube length (1.25cm). The concrete recovery is 0.29 per cent.

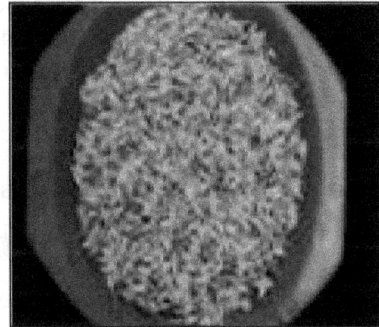

CO 1(1980)

It is a secondary clonal selection from a local type. The flower buds are white and bold with long corolla tube (1.50cm) than Parimullai. The yield is 8800 kg of flower buds per hectare in a

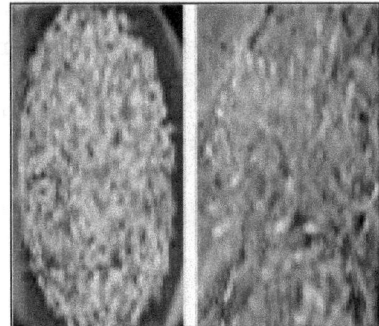

year. The jasmine concrete recovery is 0.34 per cent.

CO 2 (1988)

It is a clonal selection from progenies with desirable flower bud characters like long corolla tube and longer bud. The length of the corolla tube is 1.70 cm while it is 1.5cm in CO.1. It yields on an average of 11,198 kg of fresh flower buds as against 8,825kg in CO.1. It exhibits complete field tolerance to the phyllody disease and gall mite infestation.

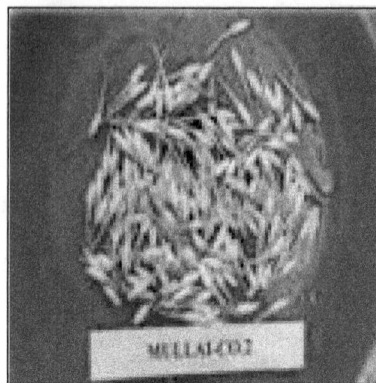

Jasmine is one of the oldest fragrant flowers cultivated by man. The flower is used for various purposes *viz.*, making garlands, bouquet, decorating hair of women, religious offering *etc.* It is also used for production of Jasmine concrete which is used in cosmetic and perfumery industries. More than 80 jasmine species are found in India, of which only three species are used for commercial cultivation. They are *Jasminum sambac* (Gundumalli/Madurai Malli), *J. auriculatum* (Mullai) and *J. grandiflorum*(Jathimalli/Pitchi). The first two species are mainly cultivated for selling as fresh flowers whereas the last one is cultivated for concrete extraction.

Tamil Nadu is the leading producer of jasmine in the country with an annual production of 77247 t from the cultivated area of 9360 ha. The flowers produced in the state are being exported to the neighbouring countries *viz.*, Sri Lanka, Singapore, Malaysia and Middle East countries. The major jasmine producing districts of Tamil Nadu are Dindigul, Salem, Madurai, Tirunelveli, Virudhunagar, Trichy, *etc.* Since the crop requires lots of manpower for harvesting and other operations, only small farmers are cultivating the crop. It is an ideal crop for small farmers whose land holdings are less than 1 acre.

The Flower Heaven on Earth

Thovalai is surrounded by flower fields, and the flower business is the main occupation of the village people. They grow many varieties of flowers and export them to different parts of the country. Thovalai is a small village located near Nagercoil in Tamil Nadu. It is a quaint village where several acres of gardens produce fresh flowers for local and export markets. The village is decorated with long garlands, some of which almost touch the ground. The village is known as a 'flowering village', and the flower market exports flowers like Malligai, Pichi, Kaakadai, Kanagamparam, Kenthi, Sampangi, Vaadamalli, roses, Kozhi poo and Chevanthi to many other places throughout the year. The village is also famous for its Murugan temple, dedicated to Lord Muruga, who is locally referred to as "Thovalai Murugan".

Location

Thovalai is located in Kanyakumari District, on the highway running between Nagercoil and Thirunelveli. It is situated near Aramboly Gap, a natural depression running through the Western Ghats range connecting Kanyakumari with rest of Tamil Nadu. Thovalai is also known as Thovalai Vadakur.

Occupation

Thovalai is surrounded by flower fields, and the flower business is the main occupation of the village people. They grow many varieties of flowers and export them to different parts of the country. Every shop in this village greets visitors with a palette of colors, including garlands in oranges and yellows, whites and pinks. People visit the flower market early in the morning, and the market remains busy with the hum of people bargaining. The village is especially famous for a fresh-looking jasmine flower called Pichchi vellai, or Pichchi poo.

Jasmine flower growers demanded that the government establish a perfume unit, so the Floriculture Research Station was set up in Thovalai for the benefit of farmers cultivating flower crops in and around the Kanyakumari District. It was set up during 2008 and 2009 as the flower producers faced difficulties in preserving the grown flowers. This facility has raised the status of farmers, the quality of the exported flowers as well as the quantity.

The main aims of this research station are:

- ☆ Breeding and cultivating flower crop varieties, such as tuberose, jasmine, scented rose, celosia, nerium, marigold, gomphrena, chrysanthemum, crossandra, and many more;
- ☆ Standardizing the technologies for disease management in flowering crops;
- ☆ Standardizing the postharvesting and packing technologies for cut and loose flowers;
- ☆ Researching and standardizing the techniques for optimal harvesting;
- ☆ Standardizing the agro-techniques;
- ☆ Developing technologies for value added products from flowers;
- ☆ Studying the possibility of flower growth in the banana and coconut cropping systems; cultivating varieties of flowers with longer lives and long stems.

Some of the farmers from nearby villages, such as Kannan Puthoor, Viswanatha Puram, Chenbagaraman Puthoor, and Aramboly, have switched to paddy cultivation due to inconsistent rainfall.

Infrastructure

The village is famous for the flower market and the Murugan temple situated nearby on a small hill. Festivals like Soora Samharam and Malar Muzhukku Vizha are celebrated here. Another Murugan temple, on Chekkargiri Hill, is also famous in Thovalai.

Andhra Pradesh

Favourable weather conditions in Araku area to start floriculture on 250 acres in Andhra Pradesh. At present 54,363 acre of land is under loose flower cultivation in AP, generating 1.26 lakh metric tonne of flowers annually. Floriculture in Andhra Pradesh is set to get a boost. At present 54,363 acre of land is under loose flower cultivation in AP, generating 1.26 lakh metric tonne of flowers annually. There are also 50 greenhouses for cultivation of cut flowers, each of which grows about 1 lakh flowers in one crop. Loose flowers are harvested from plants independently and cut flowers are harvested along with the stalk.

The AP government has decided to start floriculture in Araku on 250 acre. A horticulture department official said, "Araku has the right weather and environmental conditions because of which less maintenance will be required here than in other districts." The government has also sent a proposal to the Central government to start cultivation of dendrobium, an orchid, at Chintapalle, a hilly area in Eastern Ghats in Visakhapatnam which they say, has the right climate for orchids.

Apart from this the government has decided to give 50 per cent subsidy for development of 110 polyhouses in Kuppam, Chittoor district for promotion of Dutch Rose cultivation. Flower cultivation can fetch profits of anywhere between ₹ 40,000 - ₹ 60,000 from one hectare (nearly 2.5 acre). However, lack of training in floriculture is proving to be a major setback for those interested in cultivating flowers.

A flower cultivator from East Godavari said, "Many farmers without any technical know-how embark on floriculture and plant seeds whenever they feel like throughout the year and use whatever pesticide, fertilizer or bio-pesticides they get in the market, ending up losing their investments. Horticulture department or seed companies promote floriculture but do not provide any regular training on how to correctly go about it."

The major flowers which are harvested in Andhra Pradesh include gladiolus, lily, jasmine, aster, jasmine, gerbera, rose, African marigold and carnation. The flowers are mostly grown in Anantapur, Guntur, Prakasam, Chittoor and East Godavari districts. The major markets for flowers are Bengaluru, Kolkata, Hyderabad and Chennai. Apart from growing and selling flowers, a major business is cultivation of flower saplings, which is a big business in Kadiam, East Godavari, which has 1,500 nurseries. Mr. P. Ramakrishna, a nursery owner, said, "Every month I sell about 1 lakh saplings, mainly to Gujarat and Maharashtra. Heliconia, red ginger, chrysanthemum and anthurium are some of the flower saplings which we sell."

Small Farmers Cultivating Flowers as Allied Activity

Making better use of the available cultivable space, small and marginal farmers around the city are cultivating flowers. Though the produce is not large, these farmers are able to make a mark in the local market by taking a 20 per cent share. This allied activity is providing them financial support too. Over 100 farmers from Samalkot, Pithapuram, Peddapuram and nearby villages are into flower cultivation. Besides jasmine, they are also into the cultivation of jaji (Spanish jasmine) and kanakambaram (Crossandra/firecracker flower). The Sarpavaram flower market that registers a daily sale of 1,000 kg to 5,000 kg a day basing on the season, is mainly depending on the famous flower gardens of Kadiyam.

"The flower from the local farmer is fresh as compared to that of Kadiyam nursery. For jasmine and jaji, many of our customers prefer the local flower," says D. Nooka Raju, a flower vendor from Sarpavaram. Summer is the lone season for jasmine, whereas July to January is the peak period for jaji and kanakambaram. "The price for kanakambaram is ₹ 3 for 100 flowers during the slack season. When it comes to the peak, it may go up to ₹ 8. Since the auspicious Sravanamasam begins, we are expecting the price to go up," says Penke Satti Babu, a kanakambaram farmer from Madhavapatnam, who cultivates the flower in 10 cents of land, abutting irrigation canal.

"The expenditure is nominal for flower cultivation. But, we have to be attentive every day. In addition to agriculture works, we work in the garden during morning and evening time and the income is depending on the price," explains Mr. Satti Babu. Interestingly, none of these small and marginal farmers are taking any help from the Horticulture Department and most of them do not know of its existence.

"Initially, we bought seed from the outside market. Now we are able to make seed on our own. We can cultivate flowers round the year, as there are numerous varieties and there is no dearth of demand in the market," says Seeram Nagaraju of Samalkot.

Good Investment Potential for Floriculture

The good response to a pilot project implemented by the Horticulture Department for growing floriculture in the district has brightened the prospects for large-scale investment in exporting flowers. Floriculture is of two types — open cultivation of flowers like rose, crossandra, jasmine, tuberose (lily) and gladiolus and cut flowers under controlled conditions. Cut flowers like anthurium and orchids are raised through shade cultivation, carnation and gerbera through polyhouse and Dutch roses and Asiatic lilies through the greenhouse method.

Ideal Conditions

The climate in Anandapuram, Pendurthy and Sabbavaram mandal are said to be ideal for open cultivation. Experiments found that areas in Araku, Paderu and Chintapalle, due to hilly terrains and low temperature, are suitable for high-value roses under open cultivation. Under the pilot project grounded two years ago, 15 units were given loans for taking up polyhouse floriculture to grow carnation and gerbera by progressive farmers in Anandapuram, Sabbavaram, Anakapalle and Visakhapatanam (rural) mandals. Each unit was set up on an area of 500 square metres with individual investment of ₹5 lakhs.

Encouraging Results

The Horticulture Department gave a subsidy of ₹1.8 lakhs and the remaining amount was arranged through bank loan. "Good results during last two years has encouraged us to take up anthurium cultivation in a big way under shade houses," the Assistant Director of Horticulture, S. Ramamohan, said.

A study revealed that all the progressive farmers earned ₹2 lakh after spending ₹1 lakh on cultivation, labour and towards loan component thereby netting a profit of ₹1 lakh a year. The flowers grown in the district are sold locally due to market awareness. There is lot of demand for floral decorations for birthday and marriage parties, cultural and other meetings. "Though we have sent a few consignments to Kolkata, by and large, the total yield is consumed locally.

According to him, there is good scope for tapping the overseas market, once the work on modernisation of the airport is completed. Cold storage facility is not required as most of the flowers can stay intact for a week or two. Flowers can also be supplied to pharmaceutical companies to extract the medicinal ingredients in them. The National Horticulture Board (NHB) offers 25 per cent of the project cost or ₹20 lakhs (whichever is less) as subsidy to big investors. The Agriculture Process and Export Development Authority (APEDA) also offers sops for transportation and packaging to export-oriented units. At present, big units are confined to Bangalore, Sangli and Pune.

Karnataka

Karnataka is a major floriculture growing State in the country. The state has the highest area under modern cut flowers, and 40 flower growing and exporting units. The country's first flower auction centre is located in Karnataka. Karnataka has a land area of 22,340 hectares under floriculture in the year 2007-08 with a total production of 169, 120 metric tones of loose and 5550 Lakh number cut flowers. It plays a lead role in the cut flower exports from the Country.

Karnataka is well known for floriculture right from the 18th century onwards. During Hyder Ali and Tippu Sultan periods and also during the regime of Mysore kings, floriculture received great impetus and the colonial government also evinced lot of interest in this sector. The Lalbagh and Cubbon Parks established in Bangalore and the Brindavan Gardens established in Mysore in the early 20th century are testimony to these efforts. The flower shows are organized by the Mysore Horticultural Society, Bangalore twice a year at the historic Lalbagh Glass House.

The farmers in the state have been growing traditional flowers such as rose, chrysanthemum, tuberose, aster, jasmine, crossandra, marigold, champaka, gladiolus, and bird of paradise in the open fields. Some of these flowers are also being cultivated as cut-flowers in recent years. Rose, carnations, gerbera, and anthurium were grown under protective covers and these have gained momentum in the last 10 years. Recently, new crops like lilies, calla lily, iris, limonium, alstroemeria gypsophila, liatris, lisianthus and freesia have also emerged as potential cut-flowers in the state. Modern cut-flowers are relatively better in quality, have longer vase life and always fetch high unit price in the market. All these new trends have turned floricultural activity as an important agri-business activity in the state.

Trends in Area and Production of Commercial Flowers in Karnataka

There has been a steady increase in the area under commercial flowers in Karnataka. The area has increased from 12,000 hectares in 1990-91 to 21,000 hectares in 2004-05 with a compound growth rate of 3.72 per cent per annum. However, the area had come down to 18,000 hectares in 2002-03 from 21,000 hectares in the year 2001-02 due to drought. The production of commercial flowers increased from mere 61,000 tons to 151,000 tons during the same period with a compound growth rate of 6.59 per cent per annum. Thus, although the growth in the area under flower crops has increased at a slow pace, flower production has gone up at a fairly brisk pace.

District-wise Trends in Area and Production of Floriculture in Karnataka

Although the overall area under floriculture in the state has gone up, certain districts have witnessed a fall in the area. The annual compound growth rates of the area under floriculture are negative for many districts, the rate of decline being the highest in Raichur (ACGR = -26.52) followed by Dharwad (ACGR = -25.02). Among the districts registering positive growth rates, Dakshina Kannada stands first with the ACGR of 18.25 per cent. Among the newly formed districts, Koppal has recorded the highest ACGR of 23.34 per cent. Flower production in the state has also followed similar pattern, with the growth rates for most districts being negative, except for the districts closer to Bangalore namely Bangalore (both rural and urban), Kolar, Tumkur and Chitradurga, and Bellary among the northern region.

Bangalore is one of the largest flower trading centre in the Country catering to the marketing of flowers for southern states, especially Karnataka and Tamil Nadu. In 2003 The International Flower Auction Bangalore (IFAB), the operating company controlled by growers, has taken over the operations of the flower auction centre run by the State-owned Karnataka Agro Industries Corporation (KAIC).

The geographical area of Karnataka is 190.50 lakh ha of which, an area of 130.27 lakh ha comes under the cultivable area, constituting 68.38 per cent of the geographical area for the year 2005-06. Out of the total cultivable area, 17.25 lakh hectares were covered under horticulture (*Horticultural Crop Statistics of Karnataka State At A Glance 2006 -07*). Horticultural area in the State accounts about 13.24 per cent of the total cultivable area. Out of 17.25 lakh ha of the total horticultural cropped area, 7.65 lakh ha (44.35 per cent) come under Garden/Plantation Crops; 4.12 lakh ha (23.88 per cent) under Vegetables; 2.78 lakh ha (16.12 per cent) under Fruits; 2.45 lakh ha. (14.20 per cent under Spices and 0.25 lakh ha (1.45 per cent) under Commercial Flowers, including the area under the Medicinal and Aromatic plants.

Accordingly, the total horticultural production in the State during the year 2005-06 at 130.26 lakh tons. Detailed production figures stand at 47.36 lakh tons (36.36 per cent) with respect to Fruit Crops; 70.15 lakh tons (53.85 per cent) w.r.t., Vegetable Crops; 6.04 lakh (4.64 per cent) w.r.t. Spice Crop; 4.69 lakh tons (3.60 per cent) w.r.t. Garden/Plantation Crops and 2.02 lakh tons (1.55 per cent) w.r.t. crops coming under commercial Flowers, including the area under the Medicinal and Aromatic Plants (*Horticultural Crop Statistics of Karnataka State At A Glance 2006 -07*).

Due to the introduction of the high yielding varieties through improved technology, and also due to commercialization, the productivity of horticultural crops has improved. Recently, efforts are being made by the Government of Karnataka, to boost-up the agricultural exports, mainly of Horticultural produce like fruits, vegetables and flowers, through the effective Agricultural Policy.

Karnataka State occupied 4th place in respect of total Area in the Country contributing 14 per cent area to total Area, 2nd place in respect of total Production of loose flowers contributing 19.5 per cent production at all India level and 3rd position in production of cut flowers contributing 13 per cent in all India production. This is a clear indication that Karnataka State is in fore-front in the field of Floriculture. However, after examining the trends over last 3 years, we find a decline in both area and production compared to previous years (Indian Horticulture Data base 2008).

Floriculture, or flower farming, is a discipline of horticulture concerned with the cultivation of flowering and ornamental plants for gardens and for floristry, comprising the floral industry. The development plant breeding of new varieties is a major occupation of floriculturists.

Floriculture crops include bedding plants, flowering plants, foliage plants or houseplants, cut cultivated greens, and cut flowers. As distinguished from nursery crops, floriculture crops are generally herbaceous. Bedding and garden plants consist of young flowering plants (annuals and perennials) and vegetable plants. Flowers are mainly for export. This business is growing in the world at around 6-10 per cent per annum. In 2007 the size of the industry was $80 billion.

Floriculture, or flower farming, is a discipline of horticulture concerned with the cultivation of flowering and ornamental plants for gardens and for floristry, comprising the floral industry. The development plant breeding of new varieties is a major occupation of floriculturists. Floriculture crops include bedding plants, flowering plants, foliage plants or houseplants, cut cultivated greens, and cut flowers. As distinguished from nursery crops, floriculture crops are generally herbaceous. Bedding and garden plants consist of young flowering plants (annuals and perennials) and vegetable plants. Flowers are mainly for export. This business is growing in the world at around 6-10 per cent per annum. In 2007 the size of the industry was $80 billion.

In spite of a long tradition of Agriculture and Floriculture, India's share in the international market for these flowers is negligible. During the last ten years, taking advantage of the incentives offered by the Government, a number of Floriculture units were established in India for producing and exporting flowers to the developed countries. Most of them were located near Mumbai, Bangalore and Delhi and obtained the technical know-how from Dutch and Israeli Consultants.

Karnataka is the leader in floriculture, accounting for 75 per cent of India's total flower production. The state has the highest area under modern cut flowers, and 40 flower growing and exporting units. The country's first and only flower auction centre is located in Karnataka. In Karnataka, there are 18,000 hectares under floriculture cultivation. Karnataka is into floriculture for over 300 years. The Tigala community near Devanahalli and Chickaballapur are extremely good at growing flowers.

In 2003 The International Flower Auction Bangalore (IFAB), the operating company controlled by growers, has taken over the operations of the flower auction centre run by the State-owned Karnataka Agro Industries Corporation (KAIC).

Floriculture on the Bloom in Karnataka

Floriculture in Karnataka is growing by leaps and bounds with ₹ 7.5 crore worth flowers are expected to be auctioned this year. "We are a growing industry. In 1997-98, value of flowers auctioned was ₹ 19.63 lakh. In the next year it grew to ₹ 78.29 lakh and in 1999-2000 it was ₹ 2.05 crore. If we go by the value that has already been auctioned in 2000, total value of flowers will be ₹ 7.5 crore," said Karnataka Agro Industries Corporation Ltd., S pecial Officer Floriculture division, T V Reddy.

"As of now, KAC is not under any sort of debt and ₹ 700 crore worth interest has been waived off as the government cleared all its bills at one go. In the past two three years we have developed our own technology and except for patented plants the growers need not import anything." India's share in the world floriculture export is about one per cent of the world market at present. "But the industry has a tremendous growth potential. The climate especially in and around Bangalore is ideal for flowers. While other countries need artificial cooling in summer and artificial heating in winter and therefore have a high cost of production, we need none of that. Take for example Delhi. The cost of producing a rose there would be six rupees and we in Bangalore hardly spend one rupee.

"It is only our fault that we are not doing as well we can. The major problem is of quality. The first export samples that Indian growers send are excellent." "And then when the time comes to export, they will send 10 lots of flowers in the total say 10 million thinking that no body will bother. But the quality parameters abroad are so high that the whole 10 million gets rejected for the bad 10 lots.

Bangalore contributes 60 per cent on India's domestic floriculture trade. In the auction house, three to four batches of auction are held every day with over 20 bidders attending regularly. Of these 20 bidders, 5 or 6 are really big. Even now, in the month of May, which is the off season, they buy flowers worth ₹ 20,000 to ₹ 30,000 daily. And in the season they buy about ₹ 50,000 worth flowers every day. The auction was started in 1995 when it was done three days a week. But as growers started to come in with their produce, the bidders found it easy to procure flowers and auction was called seven days a week. From calling the bid manually, this auction house now has a digital system for bidding and the bidders can buzz their bid. Bangalore may soon get a whole new flower auction center of international standards in Hebbal.

Gujarat

What started as a hobby of few people has emerged as a huge business in south Gujarat. Floriculture has emerged as one of the most sought-after business in the region and has grown nearly seven times in the last five years. It began with plantation of gerbera and Dutch roses in a two-acre greenhouse in Kusad village near Navsari in 2004. It was in 2008-09 that commercial plantation of this two flowers came up in 40 acres of farm land. Today, these two varieties of flowers are grown in more than 400 acres of land. Close to 1,50,000 Gerbera and 10,000 Dutch roses are sent daily to different parts of the country from the region.

Farmers who have opted for greenhouse farming of these two flowers work on a bank loan but still an acre of Gerbera flowers plantation gives a return of at least ₹ 6 lakh. The flowers sell at ₹ 70-₹ 100 per bunch of 10 flowers in peak season like December-January and April- May. "South Gujarat's floriculture produce has grown remarkably in last five years. The region grows ₹ 20 crore worth Gerberas and about ₹ 5 crore worth Dutch roses," said Dinesh Pataliya, assistant director of horticulture, Surat.

According to the figures available from horticulture department, nearly 200 acres of land have plantations of flowers in Surat district alone. The rest are located in Navsari, Valsad, Tapi and Bharuch. The flowers are in huge demand in markets of Delhi, Mumbai, Bhopal and some south Indian cities too. Hitesh Patel, president, South Gujarat Flower Association told TOI: "We have huge scope to increase our production and reach more places as demand is very high. However, the lack of air cargo facility from Surat and is hampering our sales." At present, 300 boxes of 500 flowers each are sent from Surat by train and road to other states daily.

Gujarat Farmers Now bet Big on Floriculture

It's a bed of roses for floriculturalists in South Gujarat. In a span of just five years, exotic flowers from Europe like the Hybrid Tea rose, Gerbera and Carnations have emerged as big money-spinners for farmers in Gujarat. Currently about 70 growers in Navsari and Surat districts grow flowers worth crores in climate-controlled greenhouses, and a few more zealous ones have begun exporting them to countries like Japan, New Zealand, Germany and UK.

Most of the floriculturalists from this region are small growers who have set up greenhouses in about one hectare of land to grow these exotic varieties of flowers that fetch three times more returns than the traditional sugarcane and paddy crops. "Since the last three years, more and more farmers have ventured into the floriculture business, and today we have a combined turnover of about ₹ 50 crore. This figure is expected to rise three-fold in the next couple of years as the domestic market in the nearby cities of Delhi, Mumbai, Ahmedabad, Surat and Vadodara is growing at a rapid pace.

As the local black cotton soil available in this region is not suited for the growth of these flowers, these enterprising farmers have specially created mounds of red soil or Coco Peat (also known as coir pith or coir dust, it is a byproduct of extracting fibres from the husk of a coconut) to grow these flowers in the specially erected greenhouses. One of the largest exporters of exotic roses from this region, Kumar Patel of Best Roses Biotech says: "We currently grow the Hybrid Tea Rose in about six hectares of land in Kuched village of Navsari district. We annually export ₹ 5 crore of these flowers, mostly to markets in Japan and New Zealand. Besides them, we also export small amounts to Germany and UK," he said adding that India exports ₹ 250 crore of roses annually to about 30-40 countries.

The demand for good quality flowers in Japan is so huge that Best Roses has also set up two packaging houses in Tokyo and Nagoya where the perishable flower cargo from India is checked for damages and is re-packaged as per the requirements of the local market.

In order to grow some of the latest and internationally sought-after Hybrid Tea Rose varieties like Respect (Red), Royal Class (Red), Pretty Woman (Pink), Passion (Red), Aqua (Pink), Corvette (Orange), Gold Strike (Yellow), Circus (bicolor), Indian Sunset (Orange), Zaria (White), Exciting (bicolor), Kumar has set up greenhouses specially imported from Ulma Agricola, Spain. These imported units have fully automated and computer-controlled fertigation facility including for drip irrigation and misting. Apart from the roses, farmers in the region also grow 20 lakh flowers of Gerbera and 14 lakh units of Carnations.

A farmer Rajendra Bhakta who has planted 60,000 Gerbera plants and 56,000 carnations saplings in his farm at Orna village of Surat district says that the idea to take up floriculture came from one of his visits to Europe. "Initially, I have invested ₹ 1.30 crore to set up this project in 2005 and I am expecting a turnover of ₹ 35-45 lakh this year.

Kerala

Flowers play a vital role in human life from ancient times. Flowers are the inevitable part of our festive occasions. Flowers are largely used in special occasions like wedding, religious ceremonies and other functions. Women in Kerala like in other states widely use flowers to adorn themselves from ancient times onwards.

As per our heritage, our festivals and festive occasions are the festivals of flowers also. In our religious rites, flowers have very important place. Garlands, Bouquets and floral decorations are essential in our marriage occasions. In addition to this, the tendency of decorating hotels, commercial establishments and even houses with beautiful flowers is increasing now a days.

But flowers needed for these purposes are mainly coming from other states. Understanding these increased demand for flowers, steps have been taken to encourage flower cultivation in the state. Many people have shown interest in floriculture and are going ahead with this cultivation. To know the extent of cultivation and production of flowers, Economics and Statistics department had conducted a survey on floriculture along with vegetables, fruits and medicinal crops survey

As per the survey 601.97 acres of land was utilised for flower cultivation by 2945 cultivators in the state. This is only 16 per cent of the agricultural land occupied by these 2945 flower cultivators. The important flower crops cultivated are Orchids, Anthooriyam, Lotus, Banthi, Cut flowers, Ixora, Jasmine, Lilly, Marigold, Vadamulla *etc.* Orchids and Anthooriyam are cultivated in all districts except Kasargod where commercial cultivation of these flowers is not traded. Ernakulam is the largest producer of these two items followed by Thrissur. Malappuram district is in the first place in the cultivation of lotus. Among the flowers, the above three items are traded in numbers. All the other items are traded in weights. In the production of other flowers Thiruvananthapuram stood in the first place with 38.49 per cent. Thrissur is in the second place with 18.54 per cent and Ernakulam is in the third place with 13.58 per cent. In Kasaragod and Pathanamthitta districts, the production of flowers is below 1 per cent of total production of flowers.

Out of ₹1168 lakhs received by selling flowers, ₹395 lakhs (33.87 per cent) was received from Thiruvananthapuram district, 19.82 per cent received from Eranakulam and 21.5 per cent received from Thrissur district. Regarding the sales pattern of flowers it can be seen that 37 per cent of the total flowers was sold to consumers, 36 per cent to agents, 21.24 per cent sold to market within the state and 5.62 per cent to outside the state. Gross area cultivated for flowers was 667.57 acre, the income received from this cultivation is ₹116839470/- and the annual income per acre is ₹194095/- The cost of cultivation is found to be ₹13337320/- and out of this, 50.02 per cent is spent as labour charges. 10.73 per cent is spent for seeds and seedlings. The expenditure for manure is 8.09 per cent, for fertilizers 6.37 per cent, pesticides 4.5 per cent, organic manure 4.59 per cent, Hormone 0.39 per cent, transportation 4.81 per cent, irrigation 6.01 per cent, rent 0.33 per cent and other expenses 4.11 per cent.

It is found that a net income of ₹103502150/- was obtained from flower cultivation *i.e.* a net income of ₹171939/- per acre was obtained from this cultivation. Jasmine is the highest income-generating flower cultivated which has a share of 35.69 per cent of total income.

It is also understood that some assistance was received by this flower cultivators. 1137 cultivators were visited by some officers from different agencies. 287 were helped by supplying seeds or seedlings, 515 cultivators got some subsidy from few agencies, 117 cultivators got insurance protection for their crops.

It was found that an amount of ₹21985929/- was got as loan and ₹1608058/- as subsidy.

District-wise Utilisation of Area for Flower Cultivation in Kerala

Sl.No	District	Area in Cents			
		Agriculture Land Area	Non Agriculture Land Area	Total Area	Flower Cultivated Area (Net)
1	Kasargod	4393	393	4786	437
2	Kannur	1041	64	1105	159
3	Wayanad	13387	927	14314	1730
4	Kozhikkode	7070	267	7337	647
5	Malappuram	7875	689	8564	2868
6	Palakkad	90488	11250	101738	12019
7	Thrissur	39727	4551	44278	9820
8	Ernakulam	96063	14696	110759	14578

Sl.No.	District	Area in Cents			
		Agriculture Land Area	Non Agriculture Land Area	Total Area	Flower Cultivated Area (Net)
9	Idukki	4847	222	5069	446
10	Kottayam	9538	1714	11252	5009
11	Alappuzha	6857	1146	8003	1432
12	Pathanamthitta	3387	577	3964	587
13	Kollam	5271	682	5953	1490
14	Thiruvananthapuram	22835	2045	24880	8975
	Total	**312779**	**39223**	**352002**	60197

Annual Production of Flowers

Sl.No.	District	In Numbers			
		Orchids	Anthoorium	Lotus	Total
1	Kasargod	0			0
2	Kannur	400	1110		1510
3	Wayanad	12861	15577		28438
4	Kozhikkode	21633	27905		49538
5	Malappuram	10000	3980	434000	447980
6	Palakkad	1002	1197	15637	17836
7	Thrissur	86820	107407	48321	242548
8	Ernakulam	200841	442830	145	643816
9	Idukki	530	4352		4882
10	Kottayam	3515	82834	25061	111410
11	Alappuzha	2404	8965	190	11559
12	Pathanamthitta	1366	5438		6804
13	Kollam	5994	11365		17359
14	Thiruvananthapuram	85636	22735		108371
	Total	**433002**	**735695**	**523354**	**1692051**

Large quantities of flowers are coming to Kerala – major portion of its cost is for transporting and various levels of middle man involved in the business. Kerala climate is very ideal for Flower cultivation and the homestead cultivation of flowers is charming. Exotic verities of flowers shall be grown in shade houses, net houses and rain shelters. More earning flowers needs special protective structures like Poly house/greenhouse

Sl.No.	District	In Kilograms								Total
		Banthi	Cut Flower	Ixora	Jasmine	Lilly	Marigold	Vada-mulla	Other Flowers	
1	Kasargod				3021				1110	4131
2	Kannur		1730	280	600	2000	2260	190	19282	26342
3	Wayanad	998	2964	1086	1852		4902		76125	87927
4	Kozhikkode	1945	657	5750	2282	197	2201	263	23966	37261
5	Malappuram		342	600	2312	0	0	250	25850	29354
6	Palakkad	5396	787	971	216288	690	13706	3881	32918	274637
7	Thrissur	9732	61488	37388	144521	339	29814	4750	256536	544568
8	Ernakulam	4045	931	1425	336411	6939	1050	927	47042	398770
9	Idukki	1970	2200	390	2455	1380	1150	15	18118	27678
10	Kottayam	2948	305	1413	5840	264	244	2745	191539	205298
11	Alappuzha	3406	1710	274	79598	364	105	1570	31264	118291
12	Pathanamthitta	428	175	150	609		245	78	2545	4230
13	Kollam	47	2104	3934	9511	2428	2126	2179	24997	47326
14	Thiruvananthapuram	5628	6114	39786	138274	14046	36801	5846	883786	1130281
	Total	36543	81507	93447	943574	28647	94604	22694	1635078	2936094

Roses

Hybrid tea rose, polyanthas floribundas, shrub rose and miniatures are different groups of roses popular in India. The heavy rainfall in Kerala is the hardily to be overcome by special protective structure like rain shelter/polyhouse.

Gerbera

The gerbera cultivation is successfully done in poly houses in Kerala-the climate prevails at hill tract is favorable for gerbera cultivation. The flower demand is very high and hence marketing is much easier.

Anthurium

In India the most favorable climate for anthurium cultivation prevails at Kerala. The crop is rewarding from the first year onwards and the investment involved is much less.

Jasmine

The perfume of jasmine is graceful, the use of jasmine flowers is so diversify that the great demand of jasmine is never achieved local production. So many cultivars of jasmine are available for cultivation and being a shade loving plant can be cultivated easily in courtyards.

Helichornia

Wild varieties of its kind is familiar from years but new introduction of hybrid cultivars make it a profitable cultivation, The long shelf life of helichornia flowers and the shade loving nature of the crop make it more appreciative to our farmers.

Cut Flowers

Cut flowers and seasonal flower cultivation are complimentary crops cultivating in the interspaces of coconut and rubber replanting. Merry gold, aster, celosia, bachelors button, sun flower, salvia, phlox and Zinnia can be grown as complimentary crop along with cash crops during initial years.

Andaman and Nicobar

There are about 2000 species of flowering plants, of which nearly 215 are endemic to these Islands (Rao and Srivastava, 1996). Of these flowering plants, nearly 110 orchid species are naturalized here (Shrama, Singh, Sreekumar and Nair, 1998).

Despite the rich natural base of floriculture, organized cultivation of flowers and ornamentals towards commercialization is yet to begin in Andaman and Nicobar Islands. Since agro-climatic conditions of these Islands are similar to those in South-East Asia like Thailand, Singapore, Malaysia and Indonesia, there is great scope for floriculture especially orchid cultivation out here in a commercial scale. Thailand alone exports orchid spikes worth $ 39 million, followed by Singapore with $ 16 million. Besides, the indigenous wild orchids are hardy, easy to propagate and grow, resistant to many diseases and insects, and therefore good germ-plasms for breeding purpose.

A few orchids of Andaman and Nicobar Islands:

☆ *Arachne* spp.

☆ *Bulbophyllum lepidium*

☆ *Cymbidium aloifolium*

☆ *Dendrobium secundum*

☆ *Eria andamanica*

☆ *Geodorum densiflorum*

☆ *Malleola andamanica* (endangered sp.)

☆ *Rhynchostylis retusa*

☆ *Vanilla andamanica* (endangered sp.)

Exotic Dendrobium hybrids, Vanda Miss Joaquim, Cattleya, Epidendrum *etc.* have also been successfully introduced here and performing well. Plantation crop like coconut can be used for growing orchids as intercrop, providing shade as well as standards (Sharma *et al.*, 1998). Other than orchids, a number of flowers, ferns, cycads, succulents, bamboo and ornamental medicinal plants grow naturally in forests of the pristine Islands of Andaman and Nicobar, which need proper attention and perpetuation.

Meghalaya

Floriculture was a passing hobby practiced by flower lovers and enthusiasts and Meghalaya is no exception. The people of Shillong were known for their passion with flowers which adorn their verandah and in lawns of prominent houses in Pine city. However in early 2000, availability of improved planting materials, seeds and other scientific, technical inputs and the increasing market demand, has encouraged growers to exploit the commercial potentialities of this hobby. As one hobbyist in Shillong remarked "this hobby of mine has now become a self paying venture due to the increasing awareness by housewives on this art". Smt. Queency Thangkhiew, one of the earliest enthusiasts has turned this hobby into a thriving commercial enterprise. Floriculture is now here to stay and Meghalaya is set to witness a colourful revolution in the days ahead. The vision of the State Government to turn Meghalaya into a flower State of India, with the active support and participation of such brave entrepreneurs, the future is rosy indeed.

Commercial floriculture is a recent phenomenon in Meghalaya, promoted and assisted by the Government of India Scheme "Technology Mission Scheme on Horticulture". Considering the natural advantages that the State is endowed with and the varied range of agro-climatic conditions available, there is high potential for cultivation of all types of flowers. The rich flora and the many species of orchids growing wild in the State, which is the highest ever recorded in a single concentrated area, is a testimony to this fact.

Meghalaya also has a very high potential for commercial floriculture due to the many competitive advantages the State has: a favourable climate, diverse agro climatic situations suitable for tropical and temperate flowers, competitive labour cost, proximity to Guwahati and Kolkatta Airport *etc.* The State is divided into 2(two) major floriculture zones namely; the temperate zone and the sub-tropical zone. The entire East and West Khasi Hills and the upper parts of Jaintia Hills district falls under temperate zone while Ri-Bhoi and the Garo Hills falls under the sub-tropical zone.

Floriculture produce, being a non food crop, was not a thrust area in the overall agriculture development plan of the State till the late 1990's, where the thrust was on increasing food grains production and productivity. It was in 1993-94 that the Department of Agriculture started to popularize floriculture among flower growers and enthusiast through Government of India scheme. The Agri-Horticultural Society, Shillong aided by the Department of Agriculture also organizes flower show and flower competition in Shillong to promote floriculture. This became an annual event and has brought a lot of awareness to the public about modern floriculture practices and marketing prospects.

Further, with the establishment of a separate Directorate of Horticulture in the State, floriculture got a boost through the various schemes of the State Government like Development of Floriculture Scheme started in 2000-2001, Setting up of Model Floriculture Centre, Establishment of Floriculture projects at Dewlieh in Ri-Bhoi and Samgong Horticulture Farm in E.Garo Hills districts through the Technology Mission Scheme on Horticulture. The objectives of the schemes were to focus on the promotional and awareness aspects by providing incentives to the farmers and motivating them to grow traditional as well as non-traditional floral crops and houseplants for commercial purpose. The nature of assistance provided is in the form of providing the growers with diseased free planting material, organic/ inorganic fertilizers, plant protection chemicals, garden tools and implements for a minimum area of 2000 square meters, along with a package of practices for commercial production. Each unit is envisaged to serve as a demonstration model for which the Department provides technical guidelines through extension and training. Crop selection is on the basis of existing popularity and market demands. A few of the recommended ornamental crops grown are Orchids, Chrysanthemums, Gerberras, Carnations, Liliums, Strelitzia reginae, Gladiolus, Asters, Marigolds, Statice, Gomphrenas, Helichyrsums, Zinnias, Roses and different kind of house plants *etc.* The concerted effort of the Department in motivating growers as well as providing infrastructural support in the form of greenhouse, poly house, shade nets and other inputs has led to the establishment of a number of private nurseries especially in the East Khasi Hills district.

The Directorate of Horticulture has also identified floriculture clusters in the State which are suitable for growing certain high value flowers. East Khasi Hills have been identified for Orchids, Carnations and Gerbera; West Khasi Hills for Carnations; Jaintia Hills for Bird of Paradise; Ri-Bhoi for Orchid, Rose, Anthurium, Lilium and Foliage; West Garo Hills for Liliums and Bird of Paradise; East Garo Hills for Anthurium, Foliage and Bird of Paradise. At present, East Khasi Hills, East Garo Hills and Ri-Bhoi districts have established greenhouse floriculture units in the departmental farms as well as in farmer's field. The estimated area under floriculture in the State is about 500 Hectares.

The Rose Pilot project which was initially started at Dewlieh Departmental Farm in Ri-Bhoi district at an area of 0.5 hectare has been a success with a production of 2500 cut flowers per day. Today, Rose, Lilium and Anthurium cultivation has also been extended to farmer's field in village clusters through self help groups and individual farmers.

The Anthurium project at Samgong Horticulture farm is a tourist spot for farmers, high power dignitaries and the common people. This is because of the success of the project taken under the cultivation of the flower. The excellent marketing of anthurium as cut flower gives a phenomenal impact to the farmers which encouraged them to go for commercial group cultivation.

Commercially, there are several existing limitations in the floriculture development in the State, which needs to be overcome through proper planning and developmental efforts at all levels. However, from a technical point of view, the development of floriculture business enterprises in Meghalaya has a very high potential. This potential can be exploited to improve the socio-economic condition of the State. In addition, being a major export item, it can substantially contribute towards foreign exchange earning of the entire country.

The India Today ranking of States in India has shown Meghalaya's agriculture to be faring comparatively better among the Small States in North Eastern Region and the magazine ranked Meghalaya comparatively better. The State has also improved from its earlier overall 5th position during 2006 to becoming 1st among North Eastern States (except Assam which is categorized as a big State) during 2007. This goes on to show that agriculture, specially horticulture in Meghalaya is on a growth trajectory and the State only needs a boost in resources, better postharvest management, marketing linkages, organized marketing and creation of economy of scale in order to accelerate it to the take off stage. With the right policies of the State Government and adequate fund injection through the Technology Mission for Horticulture Development along with farmer's and entrepreneurs of the State coming forward to take up commercial floriculture, Meghalaya will not be lagging behind and the vision articulated by the Department of Agriculture will hopefully, flower and bloom in its full glory and beauty.

The climate of the State, particularly the areas near Shillong in East Khasi Hills District, are well suited for cut flower production. Proximity of Shillong to Guwahati and possibilities of restart of the Umroi Airport near Barapani, have strengthened the required infrastructure support for marketing of Floriculture Products from the State.

The Recommended Project Profiles are :

1. Area :
   a. Upper Shillong/Shillong
   b. Barapani
   c. Umiam
2. Products :
   a. Cymbidium, Carnation

      b. Carnation, Gladiolus, Lilium

      c. Carnation, Gladiolus, Lilium

Darjeeling

Darjeeling hills are the natural home for countless orchid species like Cymbidiums, Vandas, Dendrobiums, Paphiopedilums, Lycaste, Odontoglossum, Phaius, Arundina *etc.*the list being endless.

In the past several decades the nurseries of Kalimpong area was very much involved and buzzing with floricultural activities and developed their own techniques in tissue culture propagation of orchids and other related floricultural plants. In Kalimpong itself we have about four nurseries propagation. Exports from these hills also started 5-6 decade back. For the unlimited scope in the present multi-million dollar floriculture industry, these hills are the natural habitat for innumerable plant species and thus much has been achieved till date by our floriculturists. However, this region still has enormous potential. With the global floricultural trend these hills have limitless scope for production of Gladioli cut flowers to cater to the demand of both the domestic as well as the export market. Cut flower started trade over three decades back. Today other cut flowers, besides Gladioli are anthuriums, Orchids particularly Cymbidiums, bulbous flowers of lilies, ornithogalum and other flowers like gerberas, carnations and greens like ferns are under production.

    ☆ 56 km from Kalimpong and situated at an altitude of 5500ft. panoramic views of Kanchanjunga can be obtained from this point. Fabulous view of Sunrise over Kanchanjunga can be seen from Jhandi Dara.

Jammu and Kashmir

Ornamental Floriculture

Department is involved in maintenance and development of Gardens and Parks in the valley to boost tourism at various prime tourist destinations like Mughal Gardens (Nishat, Shalimar, Cheshmashahi, Pari Mahal, Verinag, Achabal and Jarokabagh), Botanical Garden, Kokernag, Pahalgam, Manasbal, Tulip Garden, Nehru Memorial Botanical Garden, Children Park, 98 city parks and other VIP Quarters. Besides department is involved in production of flower seedlings and other ornamental plants from its nursery known as Plant Introduction Section, Cheshmashahi. In addition maintenance and development of various Zanana/ Children Parks is also going on. Renovation, restoration and preservation of heritage gardens are some other major initiatives of the department for the last five years.

Commercial Floriculture

For the last five years department has laid major thrust on commercial floriculture under the Centrally sponsored schemes Technology Mission, RKVY for Integrated Development of Floriculture and in this field sufficient progress has been made by involving farmers of different categories particularly unemployed youth. People are showing positive response to different schemes, but several apprehensions related to the "Postharvest Management and Marketing". Postharvest

Management and Marketing is a vital component of Commercial Floriculture which is managed optimally despite shortage of manpower and infrastructure. Symmetrically, Floriculture is playing its role of creating and maintaining various parks and gardens and assets, as a supplement to tourism and recreational activities. With the inexorable adoption of floriculture as a commercial activity, not only at the national but at international level as well, department has developed link in popularizing commercial floriculture in the valley. Floriculture department has a vast scope and potential in the valley, which is evident from the fact that during 1996 an area of 80 Ha was under flower cultivation in J&K, which expanded to the level of 350 Ha, with an annual turnover of about ₹1350 lacs. Further, more than 1500 unemployed youth are directly engaged under commercial floriculture sector in Kashmir Division. Moreover, an area of about 9.297 Hac. Has been covered under protected cultivation, in private sector by revising Tubular Structure Poly Houses, High Tech Poly Houses, Shade Net Houses, *etc.* A flower mandi has also been established and inaugurated by Hon'ble Minister for Health, Horticulture and Floriculture on 15th October, 2012 at Rajbagh, Srinagar, with an idea to provide facilities like marketing and preservations of cut flowers and cold storage facilities to the flower growers on no profit no loss basis. In order to facilitate flower production, department has introduced contact farming for establishing flower-seed villages. Seeds of different flowers like, Cosmos, Zinnia and Salvia have been provided to the growers of the valley, in the buyback scheme, who have produced the flower seed of different kinds.

5

Rose

"Oh! No man knows through what wild centuries roves back the rose"; Walter de la Mare so rightly said.

A charming thought, but somewhat aggravating to rose- lovers who would like to know when the fascination of man for roses started. Even more frustrating for an Indian rose historian like me is the fact that there are no written records in India, covering periods where civilization was otherwise greatly advanced, like the Indus Valley Civilization (2500-1500 B.C.). Obviously the association of roses with India's history goes back to well over 5000 years. But, as earliest Indian history is based on oral tradition and not the written word, one has to depend on traditional myths, legends and folklore.

One such legend, widely related even today, harks back to the time when the Gods were said to live on earth. Brahma, the God of Creation, and Vishnu, the God of Protection had a debate about which flower was the most beautiful. Brahma favored the lotus while Vishnu preferred the rose. Upon being shown the arbor laden with lovely, fragrant roses in Vishnu's celestial garden, Brahma conceded the argument.

The presumption that such legends are rooted in reality is reinforced by the fact that the abode of the Gods has always been considered to be the Himalayas - this 1800 mile long sweeping chain of snow capped mountains, stretching from the north west to the north east of India, is also home to many wild rose species.

The medical treatises of the doctors, Charaka, a physician and Susruta, a surgeon — who, together are considered the founders of Ayurveda - the Indian system of medicine - ('ayurveda' literally translates as the 'science of life') and who are believed to have lived in the first century before Christ, have many references which give ample proof that roses were known in those days. The rose was known by many names, in the language of those times, Sanskrit, and each name was in fact an adjective symbolizing a particular characteristic, of its curative efficacy for a particular symptom or illness. Thus it was considered to be cooling, a laxative, an appetizer, it cured fever and boils, it lessened stomach burning.

Many other texts mention the rose in different contexts. One called 'Kashyapa's Agricultural Treatise' (c.800C.E.) lists the rose among other plants, trees and climbers that a king should plant within his kingdom. Another manuscript (Chakrapani Misra's 'Visva-vallabha', 15th century) states that the rose is an indicator of underground water.

When Alexander the Great invaded India, on its north-west border, in 327 B.C., it is said that he was amazed at the wealth of plants he beheld, and among others he sent back some rose plants to his mentor Aristotle.

Coming into the Christian era, the monk Vatsyayana, who wrote the 'Kama Sutra' (A Manual of Love) around 100 C.E., gives a list of the duties of a virtuous wife, and one of them is to tend a garden, which would have a number of flowering and fragrant plants, and the rose is one of them.

The poet Kalyana Malla's book, 'Ananga Ranga' (written about 1172 C.E.) describes the ambience of a room best suited for amorous interludes, and mentions the lavish use of rose water. Both these books have been translated by Sir Richard Burton. (19th century).

Though Avicenna (Ibn Senna) of Persia is credited with the discovery of rose water by distillation, the Ayurvedic text of Charaka, the doctor, has references to rose water, and the Buddhist sage, Nagarjuna, (1st century C.E.) who was an alchemist, describes the process of distillation in his scientific treatise, Arkaprakash.

The rose used in those bygone days would, presumably, have been what was later referred to as Rosa damascena 'Bussorah'. Even today, rose water is made in India from this rose as also with Rose Edward which is called 'panneer' rose. 'Panneer' means rose water. Trade with China existed from around 1000 B.C. as many Chinese scholars came to India, the land of the Buddha. When the Buddhist travellers Fa Hien and Hiuen Tsang (5th and 7th centuries) visited many of the sacred places associated with the Buddha, they mention in their writings the beautiful gardens, replete with roses and many other flowers and fountains that they saw and marvelled at.

The 'Gita Govinda' of poet Jayadeva (written in 1150 C.E.) and translated by Sir Edwin Arnold in 1879 as 'The Song of Songs' (who is more known for his 'Light of Asia') is replete with references to roses.

A Muslim traveller to India, Rashid-ud-din (1247-1317 C.E.) who is considered by Percy Browne to be the most reliable of Persian historians, visited the western Indian state of Gujerat in 1300 C.E. and remarked "the people are very wealthy and happy and grow no less than 70 kinds of roses". From the 14th to the 16th centuries one of the greatest and most splendiferous of dynasties was the Vijayanagar Kingdom of southern India. Two Portuguese travellers, Domingo Paes and Fernaz

Nuniz who visited this kingdom in 1537 were amazed to see that roses were integral to the daily life of both aristocrats and commoners. They saw plantations of roses in the countryside, and the bazaars of the capital city of Hampi had many shops with baskets laden with roses. Men and women adorned themselves with roses. And the king of the time, the mighty Krishna Deva Raya, a great scholar-poet-warrior, dressed in silken robes of pure white embroidered with golden roses, would, every morning, when he gave audience, shower his favorite horses and elephants and courtiers, with white roses. The king's bedchamber (and I quote) " had pillars of carved stone, the walls all of ivory, as also the pillars of the cross-timbers, which at the top had roses carved out of ivory, all beautifully executed and so rich and beautiful that you would hardly find anywhere another such".

Another traveler, Abdur Razzak who visited this kingdom earlier in 1443 writes "Roses are sold everywhere. These people could not live without roses and they look upon these as quite as necessary as food". An Italian traveller, Ludovico de Varthema, from Bologna, visited the western coast of India to a port called Calicut in 1503 and he writes that he saw innumerable flowers, and especially roses – red, white and yellow.

In the 16th century the Muslim Mughal emperors came from Persia and Afghanistan to rule India. The first emperor, Babur, (1483-1530) brought camel loads of roses- most probably musk and damask roses. His love for roses is epitomized by his poem:

'My heart, like the bud of the red, red rose,

Lies fold within fold aflame,

Would the breath of even a myriad Springs

Blow my heart's bud to a rose?'

He was so enamoured of roses that he gave all his daughters rose names – Gul chihra (rose-cheeked), Gulrukh (rose-faced), Gulbadan (rose-body) and Gul rang (rose colour). Gul means rose in Persian. Babur developed the formal geometric 'char bagh' (four quadrants) design of Mughal gardens, where many roses were planted. This style was refined by later kings and the Mughal gardens in Kashmir, called the Shalimar garden, and in Agra, in the Taj Mahal, and in Delhi, the capital, were world famous. Emperor Akbar, the 3rd Mughal Emperor, (1556-1605) invariably took camel loads of roses to give to the wives of his allies.

The next Mughal Emperor Jehangir's wife, Nur Jehan is credited with the discovery of rose oil, called attar in Persian, while she was taking a rose scented hot bath. The finest of rose liquers was distilled for the aristocracy – about 20, 000 flowers for one bottle of liquer. Manucci, the 17th century Italian traveler was amazed to see that the nobles not only used huge quantities of rose water and oil on themselves, but even rubbed their horses with these as a matter of course.

Nearly every painting of this period shows the subject of the painting holding a rose in his or her hand.

History of the Rose

Roses have enjoyed the honor of being the most popular flowers in the world for the longest time. The reason for popularity of the rose flower may be its wide variety in terms of color, size, fragrance and other attributes.

Perhaps there is no other flower that has been loved by people so much or has been as popular as the rose throughout history. In fact, roses have been in existence much before humans, who later grew their beautiful flowers, drew wonderful pictures of the flowers and even commemorated them in their lore and music. Believe it or not, the first trace of rose has been found in a fossil at Colorado's FlOdishant Fossil Beds dating back to 40 million years! Some rose fossils discovered in Montana and Oregon also date back to about 35 million years, much before the humans came into existence. In addition, rose fossils have also been discovered in Yugoslavia and Germany. In fact, there is evidence that there was a time when roses grow in the wild in the extreme northern regions, such as Norway and Alaska. Southwards, this species was also found growing in the wild in Mexico as well as North Africa. However, it is interesting to note that thus far it could not be confirmed whether any rose grew in the wild below the equator.

It is believed that the rose has its origin in Central Asia and its estimated origin dates back to anything between 60 and 70 million years - the period known as the Eocene epoch. Gradually, it spread all over the Northern Hemisphere. Available documents show that various early civilizations, such as the Egyptians, the Chinese, the Greeks, the Romans and the Phoenicians not only appreciated roses, but also cultivated them extensively as early as 5,000 years back.

Way back in 500 B.C., renowned ancient Chinese editor, politician and philosopher Confucius wrote about the roses grown in the Imperial Gardens and also mentioned that the Chinese emperor's library contained several hundred books on roses. It is interesting to note that the gardeners who cultivated rose during the Han dynasty (207 B.C. to 220 A.D.) were so passionate with roses that their parks posed a threat for the agricultural lands, often encroaching upon the latter. This led the then Chinese emperor to issue orders to plow under some rose gardens.

Rosa gallica, which is also called the French rose, has been identified as the oldest rose that exists even to this day. There was a time when Rosa gallica bloomed in the wild all over central and southern Europe, in addition to Western Asia. Even today, this rose is found growing in these places. While the precise origin of Rosa gallica is yet to be ascertained, some hints of this plant were seen way back in the 12th century B.C. In those days, it was regarded to be a symbol of love by the Persians.

The association of roses with Indian history goes back to the very beginning of time. India has had an unbroken thread of civilization stretching to well over 5000 years. The archaeological sites of the " Indus Valley Civilization " - Harappa, Mohenjodaro, Dolavira *etc.* date to around 3000 B.C. and show a very advanced culture - well planned, geometrical grid - constructed cities, with wide streets, brick houses, zoning, water supply, sewage disposal systems *etc.* which were carefully coordinated. However, earliest Indian history is based on oral not written words. The Hindu view of past has traditionally centered on myths, legends and folklore, an oral tradition prone therefore to distortions and revisions. But those legends which persist have a cachet of authenticity. One such about a discussion between the two Gods, Brahma, the creator and Vishnu, the protector, on which flower was the most beautiful. Brahma favoured the lotus and Vishnu the rose. After seeing the arbour laden with fragrant roses in Vishnu's celestial garden, Brahma acknowledged the supremacy of the rose over all the flowers, including the lotus. Another legend has it that Vishnu created his consort, the Goddess Lakshmi out of the rose. Whilst most authorities claim she was created out of the lotus, as a rosarian I would prefer to believe it was the rose.

The presumption that legends such as these have a stamp of authenticity is because the abode of the Gods. Again according to legend, were the Himalayas - the sweeping range of snowcapped mountains from the northwest to the northeast of India, which are home to more than 15 species of wild roses – roses which grew from time immortal, as substantiated by the medical treaties of Charaka and Susruta, founders of the Indian system of medicine called Ayurveda. They are believed to have lived in the 1st Century B.C. The medicinal properties of the different rose species were well known and the characteristic Sanskrit name (the language in which all ancient texts were written) given to each rose signified its curative properties for particular ailments. For example, in the Charak Samhita (*i.e.* the treatise of Charaka) the rose was named in 8 different ways - "satapatri" (having 100 petals) - centifolia perhaps?, "sivapriya" (loved by God Siva who lived on Mount Kailash in the Himalayas), "Saumyagandha" (having a pleasant smell), gandhyadya, susita, sumana, suvritta, satapatrika, - *i.e.* it is cool, bitter, laxative, an appetizer, cures fever and boils, lessens stomach burning *etc.* The rose was also called atimanjula, sevantika and tarunipushpa.

The wild roses of the Himalayas are *R. brunonii, R. sericea, R. webbiana, R. foetida, R. longicuspis, R. macrophylla, R. gigantea, R. beggariana, R. eglanteria, R. laevigata, R. baksii* and *R. bracteata.*

Going back to earlier ages, Gautama the Buddha lived in the 6th Century B.C. The Gandhara School of Sculpture (1st Century B.C. to 2nd Century A.D.) which is an amalgam of Indian and Greek traditions (Alexander the great invaded India in 327 B.C.) depicted the Buddha in many Statues. Most show him sitting on a pedestal of lotuses, but a few seem to depict him seated on a 5 petalled rose.

Coming into Christian era, the monk Vatsyayana, who wrote the Kama Sutra (Manual of Love) sometime between the 1st and 3rd Centuries A.D., mentions that one of the duties of a virtuous wife was to tend a garden, "In the garden she should plant beds of green vegetables, bunches of sugarcane and clumps of fig tree, the mustard plant, the parsley plant, the fennel plant and the xanthocymus pictorius, clusters of various flowers, such as the trapa bispinosa, the jasmine, the tabernaemontana coronaria, the CHINA ROSE and others should likewise be planted together with the fragrant grass andropogon schoenanthus and the fragrant root of the plant andropogon miricatus. She should also have seats and arbours made in the garden in the middle of which a well, tank or pool should be dug. There is a school of thought which considers that the reference to the "china rose", is to the hibiscus and not rose.

Obviously trade with China existed as early as the 2nd century B.C. as a Yue – chi chief took Buddhist scriptures to China (after Kasyapa Matanga introduced Buddhism to China). Chinese pilgrim who visited Buddhist sites talk of rose garlands being commonly used. Fa Hien and Hieun Tsang, two Chinese travelers (5th and 7th centuries A.D.) mention in the diaries that beautiful gardens with a variety of plants including roses, fountains, streams and tanks of clear water were part of garden landscapes.

Trade with China as with many other eastern and western countries was carried on over the centuries by the various kingdoms ruling different parts of India and rose products were very much a part of such commerce. A stone pillar inscription dated 1244 A.D. at Motrupalli, near Guntur on the eastern coast mentions that the King Ganapati, belonging to the Kakatiya dynasty of Warangal, waived the custom duties on various items from China including rose water. Marco Polo, the Venetian traveler, visited this kingdom around this time - 1288. It is thus obvious that roses were very much a part of the social, medical, cultural and religious fabric. India, while being largely a Hindu country, has also one of the largest populations in the World. Again, it is one of the original Christian countries - the apostle St. Thomas arrived on its western shores nearly 200 years ago in A.D. 40.

In 1300 A.D. a Muslim traveller and chronicler, Rashid – ud - din, who visited Gujarat state in Western India remarks that "the people were very wealthy and happy and grow no less than 70 kinds of roses".

One of the greatest dynasties of South India from 14th to the 16th centuries was the Vijayanagar Empire and records of many travelers show that roses were integral to the daily life of both aristocrats and commoners. Domingi Paes and Fernaz Nunis, two Portuguese travelers who came to this kingdom around 1537 mention seeing plantations of roses, bazaars where baskets laden with roses were sold, both as loose flowers and made up as garlands, and gardens of the nobility which had rose plants growing in profusion. Both men and women from all walks of life used roses in great quantities as ornamentation. And the King, dressed in robes of pure white, embroidered with golden roses and wearing lots of jewels, would offer his morning prayers, daily shower white roses on his favourite courtiers, his favourite elephants and horses, which would be bedecked with chaplets of roses. They describe the King's bedchamber " which had pillars of carved stone, the walls all of ivory as also the pillars of cross timbers, which at the top had roses carved out of ivory, all beautifully executed and so rich and beautiful that you would hardly find anywhere another such."

Abdur Razzak, a Muslim diplomat from Persia, who visited the same royal court in 1443, writes, " Roses are sold everywhere. These people could not live without roses and they look upon these as quite as necessary as food ". Rose perfume making was known and in fact was exported to other countries.

In the 16th century when the Muslim Mughal emperors came from Persia and Afghanistan to rule India, they brought camel loads of roses. In fact the first Mughal emperor, Babar, is said to have brought the damask rose into India. This royal dynasty had a great interest in art and architecture, poetry and music and the Mughal style of gardening - a very formal plan with geometrical beds, fountains, paths *etc.* all done on a grand scale, were part of the edifice they built, like the Taj Mahal, and roses were lavishly planted. The famous Shalimar gardens of Kashmir, built by Jahangir for his empress Nur Jehan are well known to this day. At a much later date this is the garden which delighted visitors with the incredible display of Marechal Neils which were planted to climb the grey - green walls of the Hall of Public Audience to hang their soft yellow globes head downwards in clusters from the carved cornices".

The same empress, Nur Jehan, celebrated for her beauty and intelligence, is credited with discovering attar (or Otto) of roses. When she was taking royal bath in hot rose scented water, she saw a scum forming which had a concentrated rose fragrance. She collected this scum. According to Emperor Jahangir "it was of such a strength in perfume that if one drop was rubbed on the palm of the hand it scented an entire assemblage, and it appeared as if many rose buds had bloomed all at once. There is no scent of equal excellence. It restores hearts that have gone and brings back withered souls."

The finest of rose liquors was distilled for emperors and royalty. About 20,000 rose flowers were distilled in order to get one bottle of rose liquor. Practically every painting of this period, of royalty and aristocracy, show the person holding a rose in their hand. Indeed, roses are considered so much a part of the Mughal period of Indian history that most Indians believe that roses came to India only with the advent of the Muslims - 10th century onwards.

When the British came to India in the 17th century, originally as a trading company, called The East India Company (later they took over the reins of the government from the Mughal kings) their ships from China carrying merchandise to England stopped for refueling at the port and capital city of Calcutta, on the east coast. Nearly every ship would carry live plants, including roses, which are becoming popular in England and France, and they would be kept in the Botanical Gardens of Howrah, a suburb of Calcutta on the banks of the river Ganges, to recover before continuing on their journey to England. Some plants from each batch would be planted down in this garden which was started by Sir William Roxburgh in 1793. One such plant was fortune's Double yellow as also other Chinas, Noisettes and Early Teas. Many of them are still available in old nurseries and gardens, though they are nameless, have names mauled beyond recognition or have names other than the ones given them after they reached their final destination - England.

Unfortunately, with modern day roses being easily available, interest in these old beauties - these heritage roses, has been lukewarm. However, with rose rustlers like my husband and myself scouting all around, searching for these old varieties, we have been able to enthuse others to collect these varieties before they disappear. We have a large collection and we have been able to tentatively name most of them, after looking at descriptions and photographs in old rose books.

Coming now to Rose products which are distinctively Indian, roses are the basis of many rose formulations, cosmetic, medicinal and dietary. In many areas of north India, especially where the soil is rich and the water abundant, like Pushkar in Rajasthan, and parts of Uttar Pradesh, Rosa Damascena as also Rose Edward are grown on a commercial scale, both for distilling Rose Oil and Rose Water. Another heritage rose widely grown is Gruss in Teplitz.

Tonnes of rose petals are dried in the shade to make Pankhuri, to be sent everyday to Middle East for use in beverages, food and medicine. When we visited Pushkar a few years ago the entire air was scented throughout the day with a heady rose fragrance.

Rose water is used as a coolant and medicinally in eye drops and lotions and as skin moisturizers. It is also used as food flavouring, especially in Indian sweets and as a welcome spray at all festive occasions, like weddings. "Arkprakash" an old Sanskrit text mentions rose water distillation and a famous Buddhist monk, Nagarjuna, who lived in the 8th/9th century A.D. has given details on how rose water is to be distilled. Around the same time the Arab historian, Ibn Khaldun also described the process.

Rose Attar or Otto (the westernized word) is Rose Oil, also called Ruh Gulab. 10 gms of rose attar are equivalent in price to 10 gms of gold! and 1 kg of oil is obtained from 4000 kgs of petals. "Attar" is a Persian word meaning perfume and is made from steam distillation of rose flowers plucked very early in the morning. The first product is rose water and from the water, over a period of days, rose oil in minute quantities is collected. Bulgaria, Morocco, France and India are the major rose oil producing countries. Rose damascene and Rosa bourboniana are the varieties generally used in India for rose oil (though R. centifolia, alba and gallica are also used elsewhere). Rose oil is used in perfumes, soaps and other cosmetics like creams and lotions, for flavouring soft drinks and alcoholic beverages, as also in the Indian system of medicine.

Gulkand is a kind of preserve made by mixing equal quantities of rose petals of R. Edward or *R. damascena* and white sugar and kept in the sun till they coalesce together. This is used as a tonic and a laxative.

Gul roghan is a hair oil made by maceration of rose flowers with warm sesame seed oil. Rose essence, rose syrup, rose sherbet, rose wine, rose liquor, rose honey - these are some of the ways in which Indian cuisine uses roses. Rose hips are high in vitamin C - they also contain vitamins A, B, K and folic acid and they do not lose their efficacy when cooked. Rose hip jam can be delicious. In the north - eastern state of Manipur, in fact, Rosa gigantean hips (which are huge) are sold in markets along with other fruits and vegetables. We have personally seen this and Nancy Steen mentions it Potpourris and sachets for perfuming rooms and linen closets are common too. In fact the ways in which roses are used in India are many and very much a part of daily life and a "must" for festive occasions.

Thus far on the past. What could be called the modern era in rose growing in India started with the advent of the pioneer Indian hybridizer, B. S. Bhatcharji in the 1940's. After Independence in 1947 rose growing received a tremendous fillip when the eminent agriculture scientist and rose lover, Dr. B. P. Pal, became the Director of the Indian Agriculture Research Institute. His enthusiasm led to the establishment of a rose society in the capital, New Delhi, He also hybridized many roses suited for the north Indian climate and started a programme of rose breeding in the I.A.R.I. His example has led to a great interest in rose hybridisation in the country.

A vast range of modern roses are cultivated in India, including classics like Crimson Glory, Ena Harkness, Mr. Lincoln, Christian Dior and many others. The latest western roses, including cut - flower roses, are well represented thanks to the enthusiasm of various rose nurseries. A landmark was the WFRS Award of Garden of Excellence to the Centenary Rose Garden, located in the southern mountains, in Ootacamund.

I would like to conclude with a quotation of Gandhi, whom we refer to as Father of the Nation. "A rose does not preach - it simply spreads its fragrance".

Roses in the Ancient World

Rosa damascena or the damask rose is a descendent of Rosa gallica. This particular rose is popular for its fragrance and since its first appearance in 900 B.C., it has been an integral part of the history of roses. In fact, some time around 50 B.C., the Romans were thrilled with a North African rose variety named Rosa damascena semperflorens, also known as the "Autumn Damask", which flowered twice every year. The Romans were actually not aware of this attribute of the "Autumn Damask" till then. This variety is thought to be a hybrid developed from Rosa gallica and Rosa moschata, also called the musk rose, and said to have its origin in the fifth century B.C. In fact, people in the West were only familiar with this repeat blooming rose till the European traders actually discovered tea and the China roses several centuries later.

In the early days, Rosa alba, also called the "White Rose of York" was another important rose variety. Rosa alba became famous in the 15th century during the War of the Roses when the House of York made this rose its emblem. However, the origin of this rose with five petals dates back to the second century A.D. or may be even earlier. It is believed that this rose has its origin in the Caucasus and was introduced to the West through Greece and Rome. It is thought that this rose variety as well as its relatives, which are known as albas, are descendants of some kind of a cross developed from Rosa corymbifera, Rosa canina, Rosa gallica and Rosa damascena.

It is interesting to note that the ancient Romans, Greeks and Phoenicians not only cultivated roses, but also traded them. These communities acquired the roses while they traveled and conquered different places. They brought these roses home and cultivated them. Consequently, roses spread rapidly and extensively all through the Middle East and other places in the region of the Mediterranean.

It is known that sometime around 300 B.C., Theophrastus, a noted ancient Greek scientist as well as writer, catalogued the roses grown during his time. According to his description of the roses, the different flowers had anything between just five to a hundred petals. In fact, he is said to be the first individual to offer a botanical description of roses. At that time, Alexander the Great was the ruler of Macedonia. He is also known to have grown roses in his garden and the credit for introducing cultivated rose varieties to Europe goes to him. In fact, it is believed that Alexander the Great may also be responsible for the spread and cultivation of roses in Egypt.

While excavating the tombs in Upper Egypt in 1888, noted English archaeologist Sir Flinders Petrie discovered vestiges of a rose garland that people had used as a funeral wreath during the second century A.D. In fact, he also identified the rose remains as being that of Rosa x richardii, a hybrid developed by crossing Rosa gallica and Rosa phoenicia. Commonly, this cross is also known as "St. John's Rose" or "Holy Rose of Abyssinia". Although the petals of the rose in the wreath had withered, they still had their pink hue and when the petals were immersed in water, they were almost restored to their natural condition. There are other researchers who discovered rose paintings on the walls of Thutmose IV's tomb. According to history, Thutmose IV was the 8th Pharaoh of the 18th dynasty of Egypt and he died in the 14th century B.C. Later, experts deciphered the references to the rose in hieroglyphics.

Almost all aristocrats in ancient Rome had rose gardens at their dwellings. In addition, people also took delight in spending their time in public rose gardens during the summer afternoons. According to available records, before the fall of Rome in 476 A.D., there were as many as 2,000 public rose gardens throughout the empire. In fact, the overload of public rose gardens in Rome led Horace, a noted poet and satirist of the time, to complain that the government was farsighted as it ought to have been used for orchards or wheat fields.

Medieval Period Roses

During the medieval period, following the fall of the Roman Empire, the Europeans were busy struggling to recover from the onslaught of various armies as well as raiders. As a result, it became very difficult to maintain rose gardens and only a few existed then. While the King of Franks Charlemagne, also known as Charles the Great (742 A.D. - 814 A.D.) also grew roses on the grounds of his palace at Aix-Ia-Chapelle, it was mainly the monks who were responsible for keeping the roses alive. These monks grew roses as well as other plants for various purposes, including therapeutic uses. In fact, the monasteries belonging to the Benedictine order developed as botanical research centers.

With the stabilization of the social conditions, the situation was once again favourable for growing roses and many private rose gardens started appearing again. During the 12th and 13th centuries, soldiers who returned from their Crusades in the Middle East carried with them stories of ostentatious rose gardens and also some sample flowers. As traveling increased all over the world, merchants, scholars and diplomats started exchanging roses as well as other plants. As a result of all these, there was a renewed interest in roses.

In the early days, herbalists also provided evidence of their increasing knowledge regarding roses. In 1597, noted English herbalist John Gerard wrote in his book Herball that people of that time were familiar with 14 types of roses. Some years later, in 1629, the apothecary (pharmacist) to James I, John Parkinson noted that as many as 24 different types of roses were grown in his herbal garden called Paradisus. By the turn of the 1700s, prominent artist Mary Lawrance not only identified about 90 different varieties of roses, but also illustrated them in her book entitled "A Collection of Roses from Nature".

Roses in the New World

Several different strains of roses appeared in North America. In fact, among the 200 different rose species that people across the globe are familiar with today, as many as 35 species are native to the United States alone. This has made the rose a native of North America much like the bald eagle, which also happens to be the emblem of the United States. The most prominent rose species native to the United States includes the *Rosa virginiana*, which was the first American rose species to be talked about in European literature. In addition, there are Rosa setigera, Rosa carolina, the "Prairie Rose", the "Pasture Rose", *Rosa woodsii, Rosa californica*, the "Swamp Rose" - named after its natural habitat, and Rosa palustris.

In his book, noted English soldier, explorer and author Captain John Smith mentions about the Indians inhabiting the James River Valley growing wild roses in their villages to beautify their locality. This is evident of the fact that even in the early days, the indigenous people in North America extensively cultivated the native roses as ornamental plants.

William Penn, who founded Pennsylvania, was a resident of Europe till the later part of the 1600s. In his writings, Penn noted that roses were a great favourites in gardens during his time. In addition, roses also had a special place in the field of arts and sciences. While returning to America in 1699, Penn carried also 18 different rose bushes along with him. Later, he talked about the beauty as well as the therapeutic value of roses in his book titled "Book of Physic". He wrote the book with a view to help the settlers in Pennsylvania to fulfill their medical requirements.

During his stay in Europe, Penn had certainly seen the exquisiteness of the cabbage rose (scientific name Rosa centifolia). This rose justified its botanical name as it has an incredible 100 petals. The petals of Rosa centifolia are so compactly arranged that they bear resemblance to small cabbages. There was a time when people believed that this rose belonged to ancient times because the Roman author and naturalist Pliny the Elder had also described a 100-petaled rose in his writings in the first century A.D. However, today many believe that the cabbage rose (Rosa centifolia) was perhaps developed by Dutch rose growers in the later part of the 17th century. Some other people are of the view that this rose was brought from

Asia sometime in 1596. Irrespective of the history of *Rosa centifolia*, it is likely that the cabbage rose is a complex cross of several ancient roses, counting the gallicas, albas and damasks.

Rosa centifolia mucosa, which is also known as the moss rose, is a well known mutation (sport) of Rosa centifolia (cabbage rose). The mutation or the moss rose was first seen way back in 1700 and many gardeners continue to grow this rose as well as hybridize it. This rose is called the moss rose because its flower buds and stems have minute, extremely fragrant hairs that bear resemblance to moss.

While European hybridizers were engaged in developing new variants of rose during this period, they mainly concentrated their introductions on a restricted gene pool. This is one reason why they failed to achieve novelty in most cases. Apart from this, the hybridizers of this period had a very poor understanding of the laws related to heredity, which eventually proved to be a handicap and it continued even after the German-speaking Moravian scientist Gregor Mendel undertook his experiments during the middle of the 1800s. Apart from this, in the initial days, European breeders were very jealous and kept their methods a close secret. They were basically concerned that their rivals may use their methods to render them useless in business.

Roses from the Orient

During the 18th and 19th centuries, Europe witnessed a sort of revolution in growing as well as breeding roses. This was the time when increased trade with the East Asian countries helped to procure the Chinese rose (Rosa chinensis), which actually attracted the attention of rose growers in Europe. The first variety of Chinese rose that reached the West is the "Old Blush", which was first introduced in Sweden way back in 1752 and to the remaining regions of Europe four decades later in 1793. On the other hand, the tea rose (Rosa x odorata) was introduced to the West sometime in 1808 or 1809. The tea rose derived its name from the fact that the aroma of its foliage bears resemblance to that of tea plants.

While the Chinese have been growing these rose varieties for several centuries, when the China rose and tea rose were introduced to the West, they really had a phenomenal impact in these alien countries. These roses had a very significant attribute - they were capable of producing repeated blooms in one season - something that the rose growers were unaware of till then. As a result, these rose varieties turned out to be a sensation instantly.

Different from these repeat-blooming rose "Autumn Damask" that was known to the European growers and only bloomed briefly two times every year, the continual blooming roses imported from the Orient produced numerous flowers over a prolonged period during each growing season. Apart from the blooming abilities of these rose varieties from China, their foliage is nearly evergreen and the foliage of the tea rose is especially mildew resistant. When the European rose breeders acquired the Chinese roses, they were eager to cross them with the existing roses to develop newer varieties possessing the good traits of both parents. Therefore, it may safely be claimed that both the China rose as well as the tea rose formed the basis for nearly all the roses we see in our times. However, it is unfortunate that most of these modern roses do not possess the cold hardiness, an important trait of the European roses, as the Chinese roses did not possess this quality.

Often the Chinese rose is also referred to as the Bengal rose, as it was initially exported from Calcutta, the capital of West Bengal in India, to the West. In fact, a large botanical garden was set up in Calcutta during the 18th century. Several varieties of roses procured by the Eastern Indian Company were grown in this famous botanical garden in the East. Sometime in 1789, a captain of the British Navy carried the flowers to England. Few years later, in 1793, the director of the East India Company Dr. William Roxburgh shipped additional rose specimens to various parts of Europe from Calcutta.

In the 18th century, trade in roses was very flourishing in the new British settlements in America. In fact, the first rose nursery in America was opened in Flushing, Long Island, by Robert Prince in 1737. This nursery began importing large number of various types of rose plants, especially new varieties. In 1746, Robert Prince marketed as many as 1,600 different rose varieties, which was definitely among the largest rose collections in the world in those days. Available documents belonging to Prince reveal that in 1791 the U.S. President Thomas Jefferson placed an order for two centifolias, a "Rosa Mundi", a "Common Moss", a musk rose, an unidentified yellow variety of rose and also a China rose. As the China roses were available in Europe only in 1793, it is believed that the rose delivered to Jefferson was bought directly from Asia by a clipper ship that traveled across the Pacific via the Cape Horn.

A new class of roses called the Portlands came into existence sometime around 1800. Probably, this new class of rose was obtained by crossing "Autumn Damask" with *Rosa gallica* and the China rose. Deriving its name from the duchess of Portland, this new rose variety was among the best quality garden rose hybrids that manifested the bond between the East and the West. Like their Chinese parent, the Portlands possessed the ability to bloom repeatedly in one growing season. The Portlands are also referred to as the damask perpetuals and people continued growing them till the variety called hybrid perpetual was eventually introduced nearly four decades later.

Josephine and Malmaison

Perhaps the contribution of Empress Josephine, Napoleon Bonaparte's wife, was the most in making the rose so popular in the early 1800s. Being very passionate about roses, Empress Josephine actually began a "rose renaissance" by trying to grow all familiar varieties of roses in her personal garden located at Malmaison, close to Paris.

In 1798, Empress Josephine started developing her rose garden and, before she died after 16 years in 1814 on her 51st birthday, the empress had already collected as many as 250 specimens of different roses. In order to encourage his wife's hobby, Napoleon I directed all his captains to search for new roses in the lands they visited and bring them home for Empress Josephine. In fact, her passion for roses was so intense that even the English, who were at war with the French at that time, not only allowed roses meant for Josephine to cross the borders, but also granted permission to her chief gardener to freely travel throughout the Channel. The esteem of the English for Josephine's love for roses is evident from this singular fact. As the repute of Empress Josephine's rose garden reached different regions of Europe, it led to a growing interest among people to grow roses. Many enthusiasts even took up hybridizing the roses, which ultimately resulted in the birth of many of the present day roses.

The presence of the celebrated rose gardens at Malmaison proved to be beneficial for France, which not only emerged as a leading rose growing country, but also started exporting large volumes of roses. By 1815, French rose growers cultivated as many as 2,000 different varieties of roses. This number increased rapidly to 5,000 varieties in just another decade. Prior to the Civil War, French growers exported their roses to New Orleans as well as other cities upstream the Mississippi River. In fact, gardeners in the south discovered that the soft China roses as well as the tea roses were most appropriate for growing in the warm climatic conditions prevailing in their region.

The Rise of the Hybrid Tea

In 1817, some people brought the Bourbon rose (scientific name Rosa x borboniana) to France from the Réunion Island (which was then known as Île de Bourbon) located close to Madagascar in the Indian Ocean. While the background of this rose is yet to be ascertained, it is believed to be a hybrid developed by crossing the "Autumn Damask" and Rosa chinensis. This belief gains ground from the fact that both these purported parents of Bourbon rose have been cultivated on the Réunion Island in the form of hedges. In the 19th century, the Bourbon rose turned out to be one of the most popular roses very quickly because the plant produced intermittent blooms. Similar to Port lands, this rose was also among the first to possess the best attributes of the Oriental and European roses. The Bourbon rose developed initially produced bright pink blooms. Unfortunately, now this rose is lost. However, we still have several hybrids developed from Bourbon rose and even to this day this rose forms a major source of reds in most modern roses.

The hybrid China class is yet another product that has been created by crossing the Oriental and European roses. These plants are quite tall growing and rather unappealing plants that also did not bloom well repeatedly. As a result, they were never popular on their own virtues. Nevertheless, they are among the ancestors of several other varieties of roses including the polyanthas, perpetuals, hybrid teas and floribundas.

On the other side of the Atlantic, American breeders also made their contribution to the history of roses in the 19th century by introducing the noisette rose (scientific name *Rosa noisettiana*). Incidentally, this is the first rose that was actually hybridized by rose growers in America. They developed the noisette rose by crossing *Rosa moschata*, *Rosa chinensis* (China rose) and the musk rose. A rice grower in South Carolina named John Champneys hybridized the noisette rose in 1812. Champneys named his new creation "Champneys' Pink Cluster". Although Champneys developed the new rose, he did not have any interest whatsoever in marketing the new rose. As a result, he asked his neighbour Philippe Noisette to market the rose in return for a cut. On his part, Philippe despatched the rose to his brother Louis, who worked as a nurseryman in Paris. Louis crossed the low-growing "Champneys' Pink Cluster" with other tall-growing roses to create a new tall rose, which he name "Blush Noisette" - a clear sign of an uncharitable snub to John Champneys, the hybridizer who originally developed the new rose.

Sometime in the middle of the 1800s, a familiar seashore rose called Rosa rugosa was introduced to the West from Japan. As this rose is not capable of hybridizing well, it has made no significant contribution in creating newer roses or to the history of roses. Nevertheless, people have been holding the Rosa rugosa in high esteem for over a thousand years now for its single blooms, creased foliage as well as its ability to produce profuse hips, which are a wonderful natural source of vitamin C.

By 1837, breeders were already working to create modern roses. This was the time when most varieties of the Chinese roses had been introduced to Europe. It was in this year that the hybrid perpetual, an intricate French hybrid, was introduced by hybridizers. This hybrid perpetual was developed by crossing China rose, Bourbon rose, Autumn Damask, Portland, Cabbage rose, noisette rose and the hybrid tea. This new rose was very resilient and produced large and aromatic flowers. The initial hybrid perpetual varieties produced pink blooms. However, when they were again crossed with Bourbon roses, they were influenced by the red hue of the Bourbon roses.

The hybrid perpetuals continued to be one of the popular roses till the end of the 19th century, and after this they were overshadowed by hybrid tea, which was undoubtedly a superior variety. However, the unfortunate part of all these is that majority of these roses does not exist any more - they have now been lost. In fact, of over 3,000 different varieties of roses that were hybridized throughout this "golden era" of roses, which ranged from Empress Josephine's gardens at Malmaison near France to the introduction as well as assimilation of roses acquired from the Orient, today you can only purchase just about 50 varieties that are still in existence.

The growth habit of the hybrid tea became more compact as a result of a cross between this rose variety and a hybrid perpetual. In addition, the hybrid tea also became additionally dependable compared to its hybrid perpetual parent, as it now inherited the continual blooming traits. Introduced in 1867, "La France" was the first hybrid tea. Eventually, the hybrid tea class was officially recognized by the National Rose Society of Great Britain in 1893. Ever since, hybridizers have been working closely on the hybrid tea and succeeded in developing many significantly improved varieties of this class. As a result, roses in the hybrid tea class are among the most popular roses even in present times.

As far as rose breeding is concerned, the development of the hybrid tea ushered in a new age in this field. After the evolution of the hybrid tea, the various classes of roses that existed prior to 1867 were considered to be old garden roses. On the other hand, all new classes of roses developed after 1867 were known as modern roses.

After trying for 13 long years, Joseph Pernet-Ducher, a French hybridizer, introduced a new rose hybrid named "Soleil d'Or", which was developed by crossing the "Persian Yellow" (scientific name *Rosa foetida persiana*) and a red hybrid perpetual rose. In 1837, the British envoy to Persia Sir Henry Willcock brought the "Persian Yellow" rose from Persia to England. The cross developed by Joseph Pernet-Ducher produced yellow blooms and possessed the aptitude to survive inter-breeding. Soon, Pernet-Ducher developed another hybrid and named it "Rayon d'Or"-a cross that produced golden yellow flowers. As a result of introducing these new varieties, an altogether new range of roses came into existence - something that was never seen before. Now the blooms of modern roses were available in a new assortment of colours, including gold, apricot, copper and salmon.

Soon, Joseph Pernet-Ducher earned the reputation of being the Wizard of Lyons - the town in east-central France where the hybridizer worked on the plants. For about 30 years of their initial existence, these roses were a part of an individual class known as the Pernetianas. Afterwards, these roses were included in the class that comprises the hybrid tea.

It is worth mentioning here that while the hybrid tea roses possessed the aptitude to resist cold weather conditions, they did not grow robustly, as their roots were thin and fragile. During the end of the 19th century, employees in several nurseries came to realize that they could make these roses grow better, provided they grafted them on the root stock of *Rosa multiflora*, which is a vigorously growing plant producing dull blooms.

Twentieth-Century Roses

In the early 1900s, Svend Poulsen, a renowned rose breeder in Denmark, produced many new hybrids by crossing polyanthas. Jean Sisley, a French nurseryman, developed a new class of roses in the later part of the 19th century that was called polyantha. This class of rose developed into low bushes densely covered with bunches of petite flowers that bloomed continually throughout the summer. Most hybridization done by Poulsen involved using rose species native to East Asia, like *Rosa wichuraiana*. These varieties passed on the winter hardiness trait to their offsprings.

During the 1920s, Poulsen developed the first floribundas by crossing the polyantha with a hybrid tea rose variety. In fact, he produced two varieties of floribundas - the red "Kirsten Poulsen" and the pink "Else Poulsen". As the name of this class of rose suggests, the floribunda bloomed in profusion - an attribute that was passed on by its polyantha parent. At the same time, floribunda inherited its height as well as the long cutting stems from its tea parent hybrid.

Even as hybridizers developed the bush roses, a new variety - the climbers, was also coming into existence. The histories as well as the ancestries of climbers are a complicated one and, hence, tracing both is quite difficult. Several climbing roses were developed from ramblers and the "Crimson Rambler" was the first among them. In 1893, this rose variety was actually imported from Japan. The other hybrid climbing roses developed from *Rosa wichuraiana* include "Blaze", "American Pillar", "Dorothy Perkins", "Dr. W. Van Fleet" and the "New Dawn". In addition, the Bourbon roses, which are large growing plants, as well as the tall growing noisettes also impacted the climbers.

A number of other climber roses were developed from bush roses and they produced elongated, supple canes; while some other climbers descended from large shrub roses. In recent times, breeders have developed several climber roses from the tall, partially climbing shrub rose called the *Rosa kordesii* - many of these were evolved by crossing the *Rosa wichuraiana* with *Rosa rugosa* in 1952. About half a century later, breeders developed the hybrid musks - which comprise both large shrubs as well as small climbers, in the 1920s by crossing *Rosa multiflora* ramblers with noisettes.

Wilhelm Kordes II, a noted German hybridizer in the 20th century who is credited with creating Rosa kordesii, also worked with Rosa spinosissima, a rose variety that has been in existence since or before the Middle Ages, to breed newer varieties of roses. He crossed the Rosa spinosissima with hybrid teas with a view to create a finer rose class of present day shrub roses known as kordesii shrubs, counting "Frülingsmorgen" and "Frülingsgold". Roses belonging to this new class produce flowers that are winter resistant as well as requiring low maintenance. These plants are usually grown in public gardens and other areas as well as the length of the roads in Europe.

As the World War II broke out, the boom in rose hybridization witnessed a significant decline, especially in Europe. However, the trend witnessed an upswing again after the war ended. Regardless of propagating roses of an assortment of forms as well as colors, blooms having a color range of pure orange to orange-red were most sought after in those days. The floribunda "Independence", which was introduced in 1951, is considered to be the first contemporary rose whose blooms belonged to the newly created orange-red range. The pigment pelargonidin was most important aspect of this exceptional coloration. In fact, it is this pigment that is responsible for giving geraniums their scarlet hue. However, rose enthusiasts had to wait till 1960 to take delight in a hybrid tea having orange-red hue. This was the time when Rosen Tantau, a noted hybridizing firm in Germany, introduced the "Tropicana" - the hybrid tea rose with orange-red blooms.

Rose breeders developed an altogether new rose class with a view to put up the rose named "Queen Elizabeth". Also known as the grandiflora, this class of rose was developed by crossing a floribunda with a hybrid tea variety. The blooms of the plants belonging to this new class of rose bore close resemblance to those of the hybrid tea. On the other hand, the flowers appeared in clusters, are similar to the blooms of the floribunda.

Taking all classes of roses, including the 11,000 hybrid teas, presently there are over 30,000 varieties of roses. Nevertheless, several of these rose varieties, particularly the older rose varieties, are not sold any more. In fact, they may only be seen growing in some private gardens. The numerous varieties of roses in existence often make it extremely difficult to classify or trace their background or history. Apart from those developed by breeders, there are many rose varieties that are considered to be natural hybrids. Therefore, it is very difficult to identify the parents of these natural hybrids. At the same time, several varieties are also commercial hybrids and the lineage of these plants has either been lost or obscured deliberately with a view to dissuade piracy. Despite this, several enthusiasts have succeeded in re-developing these commercial hybrids successfully.

Types of Roses

Rare Golden Rose Flower

The Orange rose is a rose which has had its petals artificially colored.

The method exploits the rose's natural processes by which water is drawn up the stem. By splitting the stem and dipping each part in a different colored water, the colors are drawn into the petals resulting in a multi colored rose.

History

The uncomplicated process has been known for more than a millennium. Several companies have moved to patent the process. It appears they have been partially successful It will be up to the courts to decide if the process can be patented. In 2005 the Dutch grower Peter van de Werken succeeded in using this uncomplicated process with a special color combination, using a rose, in a way that the outcome showed as a Rainbow Rose. The result was worldwide presented as Breaking News. Since then Peter van de Werken is well known as the inventor of the Rainbow Rose.

Cultivars

A lot of research was done to find the best cultivar for this unique coloring process, with the result that the Rainbow Rose is the only cultivar that absorbs all the different colorants perfectly. The Rainbow Rose is a Hybrid Tea, cream rose that grows in the Netherlands, Colombia and Ecuador. When the rose is in full bloom it has a flower diameter of 6 cm and a stem length of 40 to 100 cm. The rose isn't scented.

Other cultivars that can be used for this colouring process are Rosa La Belle and Rosa Avalanche.

We are offering imported verities as well as indigenous ones. We have taken care each and every possible ways to reach the high quality seeds to our customers. However germination of seeds depends on suitable climate, suitable temperature, and preparation of soil

Rosa

The rose has been a **symbol of love, beauty, even war and politics** from way back in time. The variety, color and even number of Roses carry symbolic meanings. The Rose is most popularly known as **the flower of love**, particularly Red Rose.

Roses have been the most popular choice of flowers for the purpose of gifting across the world. They also act as a great addition to home and office decor. A bunch of roses or even a single rose works wonders aesthetically and considerably enlivens a place. Besides fresh cut roses, artificial flowers like silk roses in different colors are also widely used as decoration.

Some Interesting Facts about Roses

☆ The birthplace of the cultivated Rose was probably Northern Persia, on the Caspian, or Faristan on the Gulf of Persia.

☆ Historically, the oldest Rose fossils have been found in Colorado, dating back to more than 35 million years ago.

☆ **Roses were considered the most sacred flowers in ancient Egypt** and were used as offerings for the Goddess Isis. Roses have also been found in Egyptian tombs, where they were formed into funeral wreaths.

☆ Confucius, 551 BC to 479 BC, reported that the Imperial Chinese library had many books on Roses.

☆ Ancient Sumerians of Mesopotamia (in the Tigris-Euphrates River Valley) mentioned Roses in a cuneiform tablet (a system of writing) written in approximately 2860 BC.

☆ The English were already cultivating and hybridizing Roses in the 15th Century when the English War of Roses took place. The winner of the war, Tudor Henry VII, created the Rose of England (Tudor Rose) by crossbreeding other Roses.

☆ While no Black Rose yet exists, there are some of such a deep Red color as to suggest Black.

☆ Roses are universal and grown across the world.

☆ The Netherlands is the world's leading exporter of Roses.

The Netherlands, with about 8000 hectares of land under Rose cultivation, is the global leader in Rose cultivation. 54 per cent (about 5000 hectares) of the cultivated land in Ecuador is under Rose cultivation!! Zambia, a small nation, had 80 per cent of its cultivated land under Roses.

Classification of Roses

Broadly, Roses are divided into three classes:

(i) Roses Species

Species Roses are often called Wild Species Roses. Species Roses often have relatively simple, 5-petaled flowers followed by very colorful hips that last well into the winter, providing food for birds and winter color.

The most popular Rose species for sale today is Rosa rugosa owing to its superior hardiness, disease resistance, and extremely easy maintenance. Species roses are widely hybridized. Wild Species Roses include many different varieties. Wild Species Roses usually bloom once in the summer.

(ii) Old Garden Roses

Old Garden Roses have a delicate beauty and wonderful perfume, not often found in modern hybrid tea roses. Old Garden Roses are a diverse group from the those with a wonderful fragrance and great winter hardiness to the tender and lovely tea roses, which are best suited for warm climates.

Old Garden Roses comprise a multifaceted group that in general are easy to grow, disease-resistant and winter-hardy. Old Garden Roses grow in several shrub and vine sizes. Although colors do vary, this class of Roses are usually white or pastel in color. These "antique Roses" are generally preferred for lawns and home gardens. Several groupings of Roses classified as Old Garden Roses are China Roses, Tea Roses, Moss Roses, Damask Roses, Bourbon Roses, *etc.*

(iii) Modern Roses

Any Rose identified after 1867, is considered a Modern Rose. Old Garden Roses are the predecessors of Modern Roses. This group of Roses are very popular. The Modern Rose is the result of crossbreeding the hybrid tea with the polyanthus (a variety of primrose).

The colors of Modern Roses are varied, rich and vibrant. The most popular roses found in the class of Modern Roses are the Hybrid Tea Roses, Floribunda Roses, and Grandiflora Roses. Although Modern Roses are adored by florists and gardeners, they do require proper care, and do not adapt well to colder environments.

Popular Hybrid Varieties of Roses

Species Involved	Hybrid Product
Hybrid Perpetual Rose and Chinese Tea Rose	Hybrid Tea Rose
Hybrid Perpetual Rose and Australian Brier Rose	Yellow Permet Rose
R. multiflora and *R. chinensis*	Hybrid/Dwarf Polyanthas or Poly Pompon roses
Hybrid Tea Rose and Floribundas	Grandifloras
R. wichuriana, *R. multiflora* and Hybrid Tea Rose	Dorothy Perkins, American Pillar, Excelsa
R. canina and *R. gallica*	Albas
R. phoenica and *R. gallica*	Damaskas Rose
R. damascena and *R. alba*	Centifolia Rose
Autumn Damask Rose and China Rose	Bourbons

Roses

☆ Roses may be grown in any well-drained soil with optimum sunlight.

☆ Most Rose varieties are grown by budding on an understock (lower portion of a plant) propagated from seeds or cuttings. Order rose seeds online and let your garden be filled with the marvellous color and fragrance of roses.

☆ Clay soils, warm temperatures are always preferred, and the rose plants grow best when not set among other plants.

☆ Cow manure is the preferred fertilizer for Rose cultivation, but other organic fertilizers, especially composts, are also used.

☆ Rose plants usually require severe pruning, which must be adapted to the intended use of the flowers.

☆ Trim off all broken and bruised roots on the Rose plant, cut top growth back to 6 to 8 inches.

☆ Dig planting holes at least 6 inches deeper to accommodate the roots of the Rose plant without crowding or bending.

☆ Mix 1 tablespoonful of fertilizer with the soil placed over the drainage material.

☆ Cover this mixture with plain soil, bringing the level to desired planting depth.

☆ Make a mound in the center to receive the Rose plant.

☆ Set Rose plant roots over this mound, spread the roots, and fill in with soil.

☆ Firm the soil tightly 2 or 3 times while filling the hole.

It is extremely easy to buy rose plants online if you do not wish to go to the trouble of actually planting one. They usually come with a care manual and some plant food. An already flowering plant in a lovely container also makes a great gift item. The blooms stay longer and after they fade there is always the next flowering, thus providing the receiver with a lasting and beautiful gift.

Noisette Roses are the only Roses that originated in the United States of America.

Rose Plant Care

☆ When watering Roses, soak the soil to a depth of 6 to 8 inches, do not merely sprinkle.

☆ When it comes to fertilizing your roses, Provide a balanced diet to your roses. See what your plant is deficient in and try to include them in the fertilizer. Timing is also an important part to maximize the benefit of your fertilizer so that the nutrients are available to the plant when it needs it most during the active growing and blooming stage. Order your rose fertilizer now to enhance the vigor of blooming in your roses.

☆ Mulching during the summer will eliminate weeds among Rose plants. Mulches should be applied 2 or 3 weeks before the Roses come into bloom.

☆ Winter mulching with straw, peat moss, or other material is advisable. This mulch regulates the soil temperature and tempers the effects of freezing and thawing on thr Roses.

☆ Pull soil up around each Rose plant to a height of about 6 inches after the first frost.

Foolproof Guide to Growing Roses by Field Roebuck is a comprehensive book on growing roses ideal for would-be growers who were always afraid of roses, as well as for gardeners who already grow these beautiful flowers and want to learn more.

Government Support

Introduction

'Blooming business' and 'rosy future'. These are the words the Indian authorities used to sing the praises of the flower industry. The central government and the state governments grant subsidies and give information to stimulate the building of greenhouses for the cultivation of roses, carnations, chrysanthemums and other cut flowers. This development takes place mainly in the countryside surrounding the metropolises Bangalore, Mumbai (Bombay) and Delhi. During the last years the acreage of flower greenhouses has tripled or even quadrupled near these cities. Hopes are running high. A growth percentage of 30 per cent is being mentioned. India hopes to get its share of the growing demand for cut flowers in the world. This trade is considered a goldmine and a panacea for India's economic and social problems. The flower trade is said to create plenty of employment opportunity. Moreover, the export of cut flowers would earn a great deal of foreign currency. This is very important to India. It has a debt of fifty thousand million US dollars and is listed third on the world rankings of debt countries - after Mexico and Brazil.

But there is quite a different story to tell about flowers. And this is the story about pesticides, pollution of water and soil, about bad labour conditions and sick labourers. The artificial cultivation of flowers in greenhouses is said to be the most polluting agricultural activity humanity ever created. The industry originates from Western Europe. Holding 60 per cent of the world trade of cut flowers, the Netherlands is by far the most important producing country. During the past two decades, however, the flower industry has gained a foothold in the Third World, the Western Europeans often holding a big finger in the pie. For instance, the floriculture in Kenya is largely in the hands of foreign investors, including the British-Dutch multinational Unilever. This means that a large part of the profits is pumped back into the Netherlands and Great Britain, leaving the Kenyans with the ecological problems. An example: the water level in the Naivasha lake falls every year by 15 cm because of the enormous water consumption by the surrounding flower industries.

Conclusion: there are plenty of reasons to take a critical look at the growth of the flower industry in India. Who are the investors in this sector, where do the profits go and who benefit by the substantial government subsidies? Does the international flower trade have as many opportunities as is often suggested? Is the demand for cut flowers growing at the same high pace as the supply? Will the market not become oversupplied and is collapse not imminent? What are the consequences of floriculture for the environment and what measures are taken to prevent environmental damage? How are labour conditions in the floral industry? How are health hazards of pesticides dealt with?

This report deals with those questions. This text is based on interviews and documents we collected in India during the first six months of 1997. We focus on the district of Bangalore, the main district for the production of flowers in India. We pay full attention to the activities of the Dutch. They play a major role in the growth of the floral industry of India.

Summary

Growing flowers in greenhouses appears to be very profitable in India. That is to say: it is profitable for foreign, mainly Dutch companies who sell planting materials, technical knowledge and cooling units. Their Indian buyers and joint venture partners still have to wait and see how profitable the sector will be for them.

Indian business people investing in the flower industry are very much dependent on their foreign partners. The Dutch and the Israelis supply the technology and the planting materials and also dominate the distributive trades. The Indians supply (cheap) labour and capital. It is characteristic of this setup that the major share (90 per cent) of the investments is at the expense of the Indians while it is exactly the opposite where the profit is concerned: approximately 90 per cent is for the foreign partners. Moreover, the risks involved in an unpredictable international flower trade are exclusively at the expense of the Indians. A surplus is imminent on the international market for roses - India's chief product. Even now a fall in price is becoming visible. If this crisis continues, the profit of the Indians will disappear like snow in summer. But the foreigners have secured their profits through the sale of raw materials and equipment. In short: the flower industry is much more profitable for foreign companies than for their Indian partners. It is characteristic that the flower industry in India began to grow after new loans by the IMF and that the sector has been stimulated by the FMO, a development bank, which represents the Dutch state and Dutch banking.

But even if the Indian flower industry becomes successful, the economic profit for the population will be minimal. The investors in the flower industry are big agricultural and industrial companies. For others investing is hardly possible: the outlay on greenhouses is between 200 and 300 thousand dollars per acre. The foreign exchange earned in the flower industry and the generous Indian state subsidies fall to the elite only. The employment results in the Indian flower industry are far from impressive. In the greenhouses there is an average of 20 workers per acre. This means that one job costs 15,000 dollars. This money can give jobs to many more people in India. It also became clear that the flower industry not only creates jobs but also costs jobs, for the land to build the greenhouses on is often purchased from (small) farmers.

Particularly in the Third World, the flower industry is a sector with a high 'casino percentage'. The risks are high, quick profits are aimed at, and there is hardly any concern about long term effects. The same goes for environmental pollution. By sketching a picture of a beautiful natural product, the government and business disguise the environmental hazards. But in reality, huge risks threaten the environment: floriculture inherently needs a high dosage of artificial fertilizer and pesticides. These developments are not only ecological hazards, but also endanger food supplies. Continuous and intensive cultivation of flowers will eventually lead to sterile soil and no cultivation of any crop will be possible at all. Another thing is that an extremely high amount of groundwater is pumped up, in areas too, where the groundwater level is critical, like in the Bangalore district.

The land used for the greenhouses for roses appears to be bought from small farmers. Since the farmers usually produce cheap food, these sales greatly affect the food supplies of the poor population.

The wages in the greenhouses are poor even by Indian standards: between 0.75 to 1 dollar per day. Moreover the fringe benefits are very bad: the workers do not get the holidays and bonuses that are provided for in other sectors of industry. Added to this they run big health risks. Protective clothing is lacking or is insufficient. The workers get hardly any or no information about or training in how to handle hazardous pesticides. Pesticides are used which the WHO classifies as 'extremely dangerous' or 'highly dangerous'. The waiting times after spraying prescribed by the WHO are not observed.

Indian Rose Federation History

In 1978, Mr. P. L. Mokashi, President and Office Bearers of 'The Bombay Rose Society' including Past President Mr. Krishnaram Kapadia, Vice Presidents, Mr. H. S. Sethi, Mr. V. S. Padhye, Mr. Sharad Palande, Secretary Mr. M. S. Pai, successfully organized the 1st 'All India Rose Convention'. An enthusiastic response from rosarians across the country made the Convention a grand success. The concept of forming an All India body was mooted during the deliberations at the convention. The MOA was drafted at Bombay for registration and dates for the Second Convention finalized, Jabalpur was selected to be the venue and December 1979 the month and year. "The Indian Rose Federation" (IRF) officially came into being. Justice K. K. Dubey and Mr. M. S. Pai were appointed the first President and Secretary respectively of the IRF. Initially the Office of the Federation was at Bombay, but was later moved to Jabalpur for many reasons. In 1983, Lt. Col. C. P. Diddi (Retd.) took over the responsibilities of the Secretary a duty he ably discharged for many years. Justice K. K. Dubey retired as President in January 1987 and was succeeded by Dr. P. S. Rao from Hyderabad. Dr. P. S. Rao retired in 1990 and Mr. P. L. Mokashi took over as the President.

Apart from functioning as the Apex Body representing Rose Societies based in different states of the Nation, the activities of IRF so far have included :

Publication of the Indian Rose Annual, with write ups contributed by Rosarians and Scientists the World over, on various aspects of the Rose. It was initially edited by Mr. V. S. Padhye and Mr. M. S. Pai. Since 1985 (Annual IV), it was edited by Mr. V. S. Padhye, Mr. M. S. Viraraghavan and Mrs. Girija Viraraghavan till 2005. From 2006 till date it is being edited by Mr. M. S. Viraraghavan and Mrs. Girija Viraraghavan. The Indian Rose Annual is revered in India and across the World. The IRF has published till 2014, '30' very informative and prestigious Annuals.

Implementation of "Quality Standard" of Rose Plants sold by IRF accredited rose nurseries.

Holding Annual Convention and All India Rose Shows, in different states of the country.

Establishing 'Guidelines' for Judging Roses in India

Rose Trial Grounds : Establishing and setting Rules and Regulations for implementation in accordance with Global standards.

IRF represents India in the World Federation of Rose Societies.

Our aim and objective is to widely spread, educate and inform the art and techniques of growing roses, both as a hobby or for commercial purposes. To encourage growers and breeders to breed new rose varieties suitable for cultivation in various climatic zones in our country. Breeders, Researchers and Rosarians doing outstanding work in spreading rose culture and hybridizing new varieties are honoured with IRF Gold Medals every year at the Convention.

It is recommended that every rosarian should enroll as a member of IRF, which in turn will enable contact with all major activities on the rose front in the country and abroad. Attending Annual Rose Conventions, meeting fellow rosarians, exchanging information, taking opinions, perspectives, suggestions and views, of fellow rosarians on the art of growing roses.

Garden Roses

Garden roses are predominantly hybrid roses that are grown as ornamental plants in private or public gardens. They are one of the most popular and widely cultivated groups of flowering plants, especially in temperate climates. Numerous cultivarshave been produced, especially over the last two centuries, though roses have been known in the garden for millennia beforehand. While most garden roses are grown for their flowers, some are also valued for other reasons, such as having ornamental fruit, providing ground cover, or for hedging.

The Hybrid Tea Rose, 'Peer Gynt'

History

It is believed that roses were grown in all the early civilisations of temperate latitudes from at least 5000 years ago. They are known to have been grown in ancientBabylon. Paintings of roses have been discovered in Egyptian pyramid tombs from the 14th century BC. Records exist of them being grown in Chinese gardensand Greek gardens from at least 500 BC.

Most of the plants grown in these early gardens are likely to have been species collected from the wild. However, there were large numbers of selected varieties being grown from early times; for instance numerous selections or cultivars of the China rose were in cultivation in China in the first millennium AD.

The significant breeding of modern times started slowly in Europe, from about the 17th century. This was encouraged by the introduction of new species, and especially by the introduction of the China rose into Europe in the 19th century. An enormous range of roses has been bred since then. A major contributor in the early 19th century was Empress Josephine of France who patronized the development of rose breeding at her gardens at Malmaison. As long ago as 1840 a collection numbering over one thousand different cultivars, varieties and species was possible when a rosarium was planted by Loddiges nursery for Abney Park Cemetery, an early Victorian garden cemetery and arboretum in England.

Features

Roses are one of the most popular garden shrubs in the world. They possess a number of general features that cause growers and gardeners to choose roses for their gardens. This includes the wide range of colours they are available in; the generally large size of flower, larger than most flowers in temperate regions; the variety of size and shape; the wide variety of species *etc.* that freely hybridize.

An Amber-Colored Rose.

Colour of Flowers

Rose flowers have always been available in a number of colours and shades; they are also available in a number of colour mixes in one flower. Breeders have been able to widen this range through all the options available with the range of pigments in the species. This gives us yellow, orange, pink, red, white and many combinations of these colours. However, they lack the blue pigment that would give a true purple or blue colour and until the 21st century all true blue flowers were created using some form of dye. Now, however, genetic modification is introducing the blue pigment.

Classification

There is no single system of classification for garden roses. In general, however, roses are placed in one of three main groups: Wild, Old Garden, and Modern Garden roses. The latter two groups are usually subdivided further according to hybrid lineage, although due to the complex ancestry of most rose hybrids, such distinctions can be imprecise. Growth habit and floral form are also used as means of classification.

Wild Roses

The wild roses (also known as species roses) include the natural species and some of their immediate hybrid descendents. The wild roses commonly grown in gardens include *Rosa moschata*, the Musk Rose; Rosa banksiae, Lady Banks' Rose; Rosa pimpinellifolia, the Scots or Burnet Rose; Rosa rubiginosa, the Sweetbriar or Eglantine; and Rosa foetida, in varieties 'Austrian Copper', 'Persian Double' and 'Harison's Yellow'. For most of these, the plants found in cultivation are often selected clones that are propagated vegetatively. Wild roses are low-maintenance shrubs in comparison to other garden roses, usually tolerating poor soil and some shade. They generally have only one flush of blooms per year, although some species have large hips in the autumn (*e.g. Rosa moyesii*) or have colourful autumn foliage (*e.g. Rosa virginiana*).

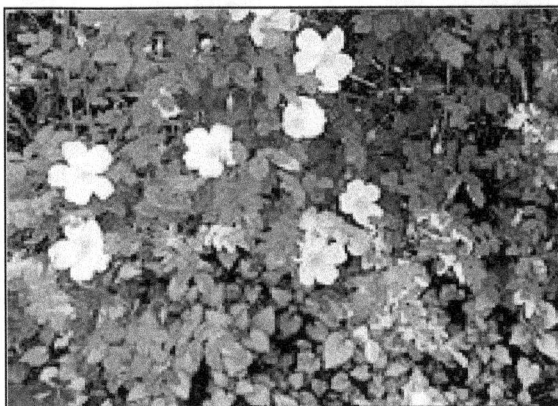

The Spring-Flowering Pimpinellifolia 'Rosa Altaica', Underplanted with Lamium.

Old Garden Roses

An old garden rose is defined as any rose belonging to a class which existed before the introduction of the first modern rose, La France, in 1867. Alternative terms for this group include heritage and historic roses. In general, Old Garden roses of European or Mediterranean origin are once-blooming woody shrubs, with notably fragrant, double-flowered blooms primarily in shades of white, pink and crimson-red. The shrubs' foliage tends to be highly disease-resistant, and they generally bloom only from canes (stems) which formed in previous years. The introduction of China and Tea roses from East Asia around 1800 led to new classes of Old Garden Roses which bloom on new growth, often repeatedly from spring to fall. Most Old Garden Roses are classified into one of the following groups.

Alba

'Maiden's Blush', an Alba Rose (before 1400).

Literally "white roses", derived from R. arvensis and the closely allied R. x alba. The latter species is a hybrid of R. gallica and R. canina. This group contains some of the oldest garden roses. The shrubs flower once yearly in the spring or early summer with scented blossoms of white or pale pink. They frequently have gray-green foliage and a vigorous or climbing habit of growth. Examples are 'Alba Semiplena', 'White Rose of York'.

Gallica

The Gallica or Rose of Provins group is a very old class developed from Rosa gallica, which is a native of central and southern Europe and western Asia. The Apothecary's Rose, R. gallica officinalis, was grown in the Middle Ages in monastic herbaria for its alleged medicinal properties, and became famous in English history

Gallica Rose 'Charles de Mills', before 1790.

as the Red Rose of Lancaster. Gallicas flower once in the summer over low shrubs rarely over 4 feet (1.25m) tall. Unlike most other once-blooming Old Garden Roses, the gallica class includes shades of red, maroon and deep purplish crimson. Examples are 'Cardinal de Richelieu', 'Charles de Mills', 'Rosa Mundi' (R. gallica versicolor).

Damask

'Autumn Damask' ('Quatre Saisons').

Named for Damascus, Damask roses (*Rosa x damascena*) originated in ancient times with a natural cross (*Rosa moschata x Rosa gallica*) x *Rosa fedtschenkoana*. Robert de Brie is given credit for bringing damask roses from the Middle East to Europe sometime between 1254 and 1276, although there is evidence from ancient Roman frescoes that at least one damask rose existed in Europe for hundreds of years before this. Summer damasks bloom once in summer. Autumn or Four Seasons damasks bloom again later, albeit less exuberantly, and these were the first remontant (repeat-flowering) Old European roses. Damask roses tend to have rangy to sprawling growth habits and strongly scented blooms. Examples: 'Ispahan', 'Madame Hardy'.

Centifolia or Provence

Centifolia roses are also known as Cabbage roses, or as Provence roses. They are derived from Rosa x centifolia, a hybrid that appeared in the 17th century in the Netherlands, related to damask roses. They are named for their "one hundred" petals; they are often called "cabbage" roses due to the globular shape of the flowers. The centifolias are all once-flowering. As a class, they are notable for their inclination to produce mutations of various sizes and forms, including moss roses and some of the first miniature roses (see below). Examples: 'Centifolia', 'Paul Ricault'.

Moss

The Moss roses are based on one or more mutations, particularly one that appeared early on Rosa centifolia, the Provence or cabbage rose. Some with Damask roses as a parent may be derived from a separate mutation. Thickly growing or branched resin-bearing hairs, particularly on the sepals, give off a pleasant woods or balsam scent when rubbed. Moss roses are cherished for this trait, but as a group they have not contributed to the development of new rose classifications. Various hybrids with other roses have yielded different forms, such as the modern miniature creeping moss rose 'Red Moss Rambler' (Ralph S. Moore, 1990). Moss roses with centifolia background are once-flowering; some moss roses exhibit repeat-blooming, indicative of Autumn Damask parentage. Examples: 'Common Moss' (centifolia-moss), 'Mousseline', also known as 'Alfred de Dalmas' (Autumn Damask moss).

Portland

The Portland roses were long thought to be the first group of crosses between China roses and European roses, and to show the influence of Rosa chinensis. Recent DNA analysis however has demonstrated that the original Portland Rose has no Chinese ancestry, but has an autumn damask/gallica lineage. This group of roses was named after the Duchess of Portland who received (from Italy about 1775) a rose then known as R. paestana or 'Scarlet Four Seasons' Rose' (now known simply as 'The Portland Rose'). The whole class of Portland roses was developed from that one rose. The first repeat-flowering class of rose with fancy European-style blossoms, the plants tend to be fairly short and shrubby, with a suckering habit, with proportionately short flower stalks. The main flowering is in the summer, but intermittent flowers continue into the autumn. Examples: 'James Veitch', 'Rose de Rescht', 'Comte de Chambord'.

China

The China roses, based on Rosa chinensis, have been cultivated in East Asia for centuries. They have been cultivated in Western Europe since the late 18th century. They contribute much to the parentage of today's hybrid roses, and they brought a change to the form of the flowers then cultivated in Europe. Compared with the older rose classes known in Europe, the Chinese roses had less fragrant, smaller blooms carried over twiggier, more cold-sensitive shrubs. However they

'Parson's Pink China' or 'Old Blush,' one of the "stud Chinas".

could bloom repeatedly throughout the summer and into late autumn, unlike their European counterparts. The flowers of China roses were also notable for their tendency to "suntan," or darken over time unlike other blooms which tended to fade after opening. This made them highly desirable for hybridisation purposes in the early 19th century. According to Graham Stuart Thomas, China roses are the class upon which modern roses are built. Today's exhibition rose owes its form to the China genes, and the China roses also brought slender buds which unfurl when opening. Tradition holds that four "stud China" roses—'Slater's Crimson China' (1792), 'Parsons' Pink China' (1793), and the Tea roses 'Hume's Blush Tea-scented China' (1809) and 'Parks' Yellow Tea-scented China' (1824)—were brought to Europe in the late 18th and early 19th centuries; in fact there were rather more, at least five Chinas not counting the Teas having been imported. This brought about the creation of the first classes of repeat-flowering Old Garden Roses, and later the Modern Garden Roses. Examples: 'Old Blush China', 'Mutabilis' (Butterfly Rose), 'Cramoisie Superieur'.

Tea

The original Tea-scented Chinas (Rosa × odorata) were Oriental cultivars thought to represent hybrids of *R. chinensis* with *R. gigantea*, a large Asian climbing rose with pale-yellow blossoms. Immediately upon their introduction in the early 19th-century breeders went to work with them, especially in France, crossing them first with China roses and then with Bourbons and Noisettes. The Tea roses are repeat-flowering roses, named for their fragrance being reminiscent of Chinese black tea (although this is not always the case). The colour range includes pastel shades of white, pink and (a novelty at the time) yellow to apricot. The individual flowers of many cultivars are semi-pendent and nodding, due to weak flower stalks. In a "typical" Tea, pointed buds produce high-centred blooms which unfurl in a spiral fashion, and the petals tend to roll back at the edges, producing a petal with a pointed tip; the Teas are thus the originators of today's "classic" florists' rose form. According to rose historian Brent Dickerson, the Tea classification owes as much to

Tea Rose 'Mrs Dudley Cross' (Paul 1907).

marketing as to botany; 19th-century nurserymen would label their Asian-based cultivars as "Teas" if they possessed the desirable Tea flower form, and "Chinas" if they did not. Like the Chinas, the Teas are not hardy in colder climates. Examples: 'Lady Hillingdon', 'Maman Cochet', 'Duchesse de Brabant', 'Mrs. Foley Hobbs'.

Bourbon

Bourbon Rose 'Climbing Souvenir de la Malmaison' (Béluze 1843/Bennett 1893).

Bourbon roses originated on the Île Bourbon (now called Réunion) off the coast of Madagascar in the Indian Ocean. They are believed to be the result of a cross between the Autumn Damask and the 'Old Blush' China rose, both of which were frequently used as hedging materials on the island. They flower repeatedly on vigorous, frequently semi-climbing shrubs with glossy foliage and purple-tinted canes. They were first Introduced in France in 1823 by Henri Antoine Jacques. Examples: 'Louise Odier', 'Mme. Pierre Oger', 'Zéphirine Drouhin' (the last example is often classified under climbing roses).

Noisette

Noisette Rose 'Desprez à fleurs jaunes' (Desprez 1830).

The first Noisette rose was raised as a hybrid seedling by a South Carolina rice planter named John Champneys. Its parents were the China rose 'Parson's Pink' and the autumn-flowering musk rose (Rosa moschata), resulting in a vigorous climbing rose producing huge clusters of small pink flowers from spring to fall. Champneys sent seedlings of his rose (called 'Champneys' Pink Cluster') to his gardening friend, Philippe Noisette, who in turn sent plants to his brother Louis in Paris, who then introduced 'Blush Noisette' in 1817. The first Noisettes were small-blossomed, fairly winter-hardy climbers, but later infusions of Tea rose genes created a Tea-Noisette subclass with larger flowers, smaller clusters, and considerably reduced winter hardiness. Examples: 'Blush Noisette', 'Lamarque' (Noisette); 'Mme. Alfred Carriere', 'Marechal Niel' (Tea-Noisette).

Hybrid Perpetual

The dominant class of roses in Victorian England, hybrid perpetuals (a misleading translation of hybrides remontants, 'reblooming hybrids') emerged

Hybrid Perpetual Rose 'La Reine' (Laffay 1844).

in 1838 as the first roses which successfully combined Asian remontancy (repeat blooming) with the old European lineages. Since re-bloom is a recessive trait, the first generation of Asian/European crosses (hybrid Chinas, hybrid bourbons, hybrid noisettes) were stubbornly once-blooming, but when these roses were recrossed with themselves or with Chinas or teas, some of their offspring flowered more than once. The hybrid perpetuals thus were something of a miscellany, a catch-all class derived to a great extent from the Bourbons but with admixtures of Chinas, teas, damasks, gallicas, and to a lesser extent Noisettes, albas and even centifolias. They became the most popular garden and florist roses of northern Europe at the time, as the tender tea roses would not thrive in cold climates, and the hybrid perpetuals' very large blooms were well-suited to the new phenomenon of competitive exhibitions. The "perpetual" in the name hints at repeat-flowering, but many varieties of this class had poor re-flowering habits; the tendency was for a massive spring bloom, followed by either scattered summer flowering, a smaller autumn burst, or sometimes nothing at all until next spring. Due to a limited colour palette (white, pink, red) and lack of reliable repeat-bloom, the hybrid perpetuals were ultimately overshadowed by their descendants, the Hybrid Teas. Examples: 'Ferdinand Pichard', 'Reine Des Violettes', 'Paul Neyron'.

Hybrid Musky

Hybrid Musk Rose 'Moonlight' (Pemberton 1913).

Although they arose too late to qualify technically as old garden roses, the hybrid musks are often informally classed with them, since their growth habits and care are much more like the old garden roses than modern roses. The hybrid musk group was mainly developed by Rev. Joseph Pemberton, a British rosarian, in the first decades of the 20th century, based upon 'Aglaia', an 1896 cross by Peter Lambert. A seedling of this rose, 'Trier', is considered to the foundation of the class. The genetics of the class are somewhat obscure, as some of the parents are unknown. Rose multiflora, however, is known to be one parent, and Rosa moschata (the musk rose) also figures in its heritage, though it is considered to be less important than the name would suggest. Hybrid musks are disease-resistant, repeat flowering and generally cluster-flowered, with a strong, characteristic "musk" scent. The stems tend to be lax and arching, with limited thorns.[8] Examples include 'Buff Beauty' and 'Penelope'.

Hybrid Rugosa

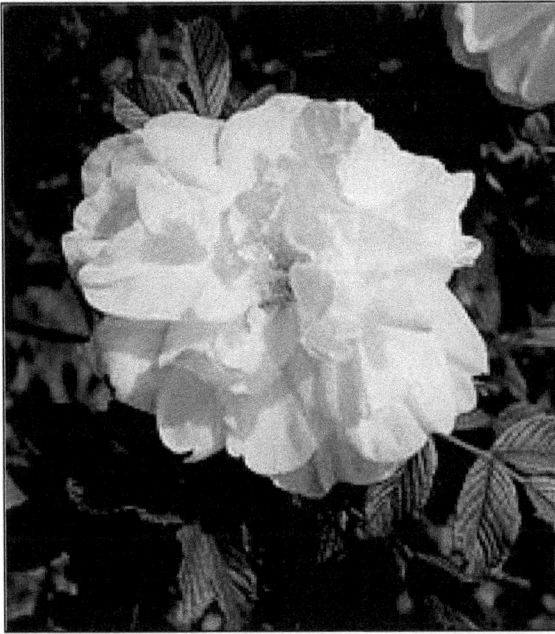

Rugosa Rose 'Blanc Double de Coubert' (Cochet 1893).

The rugosas likewise are not officially old garden roses, but tend to be grouped with them. Derived from Rosa rugosa from Japan and Korea beginning in the 1880s, these vigorous roses are extremely hardy with excellent disease resistance. Most are extremely fragrant, repeat bloomers with moderately double flat flowers. The defining characteristic of a hybrid rugosa rose is its wrinkly leaves, but some hybrids do lack this trait. These roses will often set hips. Examples include 'Hansa' and 'Roseraie de l'Häy'.

Bermuda "Mystery" Roses

This is a group of several dozen "found" roses grown in Bermuda for at least a century. The roses have significant value and interest for those growing roses in tropical and semi-tropical regions, since they are highly resistant to both nematode damage and the fungal diseases that plague rose culture in hot, humid areas. They are also capable of thriving in hot and humid weather. Most of these roses are thought to be Old Garden Rose cultivars that have otherwise dropped out of cultivation, or sports thereof. They are "mystery roses" because their "proper" historical names have been lost. Tradition dictates that they are named after the owner of the garden where they were rediscovered.

Miscellaneous

There are also a few smaller classes (such as Scots, Sweet Brier) and some climbing classes of old roses (including Ayrshire, Climbing China, Laevigata, Sempervirens, Boursault, Climbing Tea, and Climbing Bourbon). Those classes with both climbing and shrub forms are often grouped together.

Modern Garden Roses

Classification of modern roses can be quite confusing because many modern roses have old garden roses in their ancestry and their form varies so much. The classifications tend to be by growth and flowering characteristics. The following includes the most notable and popular classifications of Modern Garden Roses:

Hybrid Tea

A 'Memoriam' Hybrid Tea Rose (von Abrams 1962).

The favourite rose for much of the history of modern roses, hybrid teas were initially created by hybridising hybrid perpetuals with Tea roses in the late 19th century. 'La France', created in 1867, is universally acknowledged as the first indication of a new class of roses. Hybrid teas exhibit traits midway between both parents: hardier than the teas but less hardy than the hybrid perpetuals, and more ever-blooming than the hybrid perpetuals but less so than the teas. The flowers are well-formed with large, high-centred buds, and each flowering stem typically terminates in a single shapely bloom. The shrubs tend to be stiffly upright and sparsely foliaged, which today is often seen as a liability because it makes them more difficult to place in the garden or landscape. Hybrid teas became the single most popular garden rose of the 20th century; today, their reputation as high maintenance plants has led to a decline in popularity. The hybrid tea remains the standard rose of the floral industry, however, and is still favoured in formal situations. Examples: 'Peace' (yellow), 'Garden Party' (white), 'Mister Lincoln' (red) and 'Double Delight' (bi-colour cream and red).

Pernetiana

Pernetiana Rose 'Soleil d'Or', the First of its Class (Pernet 1900).

The French breeder Joseph Pernet-Ducher initiated the first class of roses to include genes from the old Austrian briar rose (Rosa foetida) with his 1900 introduction of 'Soleil d'Or.' This resulted in an entirely new colour range for roses: shades of deep yellow, apricot, copper, orange, true scarlet, yellow bicolours, lavender, gray, and even brown were now possible. Originally considered a separate class, the Pernetianas or Hybrid Foetidas were officially merged into the Hybrid Teas in 1930. The new colour range did much to increase hybrid tea popularity in the 20th century, but these colours came at a price: Rosa foetida also passed on a tendency toward disease-susceptibility, scentless blooms, and an intolerance of pruning to its descendants.

Polyantha

Literally "many-flowered" roses, from the Greek "poly" (many) and "anthos" (flower). Originally derived from crosses between two East Asian species (*Rosa chinensis* and *Rosa multiflora*), polyanthas first appeared in France in the late 19th century alongside the hybrid teas. They featured short plants—some compact, others spreading in habit—with tiny blooms (2.5 cm or 1 inch in diameter on average) carried in large sprays, in the typical rose colours of white, pink and red. Their main claim to fame was their prolific bloom: From spring to fall, a healthy polyantha shrub might be literally covered in flowers, creating a strong colour impact in the landscape. Polyantha roses are still regarded as low-maintenance, disease-resistant garden roses today, and remain popular for that reason. Examples: 'Cécile Brünner', 'The Fairy', 'Red Fairy', 'Pink Fairy'.

Floribunda

Rosa 'Borussia', a Modern Floribunda Rose.

Some rose breeders saw potential in crossing polyanthas with hybrid teas, to create roses that bloomed with the polyantha profusion, but with hybrid tea floral beauty and colour range. In 1907, the first polyantha/hybrid tea cross, 'Rödhätte', was introduced by Danish breeder Dines Poulsen. This had some characteristics of both parent classes, and was termed a Hybrid Polyantha or Poulsen rose. Further similar introductions followed from Poulsen, these often bearing the family name *e.g.* 'Else Poulsen' (1924). As their hybrid characteristics separated these new roses from polyanthas and hybrid teas alike, a new class was created and named floribunda, Latin for "many-flowering." Typical floribundas feature stiff shrubs, smaller and bushier than the average hybrid tea but less dense and sprawling than the average polyantha. The flowers are often smaller than hybrid teas but are carried in large sprays, giving a better floral effect in the garden. Floribundas are found in all hybrid tea colours and often with the classic hybrid tea-shaped blossom, sometimes differing from hybrid teas only in their cluster-flowering habit. Today they are still used in large bedding schemes in public parks and similar spaces. Examples: 'Anne Harkness', 'Dainty Maid','Iceberg', 'Tuscan Sun'.

Grandiflora

Grandifloras (Latin for "large-flowered") were the class of roses created in the mid-20th century to designate back-crosses between hybrid teas and floribundas that fit neither category—specifically, the 'Queen Elizabeth' rose, which was introduced in 1954. Grandiflora shrubs are typically larger than either hybrid teas or floribundas, and feature hybrid tea-style flowers borne in small clusters of three to five, similar to a floribunda. Grandifloras maintained some popularity from about the 1950s to the 1980s but today they are much less popular than either the hybrid teas or the floribundas. Examples: 'Queen Elizabeth', 'Comanche,' 'Montezuma'.

Miniature

'Meillandine' (a miniature rose) in a Terracotta Flowerpot.

Dwarf mutations of some Old Garden Roses—gallicas and centifolias—were known in Europe in the 17th century, although these were once-flowering just as their larger forms were. Miniature forms of repeat-flowering China roses were also grown and bred in China, and are depicted in 18th-century Chinese art. Modern miniature roses largely derive from such miniature China roses, especially the cultivar 'Roulettii', a chance discovery found in a pot in Switzerland.

Miniature roses are represented by twiggy, repeat-flowering shrubs ranging from 6" to 36" in height, with most falling in the 12"–24" height range. Blooms come in all the hybrid tea colours; many varieties also emulate the classic high centred hybrid tea flower shape. Miniature roses are often marketed and sold by the floral industry as houseplants, but it is important to remember that these plants are largely descended from outdoor shrubs native to temperate regions; thus, most miniature rose varieties require an annual period of cold dormancy to survive. (Examples: 'Petite de Hollande' (Miniature Centifolia, once-blooming), 'Cupcake' (Modern Miniature, repeat-blooming). Miniature garden roses only grow in the summer.

Climbing and Rambling

Rosa 'Zéphirine Drouhin', a Climbing Bourbon Rose (Bizot 1868).

All aforementioned classes of roses, both Old and Modern, have "climbing" forms, whereby the canes of the shrubs grow much longer and more flexible than the normal ("bush") forms. In the Old Garden Roses, this is often simply the natural growth habit; in many Modern roses, however, climbing roses are the results of spontaneous mutations. For example, 'Climbing Peace' is designated as a "Climbing Hybrid Tea," for it is genetically identical to the normal "shrub" form of the 'Peace' hybrid tea rose, except that its canes are long and flexible, *i.e.* "climbing." Most Climbing roses grow anywhere from 8'–20' in height and exhibit repeat-bloom. Rambler roses, although technically a separate class, are often lumped together with climbing roses. They also exhibit long, flexible canes, but are usually distinguished from true climbers in two ways: A larger overall size (20'–30' tall is common), so is a once-blooming habit. Climbing and rambling roses are not true vines such as ivy, clematis or wisteria because they lack the ability to cling to supports on their own and must be manually trained and tied over structures such as arbors and pergolas. Examples: 'Blaze' (repeat-blooming climber), 'American Pillar' (once-blooming rambler). One of the most vigorous climbing roses is the Kiftsgate rose, Rosa filipes 'Kiftsgate', named after the house garden where it was noticed by Graham Stuart Thomas in 1951. The original plant is claimed to be the largest rose in the UK, and has climbed 50 feet into a copper beech tree.

Shrub

The Shrub Rose 'Mollineux'.

This is not a precisely defined class of garden rose, but it is a description or grouping commonly used by rose reference books and catalogues. It encompasses some old single and repeat flowering cultivars, as well as modern roses that don't fit neatly into other categories. Many cultivars placed in other categories are simultaneously placed in this one. Roses classed as shrubs tend to be robust and of informal habit, making them recommended for use in a mixed shrub border or as hedging.

English/David Austin

Although not officially recognized as a separate class of roses by any established rose authority, English (aka David Austin) roses are often set aside as such by consumers and retailers alike. Development started in the 1960s by David Austin of Shropshire, England, who wanted to rekindle interest in Old Garden Roses by hybridizing them with modern hybrid teas and floribundas. The idea was to create a new group of roses that featured blooms with old-fashioned shapes and fragrances, evocative of classic gallica, alba and "damask" roses, but with modern repeat-blooming characteristics and the larger modern colour range as well. Austin mostly

Austin Rose 'Abraham Darby'(1985).

succeeded in his mission; his tribe of "English" roses, now numbering hundreds of varieties, has been warmly embraced by the gardening public and are widely available to consumers. David Austin roses are still actively developed, with new varieties released regularly. The typical winter-hardiness and disease-resistance of the classic Old Garden Roses has largely been compromised in the process; many English roses are susceptible to the same disease problems that plague modern hybrid teas and floribundas, and many are not hardy north of USDA Zone 5. Examples: 'Charles Austin', 'Graham Thomas', 'Mary Rose', 'Tamora', 'Wife of Bath'.

Canadian Hardy

Rosa **'Henry Hudson', one of the Explorer Series.**

Two main lines of roses were developed for the extreme weather conditions of Canadian winters by Agriculture Canada at the Morden Research Station in Morden, Manitoba and the Experimental Farm in Ottawa (and later at L'Assomption,

'Thérèse Bugnet', a Multi-Species Hybrid that is still Widely Available (Bugnet 1950).

Québec). They are called the Explorer series and the Parkland series. These programs have now been discontinued, the remaining plant stock has been taken over by private breeders and marketed along with the Canadian Artists roses as a single series. Derived mostly from crosses of Rosa rugosa or the native Canadian species Rosa arkansana with other species, these plants are extremely tolerant of cold weather, some down to –35C. All have repeat bloom. A wide diversity of forms and colours were achieved.

Examples of roses in the Explorer series are: 'Martin Frobisher', 'Jens Munk' (1974), 'Henry Hudson' (1976), 'John Cabot' (1978), 'David Thompson' (1979), 'John Franklin' (1980), 'Champlain' (1982), 'Charles Albanel' (1982), 'William Baffin' (1983), 'Henry Kelsey' (1984), 'Alexander Mackenzie' (1985), 'John Davis' (1986), 'J.P. Connell' (1987), 'Captain Samuel Holland' (1992), 'Frontenac' (1992), 'Louis Jolliet' (1992), 'Simon Fraser' (1992), 'George Vancouver' (1994), 'William Booth' (1999).

Roses in the Parkland series include 'Morden Centennial', 'Morden Sunrise, 'Winnipeg Parks' and 'Cuthbert Grant'. Two roses named after Canadian artists that have been added are 'Emily Carr' and 'Felix Leclerc'.

Other notable Canadian breeders include Frank Skinner, Percy Wright, Isabella Preston, Georges Bugnet and Robert Erskine.

Landscape (Ground Cover)

This type of rose was developed mainly for mass amenity planting. In the late 20th century, traditional hybrid tea and floribunda rose varieties fell out of favour with many gardeners and landscapers, as they are often labour and chemical intensive plants susceptible to pest and disease problems. So-called "landscape" roses (also known as "ground cover" roses) have thus been developed to fill the consumer desire for a garden rose that offers colour, form and fragrance, but is also low maintenance and easy to care for. Most have the following characteristics:

☆ Lower growing habit, usually under 60 cm (24 inches)

'Avon', a Ground Cover Rose Introduced by Poulson in 1992.

☆ Repeat flowering

☆ Disease and pest resistance

☆ Growing on their own roots.

☆ Minimal pruning requirements

Principal parties involved in the breeding of new varieties include: Werner Noak (Germany), Meidiland Roses (France), Boot and Co. (Netherlands), and William Radler (USA).

Patio

Chris Warner's Patio Climber 'Open Arms' (1995).

Since the 1970s many rose breeders have focused on developing compact roses (typically 1'-3' in height and spread) that are suitable for smaller gardens, terraces and containers. These combine characteristics of larger miniature roses and smaller floribundas—resulting in the rather loose classification "patio roses". Dr. D.G. Hessayon says the description "patio roses" emerged after 1996. Some rose catalogues include older polyanthas that have stood the test of time (*e.g.,* 'Nathalie Nypels', 'Baby Faurax') within their patio selection. Rose breeders, notably Chris Warner in the UK and the Danish firm of Poulson (under the name of Courtyard Climbers) have also created patio climbers, small rambler style plants that flower top-to-toe and are suitable for confined areas.

Cultivation

In the garden, roses are grown as bushes, shrubs or climbers. "Bushes" are usually comparatively low growing, often quite upright in habit, with multiple stems emerging near ground level; they are often grown formally in beds with other roses. "Shrubs" are usually larger and have a more informal or arching habit, and may additionally be placed in a mixed border or grown separately as specimens. Certain bush hybrids (and smaller shrubs) may also be grown as "standards", which are plants grafted high (typically 1 metre or more) on a rose rootstock, resulting in extra height which can make a dominant feature in a floral display. Climbing roses are usually trained to a suitable support.

Roses are commonly propagated by grafting onto a rootstock, which provides sturdiness and vigour, or (especially with Old Garden Roses) they may be propagated from hardwood cuttings and allowed to develop their own roots.

Most roses thrive in temperate climates. Those based on warm climate Asian species do well in their native sub-tropical environments. Certain species and cultivars can even flourish in tropical climates, especially when grafted onto appropriate rootstocks. Most garden roses prefer rich soil which is well-watered but well-drained, and perform best in well-lit positions which receive several hours of sun a day (although some climbers, some species and most Hybrid Musks will tolerate shade). Standard roses require staking.

Pruning

Rose pruning, sometimes regarded as a horticultural art form, is largely dependent on the type of rose to be pruned, the reason for pruning, and the time of year it is at the time of the desired pruning. Most Old Garden Roses of strict European heritage (albas, damasks, gallicas, *etc.*) are shrubs that bloom once yearly, in late spring or early summer, on two-year-old (or older) canes. Their pruning requirements are quite minimal because removal of branches will remove next year's flower buds. Hence pruning is usually restricted to just removing weak and spent branches, plus light trimming (if necessary) to reduce overall size.

Modern hybrids, including the hybrid teas, floribundas, grandifloras, modern miniatures, and English roses, have a complex genetic background that almost always includes China roses (which are descended from Rosa chinensis). China roses were evergrowing, everblooming roses from humid subtropical regions that bloomed constantly on any new vegetative growth produced during the growing season. Their modern hybrid descendants exhibit similar habits; unlike Old European Roses, modern hybrids bloom continuously (until stopped by frost) on any new canes produced during the growing season. They therefore require pruning back of any spent flowering stem in order to divert the plant's energy into producing new growth and hence new flowers.

Additionally, Modern Hybrids planted in cold winter climates will almost universally require a "hard" annual pruning (reducing all canes to 8"–12", about 30 cm in height) in early spring. Again, because of their complex China rose background, modern hybrids are typically not as cold hardy as European Old Garden Roses, and low winter temperatures often desiccate or kill exposed canes. In spring, if left unpruned, these damaged canes will often die back all the way to the shrub's root zone, resulting in a weakened, disfigured plant. The annual "hard" pruning of hybrid teas and floribundas is generally done in early spring.

Deadheading

This is the practice of removing any spent, faded, withered, or discoloured flowers. The purpose is to encourage the plant to focus its energy and resources on forming new shoots and blooms, rather than fruit production. Deadheading may also be performed for aesthetic purposes, if spent flowers are unsightly. Any roses such as Rosa glauca or Rosa moyesii that are grown for their decorative hips should not be deadheaded.

Pests and Diseases

Roses are subject to several diseases. The main fungal diseases affecting the leaves are rose black spot (Diplocarpon rosae), rose rust (*Phragmidium mucronatum*), rose powdery mildew (*Sphaerotheca pannosa*) and rose downy mildew (*Peronospora sparsa*). Stems can be affected by several canker diseases, the most commonly seen of which is stem canker (*Leptosphaeria coniothyrium*). Diseases of the root zone include honey fungus (*Armillaria* spp.), verticillium wilt, and various species of phytophthora.

Fungal leaf diseases affect some cultivars and species more than others. On susceptible plants fungicidal sprays may be necessary to prevent infection or reduce severity of attacks. Cultivation techniques may also be used, such as ensuring good air circulation around a plant. Stem cankers are best treated by pruning out infection as soon as it is noticed. Root diseases are not usually possible to treat once infection has occurred; the most practical line of defence is to ensure that growing conditions maximise plant health and thereby prevent infection. Phytophthora species are waterborne and therefore improving drainage and reducing waterlogging can help reduce infection.

The main pest affecting roses is the aphid (greenfly), which sucks the sap and weakens the plant. In areas where they are endemic Japanese beetles (Popillia japonica) take a heavy toll on rose flowers and foliage; rose blooms can also be destroyed by infestations of thrips (Thysanoptera spp). Roses are also used as food plants by the larvae of some Lepidoptera (butterfly and moth) species; see list of Lepidoptera that feed on roses. Spraying with insecticide of roses is often recommended but if this is done care is needed to minimize the loss of beneficial insects; systemic insecticides have the advantage of only affecting insects which feed on the plants.

Notable Rose Growers

☆ Some rose growers are known for their particular contributions to the field. These include:

☆ David Austin nursery, based in Shropshire, UK, is the developer of "English roses", such as 'Constance Spry', 'Mary Rose' and 'Graham Thomas'

☆ Peter Beales was a specialist in classic and species roses, preserving many old and wild roses at his Norfolk nursery and also introducing 70 new cultivars. He was also the author of several classic books on the subject of roses.

☆ Joséphine de Beauharnais (Empress Josephine) was the first great collector of roses in the modern Western world, and her horticulturalist André Dupont pioneered the development of new hybrids using controlled pollination at her Malmaison estate. She has been called the godmother of modern rosomaniacs.

☆ Cants of Colchester, in Essex, is the UK's oldest firm of commercial rose growers. Notable introductions include 'Mrs B.R. Cant' and 'Just Joey'.

☆ Conard-Pyle Co. introduced the rose 'Peace' to the US and established the marque Star Roses. 'Peace' was bred by Meilland of France (where it was introduced as 'Mme A. Meilland'); Conard-Pyle acted as Meilland's US agents, and the rose was renamed for the US market when it was introduced at the end of the Second World War.

☆ Georges Delbard of Allier, France is more famous for new varieties of fruit tree, but among his nursery's roses are 'Centenaire de Lourdes', 'Altissimo' and 'Papa Delbard'.

☆ Dickson Roses, located near Belfast introduced its first roses in 1886, focusing on breeding Hybrid Teas that could stand up to the Irish climate. Successes include 'Shot Silk' and 'Grandpa Dickson' and, more recently, 'Elina' and 'Tequila Sunrise'.

☆ Pedro Dot put Spanish rose growing on the map and is best known for the shrub 'Nevada' and his work to improve the flower shape of miniature varieties.

☆ Claude Ducher was a Lyon hybridiser and nurseryman (father-in-law of Joseph Pernet-Ducher), whose roses include the Noisette 'Reve d'Or' and the Tea rose 'Marie van Houtte'.

☆ Andre Dupont was a French horticulturalist who pioneered the creation of new rose cultivars through controlled pollination. He was employed by the Empress Josephine to use her collection of roses to create new roses.

☆ Rudolf Geschwind was an Austro-Hungarian amateur rose breeder who introduced around 140 new varieties, including 'Gruss an Teplitz'. He focused on winter hardiness and vigour.

☆ Jules Gravereaux, founder of Roseraie de L'Haÿ

☆ Harkness Roses, in Hertfordshire, UK is best known for 'Ena Harkness' (at one time reputed to be the best-selling red Hybrid Tea in the world and actually bred by amateur rosarian Albert Norman). Other famous introductions include 'Compassion' and 'Margaret Merril'.

☆ S. Reynolds Hole was Dean of Rochester Cathedral in the UK and the founder of the (Royal) National Rose Society. He organized the first specialty rose show in the UK and published books on rose cultivation, popularizing rose growing and exhibiting.

☆ Jackson and Perkins was a hugely influential American rose grower. The company's early success was 'Dorothy Perkins', but under Eugene Boerner the focus on developing Floribundas led to many All-America Rose Selection honours.

☆ W. Kordes' Sons, based in Sparrieshoop in Schleswig-Holstein, Germany, is one of the most innovative rose breeders and growers, and responsible for the early flowering "Frühlings" series, the Kordesii Hybrids and many famous Hybrid Tea and Floribunda roses, including 'Crimson Glory' and 'Iceberg' ('Schneewittchen').

☆ McGredy, of Northern Ireland, was responsible for 'Evelyn Fison', 'Dublin Bay' and also 'Regensberg', a pioneering 'handpainted' rose. Sam McGredy IV moved toNew Zealand in 1974 and focused on hybrid teas and Grandifloras, including 'Paddy Stephens' and 'Kathryn McGredy'.

☆ Meilland family made its name and fortune with 'Mme A. Meilland' ('Peace'), and has continued to be at the forefront of rose breeding, with varieties such as 'Bonica '82' and 'Swany'.

☆ Ralph S. Moore, the California-based breeder of more than 500 roses, is known as 'the father of Modern Miniatures' and was a hugely influential figure in the development of commercial approaches to rose hybridization.

☆ Werner Noack of Germany introduced the Flower Carpet (ground cover/ landscape) series.

☆ Paul was a Hertfordshire, UK nursery that involved two brothers (William and George) and their two sons (Arthur William and Paul Laing). Collectively, Paul is known today for varieties such as 'Paul's Lemon Pillar' and 'Paul's Scarlet Climber'. Experimental hybrids using species roses resulted in choice varieties such as 'Mermaid'.

☆ Joseph Pemberton was an Anglican clergyman and amateur rosarian who set out to breed 'old fashioned' roses. The resulting hybrid musks include 'Felicia' and 'Penelope'. On his death, the nursery passed to his gardener J.A. Bentall, who produced 'Buff Beauty' and the Polyantha 'The Fairy'.

☆ Jean Pernet, père was a Lyon nurseryman whose notable roses include the Moss variety 'Louis Gimard' and the hybrid perpetual 'Baronne Adolphe de Rothschild'.

☆ Joseph Pernet-Ducher was among the first rose breeders to focus on developing the new Hybrid Tea class. His introductions include 'Mme Caroline Testout' and 'Soleil d'Or'- forerunner of 20th-century yellow and orange roses.

☆ Poulson, the Danish rose dynasty, was established in 1878 and originally focused on breeding roses hardy enough to withstand the Scandinavian climate. Later introductions notable for their form and colour include 'Chinatown' (1963) and 'Ingrid Bergman' (1984). The nursery developed a number of successful ground cover (landscape) roses, including 'Kent' (1988).

☆ Rose Barni in Tuscany specialises in roses for Mediterranean climates. Notable successes include 'Castore' and 'Polluce', and striped varieties such as 'Rinascimento' and 'Missoni'.

☆ Mathias Tantau

☆ Graham Stuart Thomas is best known for reawakening interest in old garden roses, but also ensured commercial introductions in the wild rose style, including 'Bobbie James' and 'Souvenir de St Anne's'.

☆ Dr. Walter van Fleet worked for the US Department of Agriculture, focusing on crops, but also developing roses designed to thrive in the American climate. His introductions include 'American Pillar' and 'Dr W. Van Fleet'. After his death, his seedlings - including 'Mary Wallace', 'Breeze Hill' and 'Glenn Dale' - were introduced by the American Rose Society as 'dooryard climbers'.

☆ Jean-Pierre Vibert was a prolific early rose hybridizer, responsible for many older roses still found in gardens today. 'Aimee Vibert' (1828), one of his Noisettes, was named for his daughter.

☆ Weeks Roses (with Tom Carruth) is a California rose company that has focused on innovations in colour, form and vigour. Its roses include 'Night Time', 'Stainless Steel', 'Fourth of July' and 'Hot Cocoa'.

Dutch Roses

While it has been quite some time since tech-backed farming of roses began in the states of Maharashtra, Karnataka, Delhi and West Bengal, it was yet to start in Jharkhand where the every day demand of over 5,000 Dutch rose stems was until recently met by cultivatators from outside the state.

The specially designed greenhouse ensures a controlled temperature of 12-25Oc, wind velocity and humidity – all necessary for healthy growth of the flower. Unlike ordinary roses, Dutch roses have long stems of about 10-12 inches, big buds, are semi-bloomed and have a longer vase life. The plants are raised in rows of beds of earth mixed with compost. "Apart from watering the plants, drips are used to provide nutrients. Their company, which employs six workers and a technician, made an initial investment of ₹ 90 lakh, and has so far sold flowers worth more than ₹ 60 lakh. The Centre's National Horticulture Board had provided 20 per cent subsidy, Rasila is out of debt now.

Rose-nama

This summer chefs at Delhis many eateries are using the glorious gulab most gastronomically! From rose risotto, rose-petal shooters to fragrant marinades, AMIN ALI and KANIKA DHAWAN discover the luscious realm of the bloom, petal by petal.

Petal Palate

Edible flowers like lavender make for a delicious companion when added to a seasonal salad or an entree. But when it comes to a rose, not all varieties of this fragrant bloom can be added to food. Only desi gulab can be used in cooking because it has enormous nutritive value as compared to the budded variety thats available at florist shops. Nishant Choubey understand that fact pretty well. Rose doesnt just cool, it adds aroma to the meal and doubles up as an anti-oxidant. But the reason its used sparingly in the main course is because its inherently sweet and may ruin the taste of a dish. Choubey makes a decoction, brews the petals and uses them for garnishing. He also prepares pan-fried scallops with roselime glaze, rose-fresh pepper-crusted corn-fed chicken with mint glaze and rose tea-infused asparagus with couscous and parmesan shavings along with rose, chilli and kafir sorbet and rose and berry semifred do. Likewise, The Leela Palaces Vinod Saini marinates chicken in rose petals and even stuffs chicken with fresh gulkand. For highlighting the taste of vegetarian food, he uses a paste of dried rose petals for marinating vegetables. In fact, rose is even used to create out-of the-ordinary breakfast spreads. Rose marmalade is made by using the Damascus rose. Its difficult to find this bloom in Delhi. One can use desi gulab instead.It tastes really good on cakes as a glaze or even when its simply used for garnishing dry desserts.

Relishing the Rose

Summer may have kickstarted the rose revolution but there are other reasons why everyone swears by the flower. After all, the mere addition of a petal or two is known to do wonders for a dish. Ask The Claridges Neeraj Tyagi who affirms that the rose is not a seasonal flower, an attribute that works in its favour. It can be used in desserts, drinks and main course dishes right through the year. In summer it can be used to lend a pleasing aroma and a distinct taste to any dish or drink. No wonder, Tyagis ultimate rose-smorgasbord would comprise a medley of cuisines, kissed by a rose. From rose-infused kachche gosht ki biryani, rose risotto to rose granita it would be a meal, quite literally soaked in rose heaven!

Flower Food

Apart from the rose,the other kinds of edible flowers include nasturtium (its leaves and flowers are used to make pakoras and sandwiches),pumpkin and banana flowers (Bengalis make delicate fritters out of it),calendula and marigold (seasoning salads) and safflower (used for flavouring).

Flower Fascination: India Set to be Floriculture Trade Leader

India's share in global floriculture trade may not be significant but the country has, of late, shown enough potential to eventually turn itself as a favourite destination of flower importers in near future.

Surprisingly, the small land-holding pattern, considered a handicap for the country's agricultural production, comes as an advantage in floriculture due to its 'low volume high value' character. Since the sector has huge export potential, a number of small and marginal farmers have started turning towards flower production.

Increasing domestic demand for both cut and loose flowers has also attracted farmers, mainly in leading flower producing states like Tamil Nadu, Karnataka, West Bengal, Madhya Pradesh and Maharashtra, towards floriculture.

The northeastern states, especially Mizoram, have also turned towards cultivating flowers of export varieties in a big way.

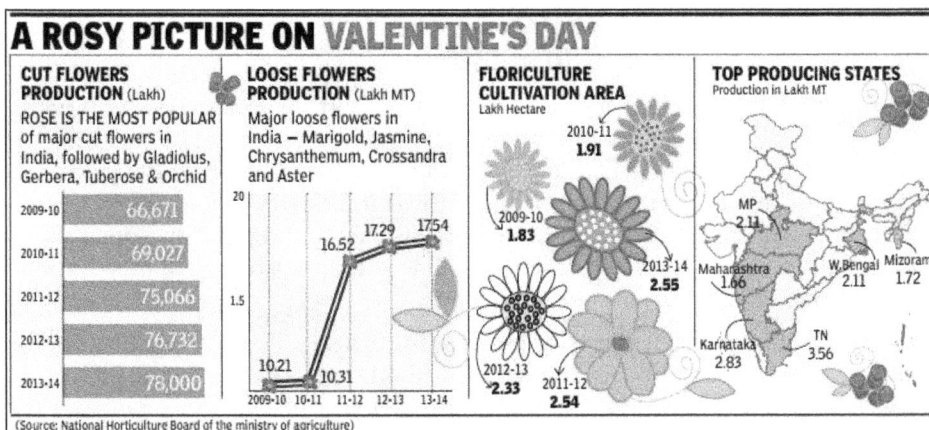

A ROSY PICTURE ON VALENTINE'S DAY

CUT FLOWERS PRODUCTION (Lakh)
ROSE IS THE MOST POPULAR of major cut flowers in India, followed by Gladiolus, Gerbera, Tuberose & Orchid

Year	Value
2009-10	66,671
2010-11	69,027
2011-12	75,066
2012-13	76,732
2013-14	78,000

LOOSE FLOWERS PRODUCTION (Lakh MT)
Major loose flowers in India – Marigold, Jasmine, Chrysanthemum, Crossandra and Aster

10.21, 10.31, 16.52, 17.29, 17.54 (2009-10 10-11 11-12 12-13 13-14)

FLORICULTURE CULTIVATION AREA Lakh Hectare
2010-11 1.91
2009-10 1.83
2013-14 2.55
2012-13 2.33
2011-12 2.54

TOP PRODUCING STATES Production in Lakh MT
MP 2.11
Maharashtra 1.66
W.Bengal 2.11
Mizoram 1.72
Karnataka 2.83
TN 3.56

(Source: National Horticulture Board of the ministry of agriculture)

Though the country had during 2014-15 exported floriculture products worth over ₹ 460 crore, its overall share is quite low in the global trade of nearly USD 40 billion or nearly ₹ 2,72,000 crore.

"There is a great demand for Indian flowers in Gulf countries. Besides the major cut flowers, the export of traditional flowers such as jasmine and marigold is also gaining momentum in recent times. How floriculture in the country moved from "dormancy to infancy", backed by growing domestic demand and policy support from the government can be explained by the fact that the country is bestowed with ideal temperature conditions for commercial floriculture throughout the year in some or other part. This has helped entrepreneurs and growers in recognising diversification into floriculture as of a commercial value.

Floriculture in India is viewed as a high growth industry and the government has already identified this sector as a sunrise industry, according it 100 per cent export-oriented status.Though major export destination of Indian cut flowers (flowers harvested in clusters or in single along with their stems) were the USA, UK, Germany, Netherlands and UAE during 2014-15, loose flowers (flowers harvested without stalk) like jasmines and marigold are generally exported to Sri Lanka, Malaysia, Singapore and West Asian countries. China, India and Peru are leading producers and exporters of marigold.

The annual world trade in marigold is currently estimated to be around ₹500 crore. Singh said, "In recent times, flower importers have been shifting their focus to India and Ethiopia for cheaper flowers in the wake of rising cost of production in Kenya - the world's largest exporter." Quoting reports from the Kenya Investment Authority (KenInvest), "India could overtake Kenya in floriculture in the near future."

The NHB managing director has reasons to believe the Ken Invest reports as he enlisted a number of factors which are conducive for the floriculture sector to grow substantially in future. "The traditional disadvantage associated with the 'small farm holding' is an advantage in floriculture ventures. Even non-arable land can be put to agricultural use since protected floriculture involves specific growing media or amendment of the existing soil.

Conclusions and Recommendations

Anyone who has observed what is happening in the Indian flower industry can draw only one conclusion: the scent of the Indian export rose is not sweet, but bitter. Just like everywhere else in the world, the Indian flower industry pollutes the environment. This form of agriculture (or is it industry?) does not fit in with the idea of a sustainable society. It is a poignant fact that the Indian flower sector is co-financed by Dutch development aid. One should wish India a more human and environmentally friendly development!

The Indian government spends many subsidies on the export flowers. But the profit from this investment is dubious. It is the Dutch entrepreneurs who profit from the Indian flower industry. The Indian workers on the other hand are the victims: they have to do hazardous work in greenhouses for moderate wages and without any social security. Of course this means more employment, but is this the kind of work the Indians really want? And in numbers this employment does not amount to much. The sector is very capital intensive. One acre needs an investment of 200 to 300 thousand US dollars. An average of 20 labourers works in one acre. This means that it costs 10,000 to 15,000 dollars to employ one labourer. This is an extremely expensive job by Indian standards. The same amount of money would provide jobs for more people in other sectors.

The most important recommendation is to stop the flower industry in India, or at least check its growth. This is certainly possible. The rise of the flower industry is not a natural process and it has little to do with a development of the market economy. The Indian and - to a lesser degree - the Dutch government play an important part in promoting and financing. This means that subsidies may be used in other ways and development may be guided into another direction. India is at an advantage as compared to other flower producers like Kenya and Colombia. Since the Indian flower sector is still small, the cut in subsidies will be less painful. Another course is still possible without too drastic consequences.

While the flower industry is still there, rules must be tightened. A good and clear registration of pesticides is needed. Moreover workers have to be trained in the use of pesticides. Protective clothing must be of the same standard as the clothing used in Dutch greenhouses. Pesticides (like DDT and captan) banned in the Netherlands and Europe must not be used. Double standards cannot be justified, particularly not in companies co-managed by Dutch people. The waiting times after spraying in greenhouses prescribed by the WHO should be observed.

6

Other Flowers in India

Flowers have always been an integral part of Indian culture. But, floriculture as a business proposition has not received the attention it deserves. This industry is still in a nascent stage and the po- tential remains untapped. Floriculture includes not just cultivation of a wide variety of flowers but also extensive research and creation of healthier, stronger seed varieties and plugs. Globally, more than 140 countries are involved in the cultivation of floriculture crops.

India's floriculture business is worth ₹9 billion. With rich and varied climatic conditions and abundance of cheap labour, it is most suit- able for floriculture. Organised floriculture in India is 15 years old. Of the 110,000 hectares under floriculture in the country, only 500 hectares come under organised floriculture. Huge projects came up in the 1990s. However, not many of the older units have survived.

The Indian floriculture industry comprises the florist trade, nursery plants, potted plants, bulb and seed production, micro propagation ma- terial and extraction of essential oils from flowers. Karnataka, Tamil Nadu, Andhra Pradesh, West Bengal, Maharashtra, Rajasthan, Uttar Pradesh, Delhi and Haryana are showing interest in floriculture. The southern states of Tamil Nadu, Karnataka and Andhra Pradesh have a major share. While Karnataka contributes a major portion of exports at ₹ 1 billion, Delhi accounts for over half of the domestic market at ₹ 4 billion.

History of Flowers

The history of flowers is older than of humans. Flowers are ubiquitous. Virtually, every non-meat food that we eat starts as a flowering plant somewhere. Even the cotton clothes we wear come from flowering plants. Flowers appeared on earth about 130 million years ago, during the Crustacean period. Humans are believed to have existed for a mere two hundred thousand years. Once flowers took firm root about 100 million years ago, they quickly diversified into some 250,000 species.

Flowering plants have the botanical name of 'angiosperm.' Angiosperms enclose their seed in fruit, and each fruit contains one or more carpels. Carpels are hollow chambers that protect and nourish the seeds. The oldest fossil of an angiosperm was discovered in China in sediments that date back to 130 million years. It was so primitive that it had not developed the lovely flower that is typical of a flowering plant. It is still considered a flowering plant since it had carpels enclosing seeds that grew into fruits.

The first angiosperms are believed to have evolved in areas where there were ecological disturbances like floodplains and volcanic regions. They were slow growing and had short life cycles. Thus, they matured and reproduced and multiplied much faster than trees. As they seeded quickly, they also evolved faster than other plants that were their competitors. Scientists believe that petals evolved 30 to 40 million years after the first an- giosperms evolved. Once this happened, the angiosperms had the distinct advantage over all the other plants, as they attracted the birds and the bees, which, in turn, polinated them.

The world must have started bustling with activity with butterflies and bees and insects swarming everywhere. When this happened some 70 to 100 million years ago, the number of flowering species on Earth exploded. By the time the first flowering plant appeared, plant-eating dinosaurs had been around for a 100 million years. Dinosaurs were eating these and when the dinosaurs became extinct, another group of animals took their place—the mammals, which dis- persed the angiosperm fruits, nuts and many vegetables. Now, the flower kingdom and the human race depended on each other.

Marigold

Introduction

Among loose flowers, marigold (*Tagetes* sp. Linn.), is an important flower which is used for floral decorations, religious offerings and also for making garlands and flower baskets. It is native to South and Central America, especially Mexico. The species such as Tagetes erecta, Tagetes patula and Tagetes minuta are commonly cultivated in India. Today, it is one of the most important commercial flowers grown worldwide and in India it accounts for more than half of the nation's loose flower production. Major marigold growing states are Karnataka, Gujarat, Maharashtra, Haryana, Andhra Pradesh, Uttar Pradesh, Chattisgarh, Odisha, Jammu and Kashmir, Puducherry, Andaman Nicobar, Arunachal Pradesh, West Bengal, Tamil Nadu, *etc.* It occupies an area of 42,880 hectares with production of 3, 60,210 metric tons loose flower (NHB database, 2012-13). Marigold has gained popularity because of its adaptability to various soil and climatic conditions, longer blooming period, beautiful flowers which have good shelf life. Marigold flowers are mainly grown in India for loose flower production and being used extensively for making garlands, for beautification of mandaps and decoration of cars in marriages and religious offerings

In India marigold is one of the most commonly grown flowers and used extensively on religious and social functions in different forms. Because of their ease in cultivation, wide adaptability to varying soil and climatic conditions, long duration of flowering and attractively coloured flowers of excellent keeping quality, the marigolds have become one of the most popular flowers in our country. Flowers are sold in the market as loose or as garlands. Due to its variable height and colour marigold is especially use for decoration and included in landscape plans.

Marigold is cultivated throughout India. It is widely grown in the Valley of Flowers, Ranthambore, National Park in India. *Calendula arvensis, Calendula bicolor, Calendula eckerleinii, Calendula lanzae, Calendula maderensis, Calendula maroccana, Calendula meuselii, Calendula stellata, Calendula suffruticosa, Calendula tripterocarpa, Calendula officinalis* are some of the other species of Marigold.

The name Marigold refers to the Virgin Mary, since Marigolds were traditionally used in Catholic celebrations concerning the mother of Jesus; and the plant received its botanical name, Calendula officinalis, Indian name : Saldbargh or Zergul, from the Romans, who noted the fact that the plants bloomed on the first days or "calends" of every month. The Calendula/Marigold was used medicinally in ancient Rome to treat scorpion bites and heal wounds, among many other applications. Some of the constituents in Marigold are essential oil, acids, carotenoid, phytosterols, calcium, vitamins C and E, saponins, flavonoids (which account for much of its anti-inflammatory activity), polysaccharides, resin and mucilage.

Marigolds were first discovered by the Portuguese in Central America probably in Mexico. in the 16th century. They introduced these flowers to Europe and India. Marigolds were well known and valued by ancient people in South Asia. Their golden colour was considered to resemble the colour of the Arya, or honourable people. It was used to demarcate special spaces like pavilions and to line sacred fire-pits or kunds in which ceremonies were performed. Today, they are naturalised in the tropics and subtropics of the Old and New Worlds. They are cultivated in India and Pakistan as a medicinal, flavouring, dye and ornamental plant.

Marigolds are now widely cultivated in the sub-continent. In Delhi, which is one of the centres of the flower trade in India, flowers are brought in from all over the country. A major centre of marigold production is the Calcutta region. The bright orange and red colours of marigolds are seen everywhere in daily life of the Indian sub-continent. They are mainly decorative plants and the flowers are used in all kinds of ceremonies including weddings. They stud cowdung balls which are used to decorate rice-powder drawings. In folk-art of eastern India they are dried and powdered to produce a yellow colour used to decorate village homes. Essential oils are also extracted from marigolds for perfumery.

It flowers from July to September. The scented flowers are hermaphrodite (have both male and female organs). Bright yellow and orange Marigold flowers are used to make garlands. They are even used to decorate the religious places. The leaves of its flowers are used as salads. Yellow dye has also been extracted from the flower, by boiling. The burning herb repels insects and flies. Pigments in the Marigold are sometimes extracted and used as the food colouring for humans and livestock. It is offered to the god and Goddess on the Durga Puja.

Plant Description

Marigold, also called Calendula, is an annual or biennial aromatic that is native to the Mediterranean countries, where it was used in early Arabic cultures and in ancient Greece and Rome as a medicinal herb, as well as a colorant for fabrics and an ingredient in food and cosmetics.

The ornamental plants bear orange or yellow flowers with dense petals and are widely grown in gardens in North America and Europe for their beauty, and the flowers are also extensively cultivated for use in herbal medicine throughout Latin America and Eastern Europe.

Medical Uses

Marigold has a long history as a superior antibacterial when used internally and externally and has been used to heal many skin irritations, wounds, and injuries, including Eczema, herpes, gingivitis, Varicose veins and athlete's foot. It is thought to be similar to Witch hazel, due to its natural iodine content, and may be used as a local application to heal all types of skin problems. Some consider Marigold to be the best tissue healer for wounds, and old herbal doctors believed that constant applications of Marigold would help or even prevent gangrene or tetanus. As a diaphoretic and febrifuge, Marigold is often used to induce perspiration and break a fever.

Marigold/Calendula is a powerful anti-inflammatory and painkilling agent that is thought to reduce inflammation of the bowel. It reduces the general tension that can promote bowel problems, relaxing the nervous constriction of the digestive muscles which will help bowel function. The herb is thought to prevent the overgrowth of yeast in the bowel and also have beneficial effects on Colitis, diverticulitis and inflammatory pelvic disease.

As an antispasmodic and effective painkiller, Marigold is an old-time remedy for menstrual cramps and for quelling the pain of an angry ulcer. Marigold is often used to soothe the digestive tract. German studies have demonstrated that Marigold prevents the hormonal reactions that produce swelling and inflammation in the stomach lining, specifically by acting on the inflammatory prostaglandin (PGE) and also has a strong bactericidal effect that may counteract infection with Helicobacter pylori, a bacterium associated with both Gastritis and peptic ulcers.

As a cholagogue, Calendula/Marigold increases the flow of bile into the intestines and is thus thought to help the gallbladder and the liver, making it useful in the treatment of hepatitis. This action further helps to promote good digestion. When taken internally, Marigold soothes and heals the tender mucous membranes and tissues within the body, improving the colon, stomach, liver, and gums after operations. When used externally, the herb provides the same soothing effects on mucous membranes that will support the skin and connective tissues.

It is thought that Marigold will support good heart health, as some recent studies indicate that the herb may reduce Blood pressure. Europeans use Marigold in numerous medicinal compounds and cosmetics. It is said to enhance the production of collagen in the skin and fill in facial Wrinkles, tonne tender skin, treat Sunburn and insect bites, and protect babies' sensitive skin (particularly when used for diaper rash).

Marigold flowers are rich source of carotenoids and being grown on commercial scale for extracting these carotenoids. Flower petals contain xanthophylls which are major carotenoids fractions and lutein forms 80-90 per cent of the total xanthophyll content. Carotenoids are the major source of pigment for poultry feed used for intensification of colour of egg yolk and boiler skin. It has been reported that dietary carotenoids are the agents for prevention and treatment of several illness such as concerned photosensitivity diseases. The purified extract of marigold petals, containing lutein dipalmitate has been marked as an ophthalmologic agent under the name "Adaptional. They also protect the eye from long term damage by light which can lead to a progressive condition known as age related macular degeneration (AMD). Some of the carotenoids have been in the market for more than 20 years and are used extensively in food colouring. Marigold petals have anti-fungal, anti-bacterial and anti-inflammatory properties that can be utilized for the production of creams. Marigold species especially T. minuta oil is the most valuable and precious for using in high grade perfumes and cosmetics. Marigold leaves and flowers possess a good insect repelling properties It has juvenile hormonal and insect repellent activities against flies, ants and mosquitoes. Therefore, marigold oil is being used on commercial scale for formulation of insect repellent. Tagetes spp. have been reported by numerous researcher's to provide a method for protecting crop plants from damage caused by various nematode and insect pests.

Varieties

There are 33 species of marigold and numerous varieties. There are two common types of marigold:

    I) The African Marigold (*Tagetes erecta*)

    II) The French Marigold (*Tagetes patula*).

Origin

African marigold: Mexico, French Marigold: Mexico and South America. Both have deeply cut foliage with a characteristic odour.

I) The African Marigold (*Tagetes erecta*)

The African Marigolds are generally tall (up to 90 cm) with large sized double globular flowers of lemon, yellow, golden yellow, primrose, orange or bright yellow colours. There are also dwarf varieties (20 to 30 cm) having large double flowers. The important varieties are: Giant Double African Orange, Giant Double African Yellow, Cracker Jack, Climax, Dubloon, Golden Age, Chrysanthemum Charm, Crown of Gold, Spun Gold.

Pusa Narangi Gainda

Pusa Narangi Gainda was developed by pedigree hybridization by crossing two exotic varieties Cracker Jack x Golden Jubilee. The plants of this variety are of medium stature, vigorous and uniform, foliage-dark green which grows to a height of 80-85 cm. It flowers in 125-136 days after sowing. The flowers are orange coloured, carnation type, double and compact in nature. The variety can yield upto 25-30 tonnes/ha and is commercially grown for loose flowers throughout the country during winter months. It is most suitable for garland making, religious offerings and carotenoid extraction.

Pusa Basanti Gainda

Pusa Basanti Gainda was developed by pedigree hybridization by crossing two exotic varieties Golden Yellow x Sun Giant. It is little late maturing variety which takes 135-145 days. The flowering duration is long (40-45 days), plants are 60 -65 cm tall, vigorous and uniform with dark green foliage. Flowers sulfur yellow coloured, carnation type, double and compact. It is ideal for loose flower production and also for growing in pots and beds in the garden.

II) French Marigold (*Tagetes patula*)

The French Marigolds are mostly dwarf, early- flowering and compact with dainty single or double blooms, borne freely and almost covering the entire plant. The colour flowers may be yellow, orange, golden yellow, primrose, mahogany, rusty red, tangerine or deep scarlet or a combination of these colours. The important varieties are: Red Borcade, Rusty Red, Butter Scotch, Valencia, Sussana. However, in the market mostly orange colour varieties are preferred and the variety which is dominating is African Giant Double Orange.

In marigold, many varieties and strains are available which vary in plant height, growth habit, flower shape and size. Mostly local varieties are being cultivated by the farmers which are generally low yielders. The Division of Floriculture and Landscaping, Indian Agricultural Research Institute, New Delhi has developed high yielding varieties of marigold which are as follows.

Pusa Arpita

The variety produces tall, bushy and vigorous plants. Plant height and spread ranges from 90-100 cm and 60-70 cm, respectively. The foliage colour is dark green and stem colour is greenish purple. The crop establishment is very good and the plants grow straight with strong stems. The variety produces medium sized light orange coloured flowers. The flowers are compact with turmeric yellow colour of petals at lower surface.

Nutrition

It was observed that fresh and dry weight of plants increased for 70-75 days until flowering and need for nitrogen lasted till this stage. Potassium was required till seed formation (90-100 days). It responds well to both macro and micro nutrients in addition to organic manures. Well rotten farmyard manure @ 30tonnes should be incorporated per hectare well in the soil at the time of soil preparation. In addition to farmyard manure, it is advisable to apply 120 kg/ha of nitrogen and 80 kg/ha each of phosphorus and potash for getting good vegetative growth and flowering. Whole quantity of phosphorus and potash should be applied at the time of land preparation. Nitrogen is to be applied in two split doses. The first dose should be applied 20 days after planting and the second dose 40 days after transplanting. It will be better if two foliar sprays of 0.2 per cent urea are done at an interval of 15 days.

Significance of Marigold Flowers in Indian Culture?

The Marigold is said to have derived its name from "Mary's Gold", taken from the fact that early Christians placed flowers instead of coins on Mary's altar as an offering. This flower is often used in festivities honoring Mary.

The marigold is likewise associated with the sun - being vibrant yellow and gold in color. The flowers are open when the sun is out. The marigold is also called the "herb of the sun", representing passion and even creativity. It is also said that marigolds symbolize cruelty, grief and jealousy. It can mean to show strong passion, being associated with the legendary brave and courageous lion. Its Victorian meaning, desire for riches, is probably consequent to the legends of the flower being Mary's gold, depicting coins.

As they are luminous and beautiful to see, they are often used as love charms. They are mostly used in weddings depicting beauty and a sign of new beginning for the married couple.

Marigold flowers are not only significant in the Indian Culture but also all over the world.

It was believed that water made from marigold was used to invoke psychic visions of fairies, if rubbed on someone's eyes.

☆ Marigold was a sacred flower to the Aztecs. It was used in many religious ceremonies and also as a medicine. The Aztecs were of the belief that marigold flowers relieved one from hiccups and it cured people who were struck by lightning.

☆ According to Mexican tradition, marigold flowers are used during Dia de los Muertos, which in English means Day of the Dead. It is a tradition which originated in Mexico, celebrated on the 1st November of every year to honor the lives of family and friends who passed away. Mexicans are of the belief that during this day, dead souls visit the living and marigold flowers guide them towards the altar.

☆ Due to its strong odor, dead souls are attracted towards the flowers. Burial sites are adorned with marigolds and even the private altars constructed for the dead are surrounded with marigold.

☆ According to the language of flowers, French marigold flower means jealousy and African marigold flower means a vulgar mind.

☆ Portuguese introduced marigolds in India, it is widely cultivated in India to make garlands, for decorative purpose in marriages and festivals.

Particularly, Dussehra where individuals adorn their vehicles and homes with marigold garlands.

Jasmine

Jasmine (taxonomic name Jasminum is a genus of shrubs and vines in the olive family (Oleaceae). It contains around 200 species native to tropical and warm temperate regions of Eurasia, Australasia and Oceania. Jasmines are widely cultivated for the characteristic fragrance of their flowers. A number of unrelated plants contain the word "Jasmine" in their common names (see Other plants called "Jasmine").

Description

Jasmine can be either deciduous (leaves falling in autumn) or evergreen (green all year round), and can be erect, spreading, or climbing shrubs and vines. Their leaves are borne, opposite or alternate. They can be simple, trifoliate, or pinnate. The flowers are typically around 2.5 cm (0.98 in) in diameter. They are white or yellow in color, although in rare instances they can be slightly reddish. The flowers are borne in cymose clusters with a minimum of three flowers, though they can also be solitary on the ends of branchlets. Each flower has about four to nine petals, two locules, and one to four ovules. They have twostamens with very short filaments. The bracts are linear or ovate. The calyx is bell-shaped. They are usually very fragrant. The fruits of jasmines are berries that turn black when ripe.

The basic chromosome number of the genus is 13, and most species are diploid (2n=26). However, natural polyploidy exists, particularly in *Jasminum sambac* (2n=39), *Jasminum flexile* (2n=52), *Jasminum mesnyi* (2n=39), and *Jasminum ngustifolium* (2n=52).

Introduction

Jasmine is one of the oldest fragrant flowers cultivated by man. The flower is used for various purposes *viz.* making garlands, bouquet, decorating hair of women, religious offering *etc*. Jasmine is also known as the "Queen of the Night", because of it's heady fragrance. It is also used for production of Jasmine concrete which is used in cosmetic and perfumery industries. More than 80 Jasmine species are found in India, of which only three species are used for commercial cultivation. They are *Jasminum sambac* (Gundumalli/Madurai Malli), *Jasminum auriculatum* (Mullai) and *J. grandiflorum* (Jathimalli/Pitchi). The first two species are mainly cultivated for selling as fresh flowers whereas the last one is cultivated for concrete extraction.

Tamil Nadu is the leading producer of jasmine in the country with an annual production of 77247 tonnes from the cultivated area of 9360 ha (2005-06). The flowers produced in the state are being exported to the neighbouring countries *viz*., Sri Lanka, Singapore, Malaysia and Middle East countries. The major jasmine producing districts of Tamil Nadu are Dindigul, Salem, Madurai, Tirunelveli, Virudhunagar, Trichy, *etc*., Since the crop requires lots of manpower for harvesting and other operations, only small farmers are cultivating the crop. It is an ideal crop for small farmers whose land holdings are less than 1 acre.

Distribution and Habitat

Jasmines are native to tropical and subtropical regions of Eurasia, Australasia and Oceania, although only one of the 200 species is native to Europe. Their center of diversity is in South Asia and Southeast Asia.

A number of jasmine species have become naturalized in Mediterranean Europe. For example, the so-called Spanish jasmine (*Jasminum grandiflorum*) was originally from Iran and western South Asia, and is now naturalized in the Iberian peninsula.

Jasminum fluminense (which is sometimes known by the inaccurate name "Brazilian Jasmine") and *Jasminum dichotomum* (Gold Coast Jasmine) are invasive speciesin Hawaii and Florida. Jasminum polyanthum, also known as White Jasmine, is an invasive weed in Australia.

Taxonomy

Species belonging to genus Jasminum are classified under the tribe Jasmineae of the olive family (Oleaceae). Jasminum is divided into five sections Alternifolia,Jasminum, Primulina, Trifoliolata, and Unifoliolata. The genus name is derived from the Persian Yasameen ("gift from God") through Arabic and Latin.

Selected Species

Jasminum Sambac 'Grand Duke of Tuscany'.

Species include:

- ☆ *J. abyssinicum* Hochst. ex DC. – forest jasmine
- ☆ *J. adenophyllum* Wall. – bluegrape jasmine, pinwheel jasmine, princess jasmine

A Double-Flowered Cultivar of *Jasminum sambac* in Flower with an unopened bud. The flower smells like the tea as it opens.

☆ *J. angulare* Vahl

☆ *J. angustifolium* (L.) Willd.

☆ *J. auriculatum* Vahl – Indian hasmine, needle-flower jasmine

☆ *J. azoricum* L.

☆ *J. beesianum* Forrest and Diels – red jasmine

☆ *J. dichotomum* Vahl – Gold Coast jasmine

☆ *J. didymum* G.Forst.

☆ *J. dispermum* Wall.

☆ *J. elegans* Knobl.

☆ *J. elongatum* (P.J.Bergius) Willd.

☆ *J. floridum* Bunge

☆ *J. fluminense* Vell.

☆ *J. fruticans* L.

☆ *J. grandiflorum* L. – Catalonian jasmine, jasmin odorant, royal jasmine, Spanish jasmine

☆ *J. humile* L. – Italian jasmine, Italian yellow jasmine

☆ *J. anceolarium* Roxb.

☆ *J. mesnyi* Hance – Japanese jasmine, primrose jasmine, yellow jasmine

☆ *J. multiflorum* (Burm.f.) Andrews – Indian jasmine, star jasmine, winter jasmine

☆ *J. multipartitum* Hochst. – starry wild jasmine

☆ *J. nervosum* Lour.

☆ *J. nobile* C.B.Clarke

☆ *J. nudiflorum* Lindl. – winter jasmine

☆ *J. odoratissimum* L. – yellow jasmine

☆ *J. officinale* L. – common jasmine, jasmine, jessamine, poet's jasmine, summer jasmine, white jasmine

☆ *J. parkeri* Dunn – dwarf jasmine

☆ *J. polyanthum* Franch.

☆ *J. sambac* (L.) Aiton – Arabian jasmine, Sambac jasmine

☆ *J. simplicifolium* G.Forst.

☆ *J. sinense* Hemsl.

☆ *J. subhumile* W.W.Sm.

☆ *J. subtriplinerve* Blume

☆ *J. tortuosum* Willd.

☆ *J. urophyllum* Hemsl.

Bunch of Jasmine Flowers.

Cultivation and Uses

The Jasmine flower is used for removing intestinal worms and is also used for jaundice and venereal diseases. The flower buds are useful in treating ulcers, vesicles, boils, skin diseases and eye disorders. The leaves extracts against breast tumours. Drinking Jasmine tea regularly helps in curing cancer. Its oil is very effective in calming and relaxing.

It is propagated through the softwood cuttings, semi hard wood cuttings and through simple layering. It needs water regularly. It is mostly propagated in the summer season. It is planted 6 inches deep inside the soil. It requires moist and well drained soil. Remove the weeds present inside the soil before planting the tree. The soil should consists of cow dung before planting the Jasmine plant. It requires frequent pruning for its fast growth. It needs warm temperature and proper watering from time to time. It grows in full Sun to partial shade. It should be fertilized in a month period. Jasmine Plant should be kept at least eight feet apart in order to save the later growth of the plant from jamming together. Tips of the plants should be pinched to stimulate lateral growth and frequent prunning. Younger plants should be tied with the stems to give a fairly heavy support.

Widely cultivated for its flowers, jasmine is enjoyed in the garden, as a house plant, and as cut flowers. The flowers are worn by women in their hair in southern and southeast Asia.

Other Uses

Jasmine oil is used used for making perfumes and incense. Its flowers are used to flavour Jasmine tea and other herbal or black tea. Its oil is also used in creams, shampoos and soaps. In India Jasmine flowers are stringed together to make garlands. Women in India wear this flower in their hair. Some communities even use this flower to cover the face of the bridegroom.

Jasmine Tea

Green Tea with Jasmine Flowers.

Jasmine tea is often consumed in China, where it is called jasmine-flower tea ; pinyin: mò lì huâ chá). Jasminum sambac flowers are also used to make jasmine tea, which often has a base of green tea or white tea, but sometimes an Oolong base is used. The flowers are put in machines that control temperature and humidity. It takes about four hours for the jasmine blossoms to absorb the fragrance and flavour. For the highest grades of jasmine tea, this process may be repeated up to seven times. It must be refired to prevent spoilage. The used flowers may be removed from the final product, as the flowers contain no more aroma. Giant fans are used to blow away and remove the petals from the denser tea leaves.

In Okinawa, Japan, jasmine tea is known as sanpin cha.

Jasmonates

Jasmine gave name to the jasmonate plant hormones, as methyl jasmonate isolated from the oil of Jasminum grandiflorumled to the discovery of the molecular structure of jasmonates.

Cultural Importance

The White Jasmine Branch, painting of ink and color on silk by Chinese artist Zhao Chang, early 12th century. Madurai, a city in Tamil Nadu is famous for its jasmine production. In the western and southern states of India, including Andhra Pradesh, Karnataka, Kerala, Maharashtra and Tamil Nadu, jasmine is cultivated in private homes. These flowers are used in worship and for hair ornaments. Jasmine is also cultivated commercially, for both the domestic and industrial uses, such as the perfume industry. It is used in rituals like marriages, religious ceremonies and festivals. In the Chandan Yatra of lord Jagannath, the deity is bathed with water flavoured with sandalwood and jasmine.

Jasmine Flower Blooming Near Hyderabad, India.

Jasmine flower vendors sell garlands of jasmine, or in the case of the thicker motiyaa (in Hindi) or mograa (in Marathi) varieties, bunches of jasmine are a common sight in many parts of India. They may be found around entrances to temples, on major thoroughfares, and in major business areas.

A change in presidency in Tunisia in 1987 and the Tunisian Revolution of 2011 are both called "Jasmine revolutions" in reference to the flower. Jasmine flowers were also used as a symbol during the 2011 Chinese pro-democracy protests in thePeople's Republic of China.

"Jasmine" is also a female forename.

Varieties

The species-wise recommended varieties are Gundumalli (*Jasminum sambac*), Co-1, and Co-2 (*J. auriculatum*) and Co-1 and Co-2 (*J. grandiflorum*).

Jasmine as a National Flower

Several countries and states consider jasmines as a national symbol. They are the following:

☆ **Hawaii:** *Jasminum sambac* ("pikake") is perhaps the most popular of flowers. It is often strung in leis and is the subject of many songs.

☆ **Indonesia:** *Jasminum sambac* is the national flower, adopted in 1990. It goes by the name "melati putih" and is the most important flower in wedding ceremonies for ethnic Indonesians, especially on the island of Java.

☆ **Pakistan:** *Jasminum officinale* is known as the "chambeli" or "yasmin", it is the national flower.

☆ **Philippines:** *Jasminum sambac* is the national flower. Adopted in 1935, it is known as "sampaguita" in the islands. It is usually strung in garlands which are then used to adorn religious images.

☆ **Syria:** The Syrian city Damascus is also called City of Jasmine and uses it as a symbol.

☆ **Thailand:** Jasmine flowers are used as a symbol of motherhood.

Other Plants Called "Jasmine"

Other names : Moghra, Kundumalligai, Arabian Jasmine. Mallika are some of the other names used for the Jasmine. In India Jasmine is called the 'Moonshine in the garden'.

☆ Brazilian Jasmine Mandevilla sanderi

☆ Cape Jasmine Gardenia,

☆ Carolina Jasmine Gelsemium

☆ Chilean Jasmine *Mandevilla laxa*

☆ New Zealand Jasmine *Parsonsia capsularis*

☆ Night-Blooming Jasmine *Cestrum nocturnum*

Bunch of Jasmine Buds.

☆ Night-Flowering Jasmine *Nyctanthes arbortristis*

☆ Red Jasmine *Plumeria rubra*

☆ Star Jasmine, Confederate *Jasmine trachelospermum*

☆ Tree Jasmine (disambiguation)

☆ Jasmine rice, a type of long-grain rice

Chrysanthemum

Chrysanthemums, sometimes called mums or chrysanths, are flowering plants of the genus Chrysanthemum in the family Asteraceae. They are native to Asia and northeastern Europe. Most species originate from East Asia and the center of diversity is in China. There are countless horticultural varieties and cultivars.

Etymology

The name "chrysanthemum" is derived from the Greek words chrysos (gold) and anthemon (flower).

Taxonomy

The genus once included more species, but was split several decades ago into several genera, putting the economically important florist's chrysanthemums in the genus Dendranthema. The naming of the genera has been contentious, but a ruling of the International Botanical Congress in 1999 changed the defining species of the genus to Chrysanthemum indicum, restoring the florist's chrysanthemums to the genus Chrysanthemum.

The other species previously included in the narrow view of the genus Chrysanthemum are now transferred to the genus Glebionis.

The other genera separate from Chrysanthemum include Argyranthemum, Leucanthemopsis, Leucanthemum, Rhodanthemum, and Tanacetum.

Description

Wild Chrysanthemum taxa are herbaceous perennial plants or subshrubs. They have alternately arranged leaves divided into leaflets with toothed or occasionally smooth edges. The compound inflorescence is an array of several flower heads, or sometimes a solitary head. The head has a base covered in layers of phyllaries. The simple row of ray florets are white, yellow or red; many horticultural specimens have been bred to bear many rows of ray florets in a great variety of colors. The disc florets of wild taxa are yellow. The fruit is a ribbed achene. Chrysanthemums, also known as 'mums', are one of the prettiest varieties of perennials that start blooming early in the fall. This is also known as favorite flower for the month of November.

History

Historical painting of chrysanthemums from the New International Encyclopedia, 1902 Chrysanthemums were first cultivated in China as a flowering herb as far back as the 15th century BC. Over 500 cultivars had been recorded by the year 1630. The plant is renowned as one of the Four Gentlemen in Chinese and East Asian art. The plant is particularly significant during the Double Ninth Festival. The flower may have been brought to Japan in the eighth century AD, and the Emperor adopted the flower as his official seal. The "Festival of Happiness" in Japan celebrates the flower.

Chrysanthemums entered American horticulture in 1798 when Colonel John Stevens imported a cultivated variety known as 'Dark Purple' from England. The introduction was part of an effort to grow attractions within Elysian Fields in Hoboken, New Jersey.

Chrysenthemum is next only to rose in importance among the flowers grown in the world. It is grown both for its aesthetic value and for commerce. Medium sized, fully double, white chrysanthemums are always in demand for making garlands and pot culture. In India it has been recognized as one among the five important commercially potent flower crops by the All-India Coordinated Floriculture Improvement Project of the Indian Council of Agricultural Research (ICAR). In the different states of the country it is grown under different names, 'guldandi' in the Hindi belt, 'chandramallika' in the eastern states, 'samanti' in the Southern states and 'shevanti' in the western states. Its share in the total flowers grown in the country is approximately 10 per cent. Chrysanthemum is versatile, it can be planted in the bed, cultured in the pot, used for garland-making and also as cut-flower for flower arrangement. Therefore, the modern techniques based on scientific research and by evolving more attractive cultivars, have immensely increased its potential as a commercial crop. The crop can be grown throughout the year.

In India the commercial cultivation of chrysanthemum is for loose flowers for worship and garland making. To a very small extent other flower is also grown for sale as cut flower with long stem and as potted plants. Some nurserymen also trade planting-material of different varieties on a small scale.

New Cultivars

Several new cultivars of the flower have been evolved at the National Botanical Research Institute, Lucknow, which naturally bloom in different seasons when planted according to a given schedule as detailed below:

Cultivar	Date of Planting	Blooming Season
`Jawala', `Jyoti'	January	Summer
`Varsha', `Meghdoot'	February	Rainy Season
`Sharada', `Sharad Shobha"	March	Sept-October
`Sharad Mala', `Megami' and `Sharad Kanti'	July	Oct. – November
All traditional CVs	July	November – December
`Vasantika', `Jaya'	August	December – January
`Illini Cascade'	August	February – March

As the growing of these cultivars does not involve any extra expenditure, their cost of production would be the same as traditional cultivars. Rather, as the flower would bloom when there is scarcity in the market, the growers would get a high price than when they grow conventional cultivars.

Problems and Potentialities

Under a fourable weather condition chrysanthemum can be grown perfectly by traditional methods and there is little chance of outbreak of diseases. But, under unfavourable weather condition, even the best growers are not confident of assured results and diseases take a heavy tall, both in terms of quantity and quality. This uncertainty, combined with short-blooming season, post – harvest spoilage and glut conditions dampen the spirit of any prospective grower whether he be a novice, professional or a commercial ones. However, use of disease –free planting material, sterilization of planting medium, precise water management, judicious feeding and protection from diseases and pests eliminate the possibilities of crop failure.

Economic Uses

Ornamental Uses

Modern cultivated chrysanthemums are showier than their wild relatives. The flower heads occur in various forms, and can be daisy-like or decorative, like pompons or buttons. This genus contains many hybrids and thousands of cultivars

Dance' 'Enbee Wedding Golden'.

'Dance' "Feeling Green'.

'King's Pleasure' – Class 1.

'Whiteout' – Class 1.

developed for horticultural purposes. In addition to the traditional yellow, other colors are available, such as white, purple, and red. The most important hybrid is *Chrysanthemum* × *morifolium* (syn. *C.* ×*grandiflorum*), derived primarily from *C. indicum*, but also involving other species.

Over 140 varieties of chrysanthemum have gained the Royal Horticultural Society's Award of Garden Merit. Chrysanthemums are divided into two basic groups, garden hardy and exhibition. Garden hardy mums are new perennials capable of wintering in most northern latitudes. Exhibition varieties are not usually as sturdy. Garden hardies are defined by their ability to produce an abundance of small blooms with little if any mechanical assistance, such as staking, and withstanding wind and rain. Exhibition varieties, though, require staking, overwintering in a relatively dry, cool environment, and sometimes the addition of night lights.

The exhibition varieties can be used to create many amazing plant forms, such as large disbudded blooms, spray forms, and many artistically trained forms, such as thousand-bloom, standard (trees), fans, hanging baskets, topiary, bonsai, and cascades.

Chrysanthemum blooms are divided into 10 different bloom forms by the US National Chrysanthemum Society, Inc., which is in keeping with the international classification system. The bloom forms are defined by the way in which the ray and disk florets are arranged. Chrysanthemum blooms are composed of many individual flowers (florets), each one capable of producing a seed. The disk florets are in the center of the bloom head, and the ray florets are on the perimeter. The ray florets are considered imperfect flowers, as they only possess the female productive organs, while the disk florets are considered perfect flowers, as they possess both male and female reproductive organs.

Irregular incurves are bred to produce a giant head called an ogiku. The disk florets are concealed in layers of curving ray florets that hang down to create a 'skirt'. Regular incurves are similar, but usually with smaller blooms and a dense, globular form. Intermediate incurve blooms may have broader florets and a less densely flowered head. In the reflex form, the disk florets are concealed and the ray florets reflex outwards to create a mop-like appearance. The decorative form is similar to reflex blooms, but the ray florets usually do not radiate at more than a 90° angle to the stem.

The pompon form is fully double, of small size, and very globular in form. Single and semidouble blooms have exposed disk florets and one to seven rows of ray florets.In the anemone form, the disk florets are prominent, often raised and overshadowing the ray florets. The spoon-form disk florets are visible and the long, tubular ray florets are spatulate. In the spider form, the disk florets are concealed, and the ray florets are tube-like with hooked or barbed ends, hanging loosely around the stem. In the brush and thistle variety, the disk florets may be visible.

Culinary Uses

Yellow or white chrysanthemum flowers of the species C. morifolium are boiled to make a sweet drink in some parts of Asia. The resulting beverage is known simply as chrysanthemum tea, pinyin: júhuâ chá, in Chinese). In Korea, a rice wineflavored with chrysanthemum flowers is called gukhwaju. Chrysanthemum leaves are steamed or boiled and used as greens, especially in Chinese cuisine. The flowers may be added to thick snakemeat soup to enhance the aroma. Small chrysanthemums are used in Japan as a sashimi garnish.

Insecticidal Uses

Pyrethrum (*Chrysanthemum* [or Tanacetum] *cinerariaefolium*) is economically important as a natural source of insecticide. The flowers are pulverized, and the active components, called pyrethrins, which occur in the achenes, are extracted and sold in the form of an oleoresin. This is applied as a suspension in water or oil, or as a powder. Pyrethrins attack the nervous systems of all insects, and inhibit female mosquitoes from biting. In sublethal doses they have an insect repellent effect. They are harmful to fish, but are far less toxic to mammals and birds than many synthetic insecticides. They are not persistent, being biodegradable, and also decomposeeasily on exposure to light. Pyrethroids such as are synthetic insecticides based on natural pyrethrum.

Environmental Uses

Chrysanthemum plants have been shown to reduce indoor air pollution by the NASA Clean Air Study.

Cultural Significance and Symbolism

In some countries of Europe (*e.g.*, France, Belgium, Italy, Spain, Poland, Hungary, Croatia), incurve chrysanthemums are symbolic of death and are used only for funerals or on graves, while other types carry no such symbolism; similarly, in China, Japan andKorea, white chrysanthemums are symbolic of lamentation and/ or grief. In some other countries, they represent honesty. In the United States, the flower is usually regarded as positive and cheerful, with New Orleans as a notable exception.

Australia

In Australia,the chrysanthemum is sometimes given to mothers for Mother's Day, which falls in May in the southern hemisphere's autumn, which is when the flower is naturally in season. Men may sometimes also wear it in their lapels to honour mothers.

China

☆ The chrysanthemum is one of the "Four Gentlemen"of China (the others being the plum blossom, the orchid, andbamboo). The chrysanthemum is said to have been favored by Tao Qian, an influential Chinese poet, and is symbolic of nobility. It is also one of the four symbolic seasonal flowers.

☆ A chrysanthemum festival is held each year in Tongxiang, near Hangzhou, China.

☆ Chrysanthemums are the topic in hundreds of poems of China.

☆ The "golden flower" referred to in the 2006 movie Curse of the Golden Flower is a chrysanthemum.

☆ "ChrysanthemumGate" (jú huâ mén), often abbreviated asChrysanthemum, is taboo slang meaning "anus" (with sexual connotations).

☆ Chrysanthemums were first cultivated in China as a flowering herb as far back as the 15th century BC.

☆ An ancient Chinese city (Xiaolan Town of Zhongshan City) was named Ju-Xian, meaning "chrysanthemum city".

☆ The plant is particularly significant during the Double Ninth Festival.

Germany

Industrial musicians Einstürzende Neubauten base their song "Blume" around the flower.

Italy

In Italy chrysanthemums traditionally represent death and are often placed on a person's tombstone as an offer, especially on All Souls' Day. Because of this, giving someone a bunch of chrysanthemums is a taboo and may be regarded as a sign of disrespect (associated with a sort of "death wish").

Japan

In Japan, the chrysanthemum is a symbol of the Emperor and the Imperial family. In particular, a "chrysanthemum crest" (kikukamonshô or kikkamonshô), *i.e.* a mon of chrysanthemum blossom design, indicates a link to the Emperor; there are more than 150 patterns of this design. Notable uses of and reference to the Imperial chrysanthemum include:

☆ The Imperial Seal of Japan, used by members of the Japanese Imperial family. In 1869, a two-layered, sixteen petal design was designated as the symbol of the Emperor. Princes used a simpler single-layer pattern.

☆ A number of formerly state-endowed shrines, kankokuheisha) have adopted a chrysanthemum crest; most notable of these is Tokyo's Yasukuni Shrine.

☆ The Chrysanthemum Throne is the name given to the position of Japanese Emperor and the throne.

☆ The Supreme Order of the Chrysanthemum is a Japanese honor awarded by the Emperor on the advice of the Japanese government.

☆ In Imperial Japan, small arms were required to be stamped with the Imperial Chrysanthemum, as they were considered the personal property of the Emperor.

☆ The city of Nihonmatsu, Japan hosts the "Nihonmatsu Chrysanthemum Dolls Exhibition" every autumn in historical ruin of Nihonmatsu Castle.

☆ The chrysanthemum is also considered to be the seasonal flower of September.

United States

☆ The founding of the chrysanthemum industry dates back to 1884, when Enomoto Brothers of Redwood City, CA (San Mateo County) grew the first chrysanthemums to be grown in America.

☆ In 1913, Sadakasu Enomoto (of San Mateo County) astounded the flower world by successfully shipping a carload of Turner Chrysanthemums to New Orleans for the famed All Saints Day Celebration.

☆ The chrysanthemum was recognized as the official flower of the city of Chicago by Mayor Richard J. Daley in 1966.

☆ The chrysanthemum is the official flower of the city of Salinas, California.

☆ The yellow chrysanthemum is the official flower of the sorority Sigma Alpha and the pharmacy fraternity Lambda Kappa Sigma.

☆ The white chrysanthemum is the official flower of Triangle Fraternity.

☆ The chrysanthemum is the official flower of Phi Mu Alpha Sinfonia. The purported color is white, per Sinfonia headquarters (likely variety is the incurved football mum species).

☆ The yellow chrysanthemum is the official flower of the fraternity Phi Kappa Sigma.

Others

☆ The term "chrysanthemum" is also used to refer to a certain type of fireworks shells that produce a pattern of trailing sparks similar to a chrysanthemum flower.

☆ The chrysanthemum is also the flower of November.

☆ Tutankhamen was buried with floral collars of chrysanthemum.

☆ Resins of the plant were used [by whom?] in incense cones used to ward off insects.

☆ In the Three Stooges short "Pop Goes the Easel"(1935), the spelling of 'chrysanthemum' is used as the foundation of a gag sequence, highlighted by Curly Howard rattling off a quick, correct spelling in a matter-of-fact tone, as though surprised that anyone would not know that.

Tuberose

Tuberose (*Polianthes tuberosa* L.) is one of the most important tropical ornamental bulbous flowering plants cultivated for production of long lasting flower spikes. It is popularly known as Rajanigandha or Nishigandha. It belongs to the family Amaryllidaceae and is native of Mexico. Tuberose is an important commercial cut as well as loose flower crop due to pleasant fragrance, longer vase-life of spikes, higher returns and wide adaptability to varied climate and soil. They are valued much by the aesthetic world for their beauty and fragrance. The flowers are attractive

and elegant in appearance with sweet fragrance. It has long been cherished for the aromatic oils extracted from its fragrant white flowers. Tuberose blooms throughout the year and its clustered spikes are rich in fragrance; florets are star shaped, waxy and loosely arranged on spike that can reach up to 30 to 45 cm in length.The flower is very popular for its strong fragrance and its essential oil is important component of high- grade perfumes. 'Single' varieties are more fragrant than 'Double' type and contain 0.08 to 0.14 per cent concrete which is used in high grade perfumes. There is high demand for tuberose concrete and absolute in international markets which fetch a very good price. Flowers of the Single type (singlerow of perianth) are commonly used for extraction of essential oil, loose flowers, making garland *etc.*, while that of Double varieties (more than two rows of perianth) are used as cut flowers, garden display and interior decoration. Fragrance of flowers is very sweet, floral and honey-like and can help give emotional strength. The flower spike of tuberose remains fresh for long time and finds a distinct place in the flower markets. Due to its immense export potential, cultivation of tuberose is gaining momentum day by day in our country.

Polianthes tuberosa, the tuberose, is a **perennial plant** related to the **agaves**, extracts of which are used as a **note** in **perfumery**.The common name derives from the **Latin** tuberosa, meaning swollen or **tuberous** in reference to its root system. Polianthes means "many flowers" in **Greek**. In Mexican**Spanish**, the flower is called nardo or vara de San José, which means "**St. Joseph**'s staff". This plant is called as rajanigandha in **India**, which means 'fragrant at night'. It is called kupaloke in Hawaiian.

The Tuberose, also known as *Polianthes tuberosa*, is a night-blooming. Because of its sweet, heavy scent, the Tuberose has a long history in the world of perfumery and has been grown in the south of France for centuries.

Origin and History

Tuberose is a native of Mexico from where it spread to different parts of the world during 16th Century. This is one of the earliest cultivated plants, and may be extinct in its natural habitat.

The Aztecs were growing it nearly 600 years ago. The Spanish found the Aztecs growing it in 1519 and took it back with them to the old World. A French missionary, returning from the Indies in the 1500's did so as well. Once introduced to Europe, it became part of the moon garden, a collection of white or pastel flowers, which release an intense fragrance after dusk. These gardens were popular among the sun-shunning Victorian ladies, who valued a milky pale complexion. The plant did fall out of favour when it became much overused at funerals. It has an intense fragrance and one or two open blossoms will fill the air of an entire garden. It is believed that tuberose was brought to India via Europe in 16th century.

From the ancient times, flowers hold a special place in our lives and special meaning and significance is assigned to different flowers and plants. Let's learn about Tuberose definition and history now. Tuberose can be defined as a prickly bush or shrub that typically bears red, pink, yellow, or white fragrant flowers, native to north temperate regions and widely grown as an ornamental Tuberose

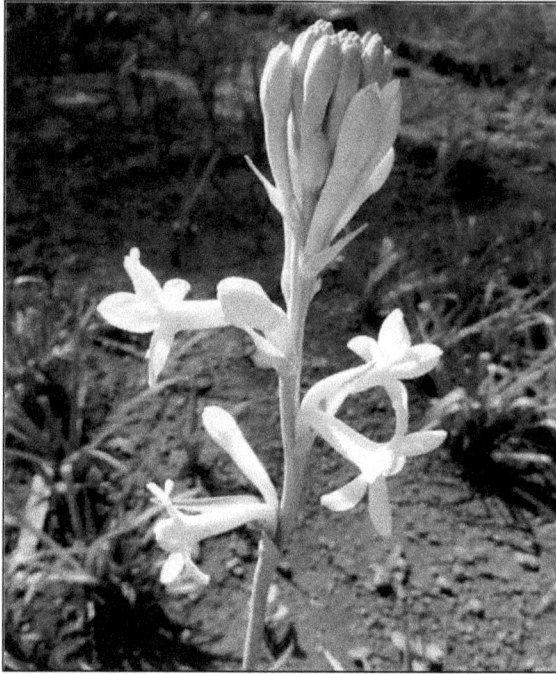

flower has a long history. Different **Types of Tuberose** are native to Mexico. After Tuberose definition and history, let's see other interesting facts about Tuberose.

Description

The tuberose is a night-blooming plant native to **Mexico**, as is every other known species of Polianthes. It grows in elongated spikes up to 45 cm (18 in) long that produce clusters of fragrant waxy white **flowers** that bloom from the bottom towards the top of the spike. It has long, bright green leaves clustered at the base of the plant and smaller, clasping leaves along the stem. Epiphyllous adhesion of stamens is seen in the flower.

Members of the closely related genus **Manfreda** are often called "tuberoses". In the **Philippines**, the plant is also known as azucena, and, while once associated with **funerals**, it is now used in floral arrangements for other occasions.

The Tuberose does not bear any relation to the rose, but its name is said to derive from the plants slender stem from which grows a tube resembling rootstock. This stem often grows 3 feet (91 cm) high and bears flowers of a pink-cream shade, each with six sword-shaped petals.

Importance and Uses

Tuberose can successfully be grown in pots, borders, beds and commercially cultivated for its various uses. The flowers of tuberose are also used for making artistic garlands, floral ornaments,bouquets, buttonholes, gajras and extraction of essential oil. It is also a popular cut flower, not only for use in arrangements, but also for the individual florets that can provide fragrance to bouquets and boutonnieres. The long flower spikes are excellent as cut flowers for table decoration. The flowers emit a delightful fragrance. Tuberose represents sensuality and is used in aromatherapy for its ability to open the heart and calm the nerves,restoring joy, peace and harmony.Tuberose flowers have long been used in perfumery as a source of essential oils and aroma compounds. Tuberose oil is used in high value perfumes and cosmetic products. Furthermore, fragrant flowers are added along with stimulants or sedatives to the favourite beverage prepared from chocolate and served either cold or hot as desired. Tuberose bulbs contain an alkaloid -lycorine, which causes vomiting. The bulbs are rubbed with turmeric and butter and applied as a paste over red pimples of infants. Dried tuberose bulbs in powdered form are used as a remedy for gonorrhoea. In Java, the flowers are eaten along with the juices of the vegetables.

Area and Distribution

Tuberose is grown commercially in a number of countries including India, Kenya, Mexico, Morocco, France, Italy, Hawaii, South Africa, Taiwan, North Carolina, USA, Egypt, China and many other tropical and subtropical areas in the world. In India, commercial cultivation of tuberose is popular in Bagnan, Kolaghat, Midnapur, Panskura, Ranaghat, Krishnanagar of West Bengal; Coimbatore and Madurai districts of Tamil Nadu; Pune, Nashik, Ahmednagar, Thane, Sangli of Maharashtra; East Godavari, Guntur, Chitoor, Krishna District of Andhra Pradesh ; Mysore, Tumkur, Kolar, Belgaum and Devanhalli taluk in Karnataka ; Guwahati and Jorhat in Assam ; Udaipur, Ajmer and Jaipur in Rajasthan; Navsari and Valsad of Gujarat and parts of Uttar Pradesh and Punjab. As per area and production statistics of National Horticulture Board (2013), the total area under tuberose cultivation in the country is about 7.95 lakh hectare. The production of loose and cut flowers is estimated to be 27.71 '000 MT and 1560.70 lakh No's respectively.

Species and Varieties

There are about fifteen species under the genus Polianthes, of which twelve species have been reported from Mexico and Central America.Of these, nine species have white flowers, one is white tinged with red and two are red. Except Polianthes tuberosa L., all the others are found growing wild.

Varieties

There are four types of tuberoses named on the basis of the number of rows of petals they bear. They are:

☆ Single
☆ Semi-double
☆ Double and
☆ Variegated

Single

They bear pure white flowers with one row/whorl of corolla segment. Flowers are highly scented and are extensively used for loose flower purpose, essential oil and concrete extraction. Single types are more fragrant than double. Concrete content has been observed to be 0.08 to 0.11 per cent.

Loose flowers are used for making floral ornaments. Its floral buds are greenish white. Also the per cent seed setting is high in single. Single Mexican, Kalyani Single, Shringar, Prajwal, Arka Nirantara, Rajat Rekha, Hyderabad Single, Calcutta Single, Phule Rajani, Kahikuchi Single, Pune Single are main varieties.

Description of some Important Single Varieties

Arka Nirantara

Arka Nirantara is released by ICAR-Indian Institute of Horticultural Research (IIHR), Bangalore. It has white, single flowers with prolonged blooming.

Shringar

This tuberose hybrid has been developed from a cross between 'Single x Double' and was released by ICAR-Indian Institute of Horticultural Research (IIHR), Bangalore. It bears single type fragrant flowers on strong and sturdy, medium spikes. The flower buds are attractive with slightly pinkish tinge. The spikes have more number of flowers and the individual florets are larger and appealing compared to the local 'single' cultivar. Loose flowers of this hybrid can be used for garlands and for extraction of tuberose concrete.The spikes can also be used as cut flower.

The loose flower yield of this hybrid is about 36 per cent higher than the existing local single variety.Yield of loose flowers is about 15,000 kg/ha per year, which is 40 per cent higher than 'Calcutta or Mexican Single' and the concrete content of the hybrid is at par with Mexican Single. Shringar is preferred by farmers and perfumery industries. This hybrid is tolerant to root knot nematodes (*Meloidogyne incognita*).

Prajwal

This hybrid which bears single type flowers on tall stiff spikes is a cross between 'Shringar' x 'Mexican Single'.The hybrid was released by Indian Institute of Horticultural Research (IIHR), Bangalore. The flower buds are slightly pinkish in colour, while the flowers are white. The individual florets are large in size, compared to 'Local Single'. It yields twenty per cent more loose flowers than 'Shringar'. It is recommended both for loose flower and cut flower purpose.

Single Mexican

It is a single flowered variety. This variety produces maximum flowers is considered as lean months for tuberose flowers yield.

Semi-double

They consist of flowers bearing 2-3 rows of corolla segments on straight spikes.

Double

They bear flowers having more than three rows of corolla segments It is a single flowered variety. This on straight spikes. Flower colour is white and also tinged with pinkish red. The main varieties are Pearl for Double, Kalyani Double, Swarna Rekha, Hyderabad Double, Culcutta Double, Vaibhav and Suvasini.

Description of some Important Double Varieties

Suvasini

It is a double flowered multi whorled variety released by ICAR-Indian Institute of Horticultural Research (IIHR), Bangalore. It is a cross between 'Single' and 'Double'. This variety produces more number of flowers per spike. The spikes are best suited for cut flowers. This tuberose hybrid is multi-whorled with bold,large, pure white fragrant flowers borne on long spikes in contrast to off-white flowers of local cv. 'Double'. The number of flowers per spike is more and flower opening is uniform in this hybrid as compared to the local 'Double' cultivar. Spike yield is 26 per cent more compared to the local Pearl Double 'Double' cultivar. Spikes are best suited for cut flower purpose.

Vaibhav

This hybrid which bears double flowers on medium spikes is from the cross 'Mexican Single' x IIHR – 2' and was released by ICAR-Indian Institute of Horticultural Research (IIHR), Bangalore. The flower buds are greenish in colour in contrast to pinkish buds in 'Suvasini' and 'Local Double'. Flowers are white. Spike yield is 50 per cent higher compared to 'Suvasini'. Hence, recommended for cut flower purpose.

Pearl Double

The flowers tinged with red in the 'Double' type are known as 'Pearl'. Pearl Double is high flower yielder with quality flowers. They are mainly used for cut flower and bouquet purpose as well as loose flower and for extraction of essential oil. Concrete recovery has been found to be 0.06 per cent. It does not open well and is not commercially viable as the single cultivar.

Variegated

These are some streaked leaf-forms, known as 'variegated'. In these varieties, silvery white or golden yellow streaks are visible on leaves. CSIR-National Botanical Research Institute, Lucknow has developed two variegated varieties Rajat Rekha and Swarna Rekha by gamma.

Rajat Rekha

It is a double flowered variety released by National Botanical Research Institute (NBRI), Lucknow.The flowers are double with golden yellow steaks along the margins of leaf. It is a gamma ray induced mutant, in which mutation occurred in chlorophyll synthesis resulting in change in leaf colour. Concrete content has been found to be 0.062 per cent.

Swarna Rekha

It is a double flowered variety released by CSIR-National Botanical Research Institute (NBRI), Lucknow.The flowers are double with golden yellow steaks along the margins of leaf. It is a gamma ray induced mutant, in which mutation occurred in chlorophyll synthesis resulting in change in leaf colour. Concrete content has been found to be 0.062 per cent

Tuberose Lifespan

Get the interesting facts about Tuberose like its lifespan and habit in this section.Tuberose lifespan is the maximum length of time for which Tuberose lives. Tuberose is a Perennials - a plant that lives for three or more years, growing habits of Tuberose is Shrubs.

Tuberose Symbolism

Learn about Tuberose symbolism and other interesting facts about Tuberose here. Tuberoses is used as a symbol of Dangerous and forbidden pleasures, Sensuality, Voluptuousness. Get Tuberose facts as birth flower in the upcoming section.

What is your Birth Flower?

What is your birth flower? Knowing birth flower is one of the unique interesting facts about Tuberose. It is believed that flowers characteristics may be inherited by whoever is born in that month. Tuberose is NA's birth flower. Gifting a Tuberose will be a good idea for those who have birthdays in the month of NA. Also check out **Flowers for Occasions** and never miss a single occasion to gift these beautiful flowers.

Habitat

Believed to have originated in Mexico, the Tuberose is today found in many parts of the world, although predominantly in the southern hemisphere.

Availability

The Tuberose is readily available in most seasons to buy in the UK as both a potted plant and as cut flowers, though it's natural blooming season is in July and August.

Care Tips

For greater growing success try planting in pots outdoors. In colder weather, move them indoors to avoid the cold and frosts. If growing in garden soils, make sure you dig up the plant in winter and transfer to a warmer environment. If you choose to buy in the Tuberose, ensure you use cut-flower or bulb-flower food. With care, this flower will last up to fourteen days so such feed is well worth using. Trim off the very top buds to maintain the flower's straight length and to avoid the stem bending towards the light.

Tulip

The tulip is a Eurasian and North African genus of perennial, bulbous plants in the lily family. It is a herbaceous herb with showy flowers, of which around 75 wild species are currently accepted.

The genus's native range extends west to the Iberian Peninsula, through North Africa to Greece, the Balkans, Turkey, throughout the Levant (Syria, Israel, Palestinian Territories, Lebanon, Jordan) and Iran, north to Ukraine, southern Siberiaand Mongolia, and east to the Northwest of China. The tulip's centre of diversity is in the Pamir, Hindu Kush, and Tien Shan mountains. It is a common element of steppe and winter-rain Mediterranean vegetation.

A number of species and many hybrid cultivars are grown in gardens or as potted plants.

Description

Tulips are spring-blooming perennials that grow from bulbs. Depending on the species, tulip plants can be between 4 inches (10 cm) and 28 inches (71 cm) high. The tulip's large flowers usually bloom on scapes with leaves in a rosette at ground level and a single flowering stalk arising from amongst the leaves. Tulip stems have few leaves. Larger species tend to have multiple leaves. Plants typically have two to six leaves, some species up to 12. The tulip's leaf is strap-shaped, with a waxy coating, and the leaves are alternately arranged on the stem; these fleshy blades are often bluish green in color. Most tulips produce only one flower per stem, but a few species bear multiple flowers on their scapes (*e.g. Tulipa turkestanica*). The generally cup or star-shaped tulip flower has three petals and three sepals, which are often termed tepals because they are nearly identical. These six tepals are often marked on the interior surface near the bases with darker colorings. Tulip flowers come in a wide variety of colors, except pure blue (several tulips with "blue" in the name have a faint violet hue).

Tulipa Agenensis in Israel (details).

Tip of a Tulip Stamen. Note the Pollen Grains

The flowers have six distinct, basifixed stamens with filaments shorter than the tepals. Each stigma has three distinct lobes, and the ovaries are superior, with three chambers. The tulip's seed is a capsule with a leathery covering and an ellipsoid to globe shape. Each capsule contains numerous flat, disc-shaped seeds in two rows per chamber. These light to dark brown seeds have very thin seed coats and endosperm that does not normally fill the entire seed.

Phytochemistry

Tulipanin is an anthocyanin found in tulips. It is the 3-rutinoside of delphinidin. The chemical compounds named tuliposides andtulipalins can also be found in tulips and are responsible for allergies. Tulipalin A, or α-methylene-dakua-butyrolactone, is a commonallergen, generated by hydrolysis of the glucoside tuliposide A. It induces a dermatitis that is mostly occupational and affects tulip bulb sorters and florists who cut the stems and leaves. Tulipanin A and B are toxic to horses, cats and dogs.

Taxonomy

The genus Tulipa was traditionally divided into two sections, Eriostemones and Tulipa (as Leiostemones), and comprises ca. 76 species. In 1997, the two sections were raised to subgenera and subgenus Tulipa was divided into five sections:

☆ Clusianae

☆ Eichleres

   ❏ subdivided into eight series

☆ Kopalkowskiana

☆ Tulipanum

☆ Tulipa

Subgenus Eriostemones was divided into the sections:

☆ Biflores

☆ Sylvestres

☆ Saxatiles

In 2009, two other subgenera were proposed, Clusianae and Orithyia, and this total of four subgenera was corroborated by a recent study (Christenhusz *et al.*, 2013). That study did not find support for any of the previous sections proposed, and since hybridisation is relatively common, it is probably better to refrain from subdividing the subgenera any further. Some species formerly classified as Tulipa are now considered as the separate genus Amana, including *Amana edulis* (*Tulipa edulis*). These species are more closely allied to Erythronium.

Species

The classification into four subgenera below is based on Christenhusz *et al.*

(2013). This list was used as the basis for The Genus Tulipa. Tulips of the World (D. Everett, 2013, Kew Publishing).

Subgenus Clusianae

Tulipa clusiana Redouté (Lady Tulip) - Iran, Iraq, Afghanistan, Pakistan, W Himalayas Tulipa harazensis Rech.f – Iran Tulipa linifolia Regel (Bokhara Tulip) - Iran, Afghanistan, Tajikistan, Uzbekistan Tulipa montana Lindl - Turkmenistan, Iran.

Subgenus Orithyia

Tulipa heteropetala Ledeb. - Altay Krai, Kazakhstan, XinjiangTulipa heterophylla (Regel) Baker - Kazakhstan, Xinjiang, Kyrgyzstan Tulipa sinkiangensis Z.M.Mao – Xinjiang *Tulipa uniflora* (L.) Besser ex Baker (Siberian Tulip) - Siberia, Mongolia, Xinjiang, Inner Mongolia, Kazakhstan

Tulipa Agenensis Sharonensis in HaSharon, Israel

- ☆ *Tulipa agenensis* Redouté (Eyed Tulip) - Middle East
- ☆ *Tulipa albanica* Kit Tan and Shuka (Albanian Tulip) - Albania
- ☆ *Tulipa alberti* Regel (Albert's Tulip) - Kazakhstan, Kyrgyzstan
- ☆ *Tulipa aleppensis* Boiss. ex Regel (Aleppo Tulip)- Turkey, Syria, Lebanon
- ☆ *Tulipa altaica* Pall. ex Spreng. (Altai Tulip) - Altai Krai, Western Siberia, Kazakhstan, Xinjiang
- ☆ *Tulipa anisophylla* Vved. - Tajikistan
- ☆ *Tulipa armena* Boiss. (Armenian Tulip) - Turkey, Iran, South Caucasus
- ☆ *Tulipa banuensis* Grey-Wilson (Afghan Tulip) - Afghanistan
- ☆ *Tulipa borszczowii* Regel - Kazakhstan
- ☆ *Tulipa botschantzevae* S.N.Abramova and Zakal. - Turkmenistan, Iran
- ☆ Tulipa butkovii Botschantz. - Uzbekistan
- ☆ *Tulipa carinata* Vved. (Pamir Tulip) - Tajikistan, Uzbekistan, Afghanistan
- ☆ *Tulipa cypria* Stapf ex Turrill (Cyprian Tulip) - Cyprus
- ☆ *Tulipa dubia* Vved. - Uzbekistan, Kyrgyzstan, Kazakhstan
- ☆ *Tulipa faribae* Ghahr., Attar and Ghahrem.-Nejad - Iran
- ☆ *Tulipa ferganica* Vved. - Uzbekistan, Kyrgyzstan
- ☆ *Tulipa foliosa* - Turkey
- ☆ *Tulipa fosteriana* W.Irving - Afghanistan, Central Asia
- ☆ *Tulipa gesneriana* L. (Garden Tulip) - Turkey
- ☆ *Tulipa greigii* Regel (Maculate Tulip) - Iran, Central Asia
- ☆ *Tulipa heweri* Raamsd. - Afghanistan
- ☆ *Tulipa hissarica* Popov and Vved. - Tajikistan, Uzbekistan
- ☆ *Tulipa hoogiana* B.Fedtsch. - Turkmenistan, Iran

☆ *Tulipa hungarica* Borbás (Rhodope Tulip) - Hungary, Serbia, Bulgaria

☆ *Tulipa iliensis* Regel - Kazakhstan, Kyrgyzstan, Xinjiang

☆ *Tulipa ingens* (Tubergen's Tulip) - Tajikistan, Uzbekistan

☆ *Tulipa julia* K.Koch (Julia Tulip) - Turkey, South Caucasus, Syria, Lebanon

☆ *Tulipa kaufmanniana* Regel (Waterlily Tulip) - Central Asia

☆ *Tulipa kolpakowskiana* Regel - Kazakhstan, Kyrgyzstan, Xinjiang, Afghanistan

☆ *Tulipa korolkowii* Regel - Central Asia

☆ *Tulipa kosovarica* Kit Tan, Shuka and Krasniqi - Kosovo

☆ *Tulipa kuschkensis* B.Fedtsch. - Turkmenistan, Afghanistan, Iran

☆ *Tulipa lanata* Regel - Tajikistan, Afghanistan, Pakistan, W Himalayas

☆ *Tulipa lehmanniana* Merckl. (Lehmann's Tulip) - Afghanistan, Iran, Central Asia

☆ *Tulipa lemmersii* - Kazakhstan

☆ *Tulipa ostrowskiana* Regel - Kazakhstan, Kyrgyzstan

☆ *Tulipa persica* (Lindl.) Sweet (Persian Tulip) - Iran

☆ *Tulipa platystemon* Vved. - Kyrgyzstan

☆ *Tulipa praestans* H.B.May (Multiflowered Tulip) - Tajikistan

☆ *Tulipa scardica* Bornm. (Balkan Tulip) - Kosovo, Greece

☆ Tulipa scharipovii Tojibaev - Kyrgyzstan, Uzbekistan

☆ *Tulipa schmidtii* Fomin - Iran, South Caucasus

☆ *Tulipa serbica* Tatic and Krivošej - Kosovo, Serbia

☆ *Tulipa sosnowskyi* Achv. and Mirzoeva - South Caucasus

☆ *Tulipa suaveolens* Roth (Scented or Crimean Tulip) - Crimea, Caucasus, Turkey, Iran, Kazakhastan

☆ *Tulipa subquinquefolia Vved.* - Tajikistan, Uzbekistan

☆ *Tulipa systola* Stapf - Middle East

☆ *Tulipa talassica* Lazkov - Kyrgyzstan

☆ *Tulipa tetraphylla* Regel - Xinjiang, Kazakhstan, Kyrgyzstan

☆ *Tulipa × tschimganica* Botschantz. - Kazakhstan, Uzbekistan

☆ *Tulipa ulophylla* Wendelbo - Iran

☆ *Tulipa undulatifolia* Boiss. (Eichler's Tulip) - Greece, Balkans, Caucasus, Middle East, Iran, Central Asia

☆ *Tulipa micheliana* Hoog – Central Asia to N.E. Iran, accepted by the World Checklist of Selected Plant Families as of May 2015, but regarded as a synonym of T. undulatifolia by others

☆ *Tulipa uzbekistanica* Botschantz. and Sharipov – Uzbekistan

☆ *Tulipa vvedenskyi* Botschantz. – Tajikistan.

Subgenus Eriostemones

☆ *Tulipa biflora* Pall. (Two-flowered Tulip) - Macedonia, Egypt, Crimea, Russia, Asia from Saudi Arabia to Xinjiang + Western Siberia

☆ *Tulipa bifloriformis* Vved. - Central Asia

☆ *Tulipa cinnabarina* K.Perss. - Turkey

☆ *Tulipa cretica* Boiss. and Heldr. (Cretan Tulip) - Crete

☆ *Tulipa dasystemon* (Regel) Regel - Central Asia, Xinjiang

☆ *Tulipa humilis* Herb. (Rainbow Tulip) - Caucasus, Middle East

As of May 2015, the World Checklist of Selected Plant Families accepts some species which other sources include in T. humilis:

☆ *Tulipa aucheriana* Baker – E. Turkey to Afghanistan

☆ *Tulipa kurdica* Wendelbo – N. Iraq

☆ *Tulipa pulchella* (Regel) Baker – S. and S.E. Turkey to N. Iran

☆ *Tulipa violacea* Boiss. and Buhse – S.E. Transcaucasus

☆ *Tulipa kolbintsevii* Zonn. - Kazakhstan

☆ *Tulipa koyuncui* Eker and Babaç - Turkey

☆ *Tulipa orithyioides* Vved. - Central Asia

☆ *Tulipa orphanidea* Boiss. and Heldr. (Green or Orange Wild Tulip) - Greece, Bulgaria, Turkey

☆ *Tulipa regelii* Krassn. (Plicate or Regel's Tulip) - Kazakhstan

☆ *Tulipa saxatilis* Sieber ex Spreng. (Rock Tulip) - Greece, Turkey

☆ *Tulipa sprengeri* Baker - Turkey

☆ *Tulipa sylvestris* L. (Wild Tulip) - Eurasia from Portugal to Xinjiang

☆ *Tulipa turkestanica* (Regel) Regel - Central Asia, Xinjiang

☆ *Tulipa urumiensis* Stapf (Tarda Tulip) - Kazakhstan, Kyrgyzstan, Iran

Unplaced

The horned tulip is often offered in the trade as *"Tulipa acuminata"*, but is in fact a cultivar, unknown from the wild, and should be distributed under its correct cultivar name: Tulipa 'Cornuta'.

Tulipa boettgeri Regel – Central Asia; accepted by the World Checklist of Selected Plant Families as of May 2015, but regarded as unplaced by Christenhusz *et al.*

Species not Belonging to the Genus Tulipa; classified in other genera[edit]

☆ *Tulipa anhuiensis* X.S.Shen, now: *Amana anhuiensis* (X.S.Shen) Christenh.

☆ *Tulipa breyniana* L., now: *Moraea collina* Thunb. (Iridaceae).

☆ *Tulipa edulis* (Miq.) Baker, now: *Amana edulis* (Miq.) Honda.

☆ *Tulipa erythronioides* Baker, now: *Amana erythronioides* (Baker) D.Y.Tan and D.Y.Hong.

☆ *Tulipa graminifolia* Baker ex S.Moore, now: *Amana edulis* (Miq.) Honda.

☆ *Tulipa latifolia* (Makino) Makino, now: *Amana erythronioides* (Baker) D.Y.Tan and D.Y.Hong

☆ *Tulipa ornithogaloides* Fisch. ex Besser, now: *Gagea triflora* (Ledeb.) Schult. and Schult.f.

☆ *Tulipa pudica* (Pursh) Raf., now: *Fritillaria pudica* (Pursh) Spreng.

☆ *Tulipa sibthorpiana* Sm., now: *Fritillaria sibthorpiana* (Sm.) Baker.

Etymology

The word tulip, first mentioned in western Europe in or around 1554 and seemingly derived from the "Turkish Letters" of diplomat Ogier Ghiselin de Busbecq, first appeared in English as tulipa or tulipant, entering the language by way of French: tulipe and its obsolete form tulipan or by way of Modern Latin tulîpa, from Ottoman Turkish tülbend ("muslin" or "gauze"), and may be ultimately derived from the Persian: delband ("Turban"), this name being applied because of a perceived resemblance of the shape of a tulip flower to that of a turban. This may have been due to a translation error in early times, when it was fashionable in the Ottoman Empire to wear tulips on turbans. The translator possibly confused the flower for the turban.

Distribution and Habitat

Tulips are indigenous to mountainous areas with temperate climates and need a period of cool dormancy, known as vernalization. They thrive in climates with long, cool springs and dry summers. Tulip bulbs imported to warm-winter areas are often planted in autumn to be treated as annuals.

Ecology

Botrytis tulipae is a major fungal disease affecting tulips, causing cell death and eventually the rotting of the plant. Other pathogens include anthracnose, bacterial soft rot, blight caused by *Sclerotium rolfsii*, bulb nematodes, other rots including blue molds, black molds and mushy rot.

The fungus *Trichoderma viride* can infect tulips, producing dried leaf tips and reduced growth, although symptoms are usually mild and only present on bulbs growing in glasshouses.

Variegated tulips admired during the Dutch tulipomania gained their delicately feathered patterns from an infection with the tulip breaking virus, a mosaic virus

Variegated Colors Produced by Tulip Breaking Virus.

that was carried by the green peach aphid, Myzus persicae. While the virus produces fantastically streaked flowers, it also weakens plants and reduces the number of offsets produced. Tulips affected by mosaic virus are called "broken"; while such plants can occasionally revert to a plain or solid colouring, they will remain infected and have to be destroyed. Today the virus is almost eradicated from tulip growers' fields. The multicoloured patterns or modern varieties result from breeding; they normally have solid, not feathered borders between the colours. Tulips are called lale in Turkish (from Persian: "lale"). When written in Arabic letters, "lale" has the same letters as Allah, which is why the flower became a holy symbol. It was also associated with the House of Osman, resulting in tulips being widely used in decorative motifs on tiles, mosques, fabrics, crockery, *etc.* in the Ottoman Empire.

Cultivation

History

Islamic World

Cultivation of the tulip began in Persia, probably in the 10th century. Early cultivars must have emerged from hybridisation in gardens from wild collected plants, which were then favoured, possibly due to flower size or growth vigour. The tulip is not mentioned by any writer from antiquity, therefore it seems probable that tulips were introduced into Anatolia only with the advance of the Seljuks. In the Ottoman Empire, numerous types of tulips were cultivated and bred, and today, 14 species can still be found in Turkey. Tulips are mentioned by Omar Kayam and Celaleddin Rûmi.

In 1574, Sultan Selim II. ordered the Kadi of A'azâz in Syria to send him 50,000 tulip bulbs. However, Harvey points out several problems with this source, and there is also the possibility that tulips and hyacinth (sümbüll, originally Indian spikenard (*Nardostachys jatamansi*) have been confused. Sultan Selim also imported 300,000 bulbs of Kefe Lale (also known as Cafe-Lale, from the medieval name Kaffa, probably *Tulipa schrenkii*) from Kefe for his gardens in the Topkapý Sarayý inIstanbul. Sultan Ahmet III maintained famous tulip gardens in the summer highland pastures (Yayla) at Spil Daðý above the town of Manisa. They seem to have consisted of wild tulips. However, from the 14 tulip species known from Turkey, only four are considered to be of local origin, so wild tulips from Iran and Central Asia may have been brought into Turkey during the Seljuk and especially Ottoman periods. Sultan Ahmet also imported domestic tulip bulbs from the Netherlands.

The gardening book Revnak'ý Bostan (Beauty of the Garden) by Sahibül Reis ülhaç Ibrahim Ibn ülhaç Mehmet, written in 1660 does not mention the tulip at all, but contains advice on growing hyacinths and lilies. However, there is considerable confusion of terminology, and tulips may have been subsumed under hyacinth, a mistake several European botanists were to perpetuate. In 1515, the scholar Qasim from Herat in contrast had identified both wild and garden tulips (lale) asanemones (shaqayq al-nu'man), but described the crown imperial as laleh kakli.

In a Turkic text written before 1495, the Chagatay Husayn Bayqarah mentions tulips (lale). Babur, the founder of the Mughal Empire, also names tulips in the Baburnama. He may actually have introduced them from Afghanistan to the plains of India, as he did with other plants like melons and grapes.

In Moorish Andalus, a "Makedonian bulb" (basal al-maqdunis) or "bucket-Narcissus" (naryis qadusi) was cultivated as an ornamental plant in gardens. It was supposed to have come from Alexandria and may have been Tulipa sylvestris, but the identification is not wholly secure.

Introduction to Western Europe

Although it is unknown who first brought the tulip to Northwestern Europe, the most widely accepted story is that it was Oghier Ghislain de Busbecq, an ambassador for Emperor Ferdinand I to Suleyman the Magnificent. According to a letter, he saw "an abundance of flowers everywhere; Narcissus, hyacinths and those in Turkish called Lale, much to our astonishment because it was almost midwinter, a season unfriendly to flowers." However, in 1559, an account by Conrad Gessner describes tulips flowering in Augsburg, Swabia in the garden of Councillor Heinrich Herwart. In Central and Northern Europe, tulip bulbs are generally removed from the ground in June and must be replanted by September for the winter. It is doubtful that Busbecq could have had the tulip bulbs harvested, shipped to Germany and replanted between March 1558 and Gessner's description the following year. Pietro Andrea Mattioli illustrated a tulip in 1565 but identified it as a narcissus, however.

Carolus Clusius planted tulips at the Vienna Imperial Botanical Gardens in 1573. After he was appointed director of theLeiden University's newly established Hortus Botanicus, he planted some of his tulip bulbs here in late 1593. Thus, 1594 is considered the date of the tulip's first flowering in the Netherlands, despite reports of the cultivation of tulips in private gardens in Antwerp and Amsterdam two or three decades earlier. These tulips at Leiden would eventually lead to both the Tulip mania and the tulip industry in the Netherlands.

The Reproductive Organs of a Tulip.

Carolus Clusius is largely responsible for the spread of tulip bulbs in the final years of the sixteenth century. He finished the first major work on tulips in 1592, and made note of the variations in colour. While a faculty member in the school of medicine at the University of Leiden, Clusius planted both a teaching garden and his private garden with tulips. In 1596 and 1598, over a hundred bulbs were stolen from his garden in a single raid.

Between 1634 and 1637, the enthusiasm for the new flowers triggered a speculative frenzy now known as the tulip mania. Tulip bulbs became so expensive that they were treated as a form of currency, or rather, as futures. Around this time, the ceramic tulipiere was devised for the display of cut flowers stem by stem. Vases and bouquets, usually including tulips, often appeared in Dutch still-life painting. To this day, tulips are associated with the Netherlands, and the cultivated forms of the tulip are often called "Dutch tulips." The Netherlands have the world's largest permanent display of tulips at the Keukenhof.

Introduction to the United States

It is believed the first tulips in the United States were grown near Spring Pond at the Fay Estate in Lynn and Salem,Massachusetts. From 1847 to 1865, Richard Sullivan Fay, Esq., one of Lynn's wealthiest men, settled on 500 acres (2.0 km²) located partly in present-day Lynn and partly in present-day Salem. Mr. Fay imported many different trees and plants from all parts of the world and planted them among the meadows of the Fay Estate.

Propagation

Tulips can be propagated through bulb offsets, seeds or micro propagation. Offsets and tissue culture methods are means of asexual propagation for producing genetic clones of the parent plant, which maintains cultivar genetic integrity. Seeds are most often used to propagate species and subspecies or to create new hybrids. Many tulip species can cross-pollinate with each other, and when wild tulip populations overlap geographically with other tulip species or subspecies, they often hybridize and create mixed populations. Most commercial tulip cultivars are complex hybrids, and often sterile.

Offsets require a year or more of growth before plants are large enough to flower. Tulips grown from seeds often need five to eight years before plants are of flowering size. Commercial growers usually harvest the tulip bulbs in late summer and grade them into sizes; bulbs large enough to flower are sorted and sold, while smaller bulbs are sorted into sizes and replanted for sale in the future. The Netherlands are the world's main producer of commercial tulip plants, producing as many as 3 billion bulbs annually, the majority for export.

Horticultural Classification

In horticulture, tulips are divided up into fifteen groups (Divisions) mostly based on flower morphology and plant size.

> ☆ Div. 1: Single early – with cup-shaped single flowers, no larger than 8 cm across (3 inches). They bloom early to mid season. Growing 15 to 45 cm tall.

'Yonina' is a Division 6 Cultivar.

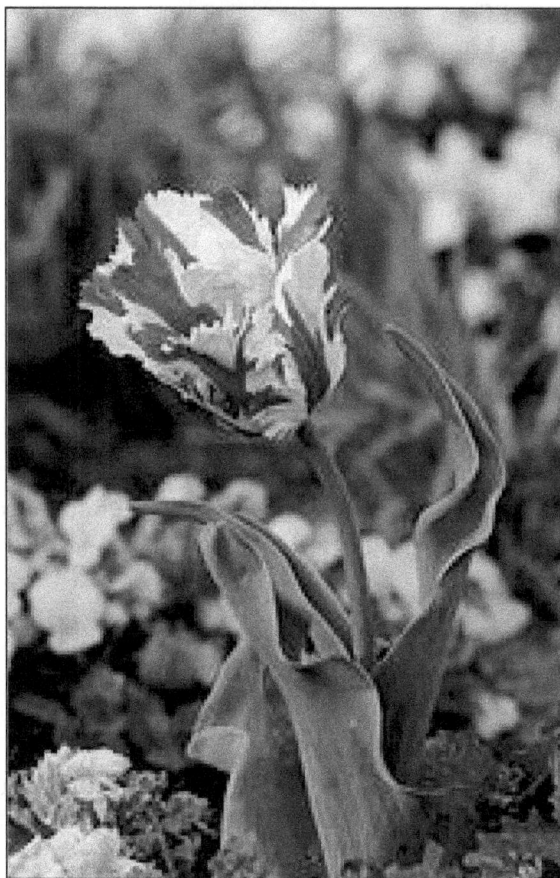

'Flaming Parrot' is a Division 10 Cultivar.

☆ Div. 2: Double early – with fully double flowers, bowl shaped to 8 cm across. Plants typically grow from 30–40 cm tall.

☆ Div. 3: Triumph – single, cup shaped flowers up to 6 cm wide. Plants grow 35–60 cm tall and bloom mid to late season.

☆ Div. 4: Darwin hybrid – single flowers are ovoid in shape and up to 8 cm wide. Plants grow 50–70 cm tall and bloom mid to late season. This group should not be confused with older Darwin tulips, which belong in the Single Late Group below.

☆ Div. 5: Single late – cup or goblet-shaded flowers up to 8 cm wide, some plants produce multi-flowering stems. Plants grow 45–75 cm tall and bloom late season.

☆ Div. 6: Lily-flowered – the flowers possess a distinct narrow 'waist' with pointed and reflexed petals. Previously included with the old Darwins, only becoming a group in their own right in 1958.

☆ Div. 7: Fringed (Crispa)

☆ Div. 8: Viridiflora

☆ Div. 9: Rembrandt

☆ Div. 10: Parrot

☆ Div. 11: Double late – Large, heavy blooms. They range from 18 to 22 inches tall.

☆ Div. 12: Kaufmanniana – Waterlily tulip. Medium-large creamy yellow flowers marked red on the outside and yellow at the center. Stems 6 in. tall.

☆ Div. 13: Fosteriana (Emperor)

☆ Div. 14: Greigii – Scarlet flowers 6 in. across, on 10 in. stems. Foliage mottled with brown.

☆ Div. 15: Species (Botanical)

☆ Div. 16: Multiflowering – not an official division, these tulips belong in the first 15 divisions but are often listed separately because they have multiple blooms per bulb.

They may also be classified by their flowering season:

☆ Early flowering: Single Early Tulips, Double Early Tulips, Greigii Tulips, Kaufmanniana Tulips, Fosteriana Tulips, Species Tulips

☆ Mid-season flowering: Darwin Hybrid Tulips, Triumph Tulips, Parrot Tulips

☆ Late season flowering: Single Late Tulips, Double Late Tulips, Viridiflora Tulips, Lily-flowering Tulips, Fringed Tulips, Rembrandt Tulips

Neo-tulipae

A number of names are based on naturalised garden tulips, and are usually referred to as neo-tulipae. These are often difficult to trace back to their original cultivar, and in some cases have been occurring in the wild for many centuries. The history of naturalisation is unknown, but populations are usually associated with agricultural practices and are possibly linked to saffron cultivation. Some neo-tulipae have been brought into cultivation, and are often offered as botanical tulips. These cultivated plants can be classified into two Cultivar Groups: 'Grengiolensis Group', with picotee tepals, and the 'Didieri Group' with unicolourous tepals.

Horticulture

Tulip bulbs are typically planted around late summer and fall, in well-drained soils, normally from 4 to 8 inches (10 to 20 cm) deep, depending on the type. Species tulips are normally planted deeper.

The tulip was a topic for Persian poets from the thirteenth century. In the poem Gulistan by Musharrifu'd-din Saadi, described a visionary garden paradise with 'The murmur of a cool stream/bird song, ripe fruit in plenty/bright multi coloured tulips and fragrant roses. The gift of a red or yellow tulip was a declaration of love, the flower's black center representing a heart burned by passion. In recent times, tulips have featured in the poems of Simin Behbahani. During the Ottoman Empire, the tulip was seen as a symbol of abundance and indulgence. The era during which the Ottoman Empire was wealthiest is often called the Tulip era or Lale Devri in Turkish.

Tulipa sylvestris subsp. *australis* with Seedpod (Right), 1804.

The shape of emblem of Iran is chosen to resemble a tulip, in memory of the people who died for Iran. Tulips became popular garden plants in east and west, but, whereas the tulip in Turkish culture was a symbol of paradise on earth and had almost a divine status, in the Netherlands it represented the briefness of life.

The Black Tulip is a historical romance by Alexandre Dumas, père. The story takes place in the Dutch city of Haarlem, where a reward is offered to the first grower who can produce a truly black tulip.The Tulip is also viewed prominently in a number of the Major Arcana cards of the Oswald Wirth Tarot deck. Specifically: Arcanas Zero,One, Four, and Fifteen.

Today, Tulip festivals are held around the world, for example in the Netherlands and Spalding, England. There is also a popular festival in Morges, Switzerland. Every spring, there are tulip festivals in North America, including the Tulip Time Festival in Holland, Michigan, the Skagit Valley Tulip Festival in Skagit Valley, Washington, the Tulip Time Festival in Orange City and Pella, Iowa, and the Canadian Tulip Festival in Ottawa, Canada. Tulips are also popular in Australia and several festivals are held in September and October, during the Southern Hemisphere's spring.

Tulip cultivars have usually several species in their direct background, but most have been derived from Tulipa suaveolens, often erroneously listed as Tulipa schrenkii. Tulipa gesneriana is in itself an early hybrid of complex origin and is probably not the same taxon as was described by Conrad Gesner in the 16th century.

Anthurium

Anthurium, is a genus of about 1000 species of flowering plants, the largest genus of the arum family, Araceae. General common names include anthurium, tailflower, flamingo flower, and laceleaf. The genus is native to the Americas, where it is distributed from northern Mexico to northern Argentina and parts of the Caribbean.

Description and Biology

Anthurium is a genus of herbs often growing as epiphytes on other plants. Some are terrestrial. The leaves are often clustered and are variable in shape. The inflorescence bears small flowers which are perfect, containing male and female structures. The flowers are contained in dense spirals on the spadix. The spadix is often elongated into a spike shape, but it can be globe-shaped or club-shaped. Beneath the spadix is the spathe, a type of bract. This is variable in shape, as well, but it is lance-shaped in many species. It may extend out flat or in a curve. Sometimes it covers the spadix like a hood. The fruits develop from the flowers on the spadix. They are juicy berries varying in color, usually containing two seeds.

The spadix and spathe are a main focus of Anthuirium breeders, who develop cultivars in bright colors and unique shapes.Anthurium scherzerianum and A. andraeanum, two of the most common taxa in cultivation, are the only species that grow bright red spathes. They have also been bred to produce spathes in many other colors and patterns.

Flamingo Flower Orchid.

Anthurium plants are poisonous due to calcium oxalate crystals. The sap is irritating to the skin and eyes.

Cultivation

Anthurium andraeanum **Princess Amalia Elegance.**

Like other aroids, many species of Anthurium can be grown as houseplants, or outdoors in mild climates in shady spots. They include forms such as A. crystallinum f peltifolium with its large, velvety, dark green leaves and silvery white venation. Manyhybrids are derived from A. andreanum or A. scherzerianum because of their colorful spathes. They thrive in moist soils with high organic matter. In milder climates the plants can be grown in pots of soil. Indoors plants thrive at temperatures between 16°C-22°C (60°F-72°F) and at lower light than other house plants. Wiping the leaves off with water will remove any dust and insects. Plant in pots with good root systems will benefit from a weak fertilizer solution every other week. In the case of vining or climbing Anthuriums, the plants benefit from being provided with a totem to climb.

Propagation

Anthurium can be propagated by seed or vegetatively by cuttings. In the commercial Anthurium trade, most propagation is via tissue culture.

Species

For a full list, see the List of *Anthurium* species.

***Anthurium scherzerianum* Inflorescence.**

In 1860 there were 183 species known to science, and Heinrich Wilhelm Schott defined them in 28 sections in the bookProdromus Systematis Aroidearum. In 1905 the genus was revised with a description of 18 sections. In 1983 the genus was divided into the following sections:

☆ Belolonchium

☆ Calomystrium

☆ Cardiolonchium

☆ Chamaerepium

☆ Cordatopunctatum

☆ Dactylophyllium

☆ Decurrentia

☆ Digitinervium

☆ Gymnopodium

☆ Leptanthurium

☆ Pachyneurium

☆ Polyphyllium

☆ Polyneurium

☆ Porphyrochitonium

☆ Schizoplacium

☆ Semaeophyllium

☆ Tetraspermium

☆ Urospadix

☆ Xialophyllium

Anthurium Information

Anthurium is one of the most popular of the tropical cut flowers and pot plants which are being grown commercially for export as well as for the local market. The anthurium belongs to the complex family Araceae. Within the family Araceae, Anthurium is the largest genus, which comprises of some 900 varieties, including well-known cultivated varieties *Anthuriurn andrenum* and *Anthurium scherzerianurn*. The inflorescence of Anthurium andreanum is comprised of a spathe with a straight spadix bearing the flowers. Anthurium scherzerianum has a coiled spadix. An inflorescence can develop from each leaf axil. The foliage of *Anthurium scherzerianum* is somewhat more leathery than that of *Anthurium andreanum*.

Anthurium varieties are common throughout South and Middle America. The northern boundary of distribution area lies near the Mexican town. Anthurium varieties are found in areas with widely different climatic conditions; from the dry regions of Mexico to the tropical rain forests of South America. The location altitude is species-related, varying from sea level to heights of 3000 meters.

Anthurium grows in many forms, mostly evergreen, bushy or climbing epiphytes with roots that can hang from the canopy all the way to the floor of the rain forest. There are also many terrestrial forms which are found as understory plants, as well as hemiepiphytic forms. A hemiepiphyte is a plant capable of beginning life as a seed and sending roots to the soil, or beginning as a terrestrial plant that climbs a tree and then sends roots back to the soil. They occur also as lithophytes.

Anthuriums are cultivated for its attractive long lasting 'flower' which is not really a flower but an inflorescence rising from the base of a bract. Anthurium flowers are small (about 3 mm) and develop crowded in a spike on a fleshy axis, called a spadix, a characteristic of the Araceae. A common feature of the Araceae is the typical, cup-shaped inflorescence. It consists of numerous flowers closely arranged in a spadix together with an outer colourful heart shaped sheath called spathe. Flowers are inconspicuous, hermoproditic with two carpel ovary and four anthers. The sepals and petals are rudimentary; stigma appears as a rounded protuberance on the spadix when it is mature. Pollen matures week to ten days after stigma becomes receptive to prevent self pollination. The large variations in inflorescences are due to the variations in shape, color and the size of the spathe and spadix.

Like other Aroids, many species of Anthurium can be grown as houseplants, or outdoors in mild climates in shady spots. They thrive in moist soils with high organic matter. In milder climates the plants can be grown in pots of soil. Indoors plants thrive at temperatures between 60-72 ° F and at lower light than other house plants.

Fertilizer

Both chemical and organic fertilizers may be used in Anthurium. Complete fertilizers such as 17-17-17, 27-13-13 and 13-27-27 at the rate of 1 gram per liter of water sprayed or drenched on the medium, though it is better to supply Anthurium with fertilizer via irrigation water at the roots rather than as a foliar spray. Since Anthurium leaves are covered with a thick layer of wax, absorption through leaf is poor. Another advantage of fertilization via irrigation water at the roots is that, the leaves and flowers remain clean.

Gladiolus

Gladiolus (from Latin, the diminutive of gladius, a sword) is a genus of perennial cormous flowering plants in the iris family(Iridaceae). It is sometimes called the 'sword lily', but usually by its generic name (plural gladioli). The genus occurs in Asia, Mediterranean Europe, South Africa, and tropical Africa. The center of diversity is in the Cape Floristic Region. The genera Acidanthera, Anomalesia, Homoglossum, and Oenostachys, formerly considered distinct, are now included in Gladiolus.

Gladiolus Growing in the Foot Hills Of Himalayas Gladiolus flowers are in demand for their elegant attractive spikes of different hues and good keeping quality. Gladiolus can be easily grown with a little care and attention in beds for garden decoration and cut flower production and also in pots for interior and outdoor decoration. In India, gladiolus has become a very popular flower and millions and millions of spikes are being sold every year. Gladiolus can be grown outdoor and under greenhouse conditions. In the plains, gladiolus flowers are available during the winter months only. Very high temperature during the summer days may adversely affect the flower spikes. In the hilly areas (upto 2000 meters) with moderate summer, the gladiolus can be grown for almost for throughout the year. In places where winter is severe, the gladiolus crop may affect by frost. Gladiolus Cultivars can be classified into five groups based on the flower size *viz.* i) Miniature (floret diameter < 6.4 cm), ii) Small (6.4 – 8.9 cm) iii) Decorative (8.9 - 1.4 cm), iv) Standard or Large (11.4-14.0 cm) and v) Giant (>14.0 cm). Gladiolus can also be classified into three groups on the basis of time to flower from the date of planting of corms – a) Early (flowering within 60 days), b) Medium (between 60-90 days) and c) are (> 90 days).

Description

The genus Gladiolus contains about 260 species, of which 250 are native to sub-Saharan Africa, mostly South Africa. About 10 species are native to Eurasia. There are 160 species of Gladiolus endemic in southern Africa and 76 in tropical Africa. The flowers of unmodified wild species vary from very small to perhaps 40 mm across, and inflorescences bearing anything from one to several flowers. The spectacular giant flower spikes in commerce are the products of centuries of hybridisation, selection, and perhaps more drastic manipulation.

Gladioli are half-hardy in temperate climates. They grow from rounded, symmetrical corms, that are enveloped in several layers of brownish, fibrous tunics.

Their stems are generally unbranched, producing 1 to 9 narrow, sword-shaped, longitudinal grooved leaves, enclosed in a sheath. The lowest leaf is shortened to a cataphyll. The leaf blades can be plane or cruciform in cross section.

The flower spikes are large and one-sided, with secund, bisexual flowers, each subtended by 2 leathery, green bracts. Thesepals and the petals are almost identical in appearance, and are termed tepals. They are united at their base into a tube-shaped structure. The dorsal tepal is the largest, arching over the three stamens. The outer three tepals are narrower. Theperianth is funnel-shaped, with the stamens attached to its base. The style has three filiform, spoon-shaped branches, each expanding towards the apex.

The ovary is 3-locular with oblong or globose capsules, containing many, winged brown, longitudinally dehiscent seeds. In their center must be noticeable the specific pellet-like structure which is the real seed without the fine coat. In some seeds this feature is wrinkled with black color. These seeds are unable to germinate. These flowers are variously colored, pink to reddish or light purple with white, contrasting markings, or white to cream or orange to red.

The South African species were originally pollinated by long-tongued anthrophorine bees, but some changes in the pollination system have occurred, allowing pollination by sunbirds, noctuid and Hawk-moths, long-tongued flies and several others. In the temperate zones of Europe many of the hybrid large flowering sorts of gladiolus can be pollinated by small well-known wasps. Actually, they are not very good pollinators because of the large flowers of the plants and the small size of the wasps. Another insect in this zone which can try some of the nectar of the gladioli is the best-known European Hawk-moth Macroglossum stellatarum which usually pollinates many popular garden flowers like Petunia, Zinnia, Dianthus and others.

Gladioli are used as food plants by the larvae of some Lepidoptera species including the Large Yellow Underwing. The Gladioli is also the official flower of Elmira Ontario in Canada adopted by the council on March 15, 1926. Gladioli have been extensively hybridized and a wide range of ornamental flower colours are available from the many varieties. The main hybrid groups have been obtained by crossing between four or five species, followed by selection: Grandiflorus, Primulines and Nanus. They make very good cut flowers.

The majority of the species in this genus are diploid with 30 chromosomes but the Grandiflora hybrids are tetraploid and possess 60 chromosomes. This is because the main parental species of these hybrids is Gladiolus dalenii which is also tetraploid and includes a wide range of varieties (like the Grandiflora hybrids).

Varieties

A number of gladiolus varieties have been found to grow well under sub-temperate conditions *viz.* Aldebaran, American Beauty Eurovision, Friendship, Friendship Pink, Gold Field, Green Wood- pecker, Her Majesty, Jacksonvilla Gold, Jestsr, Oranqe Emperor, Oscar, Peter Pears, Priscilla, Red Beauty, Spic and Span, Top Brass, Traderhorn, Tropic Seas, Victor Borge, Video, White Goddess, White Friendship, White Prosperity.

Species

The genus Gladiolus has been divided into many sections. Most species, however, are only tentatively placed. As of June 2014, the World Checklist of Selected Plant Families accepts 276 species:

* ☆ *Gladiolus abbreviatus* Andrews
* ☆ *Gladiolus abyssinicus* (Brongn. ex Lem.) B.D.Jacks.
* ☆ *Gladiolus actinomorphanthus*P.A.Duvign. and Van Bockstal
* ☆ *Gladiolus acuminatus* F.Bol.
* ☆ *Gladiolus aequinoctialis* Herb.
* ☆ *Gladiolus alatus* L. (sect. Hebea)
* ☆ *Gladiolus albens* Goldblatt and J.C.Manning
* ☆ *Gladiolus amplifolius* Goldblatt
* ☆ *Gladiolus anatolicus* (Boiss.) Stapf
* ☆ *Gladiolus andringitrae* Goldblatt
* ☆ *Gladiolus angustus* L. (sect. Blandus) - long-tubed painted lady
* ☆ *Gladiolus antakiensis* A.P.Ham.
* ☆ *Gladiolus antandroyi* Goldblatt
* ☆ *Gladiolus appendiculatus* G.Lewis
* ☆ *Gladiolus aquamontanus* Goldblatt and Vlok
* ☆ *Gladiolus arcuatus* Klatt
* ☆ *Gladiolus atropictus* Goldblatt and J.C.Manning
* ☆ *Gladiolus atropurpureus* Baker
* ☆ *Gladiolus atroviolaceus* Boiss.
* ☆ *Gladiolus attilae* Kit Tan
* ☆ *Gladiolus aurantiacus* Klatt
* ☆ *Gladiolus aureus* Baker - golden gladiolus
* ☆ *Gladiolus balensis* Goldblatt
* ☆ *Gladiolus baumii* Harms
* ☆ *Gladiolus bellus* C. H. Wright
* ☆ *Gladiolus benguellensis* Baker (sect. Ophiolyza)
* ☆ *Gladiolus bilineatus* G. J. Lewis
* ☆ *Gladiolus blommesteinii* L.Bolus
* ☆ *Gladiolus bojeri* (Baker) Goldblatt
* ☆ *Gladiolus bonaespei* Goldblatt and M.P.de Vos
* ☆ *Gladiolus boranensis* Goldblatt
* ☆ *Gladiolus brachyphyllus* Bolus f.

☆ *Gladiolus brevifolius* Jacq. (sect. Linearifolius)

☆ *Gladiolus brevitubus* G. Lewis

☆ *Gladiolus buckerveldii* (L. Bolus) Goldblatt

☆ *Gladiolus bullatus* Thunb. ex G. Lewis - Caledon bluebell

☆ *Gladiolus caeruleus* Goldblatt and J.C. Manning

☆ *Gladiolus calcaratus* G. Lewis

☆ *Gladiolus calcicola* Goldblatt

☆ *Gladiolus canaliculatus* Goldblatt

☆ *Gladiolus candidus* (Rendle) Goldblatt

☆ *Gladiolus cardinalis* Curtis (sect. Blandus)

☆ *Gladiolus carinatus* Aiton

☆ *Gladiolus carmineus* C. H. Wright (sect. Blandus) - cliff lily

☆ *Gladiolus carneus* F. Delaroche (sect. Blandus) - large painted lady

☆ *Gladiolus caryophyllaceus* (Burm. f.) Poiret

☆ *Gladiolus cataractarum* Oberm.

☆ *Gladiolus caucasicus* Herb.

☆ *Gladiolus ceresianus* L. Bolus

☆ *Gladiolus chelamontanus* Goldblatt

☆ *Gladiolus chevalierianus* Marais

☆ *Gladiolus communis* L. (sect. Gladiolus) - cornflag (type species)

☆ *Gladiolus comptonii* G.J.Lewis

☆ *Gladiolus crassifolius* Baker

☆ *Gladiolus crispulatus* L. Bolus

☆ *Gladiolus cruentus* T. Moore (sect. Ophiolyza)

☆ *Gladiolus cunonius* (L.) Gaertn.

☆ *Gladiolus curtifolius* Marais

☆ *Gladiolus curtilimbus* P.A.Duvign. and Van Bockstal ex S.Córdova

☆ *Gladiolus cylindraceus* G. Lewis

☆ *Gladiolus dalenii* Van Geel (sect. Ophiolyza)

☆ *Gladiolus davisoniae* F.Bolus

☆ *Gladiolus debeerstii* De Wild.

☆ *Gladiolus debilis* Ker Gawler (sect. Homoglossum) - small painted lady

☆ *Gladiolus decaryi* Goldblatt

☆ *Gladiolus decoratus* Baker

☆ *Gladiolus delpierrei* Goldblatt

☆ *Gladiolus densiflorus* Baker

☆ *Gladiolus deserticola* Goldblatt

☆ *Gladiolus dichrous* (Bullock) Goldblatt

☆ *Gladiolus diluvialis* Goldblatt and J.C.Manning

☆ *Gladiolus dolichosiphon* Goldblatt and J.C.Manning

☆ *Gladiolus dolomiticus* Oberm.

☆ *Gladiolus dzavakheticus* Eristavi

☆ *Gladiolus ecklonii* Lehm.

☆ *Gladiolus elliotii* Baker (sect. Ophiolyza)

☆ *Gladiolus emiliae* L. Bolus

☆ *Gladiolus engysiphon* G. Lewis

☆ *Gladiolus equitans* Thunb. (sect. Hebea)

☆ *Gladiolus erectiflorus* Baker

☆ *Gladiolus exiguus* G. Lewis

☆ *Gladiolus exilis* G.J. Lewis

☆ *Gladiolus fenestratus* Goldblatt

☆ *Gladiolus ferrugineus* Goldblatt and J.C.Manning

☆ *Gladiolus filiformis* Goldblatt and J.C. Manning

☆ *Gladiolus flanaganii* Baker - suicide gladiolus

☆ *Gladiolus flavoviridis* Goldblatt

☆ *Gladiolus floribundus* Jacq.

☆ *Gladiolus fourcadei* (L.Bolus) Goldblatt and M.P.de Vos

☆ *Gladiolus geardii* L. Bolus

☆ *Gladiolus goldblattianus* Geerinck

☆ *Gladiolus gracilis* Jacq. (sect. Homoglossum) - reed bells

☆ *Gladiolus gracillimus* Baker

☆ *Gladiolus grandiflorus* Andrews (sect. Blandus)

☆ *Gladiolus grantii* Baker

☆ *Gladiolus gregarius* Welw. ex Baker(sect. Densiflorus)

☆ *Gladiolus griseus* Goldblatt and J.C. Manning

☆ *Gladiolus gueinzii* Kunze

☆ *Gladiolus gunnisii* (Rendle) Marais

☆ *Gladiolus guthriei* F. Bol. (sect. Linearifolius)

☆ *Gladiolus hajastanicus* Gabrieljan

☆ *Gladiolus halophilus* Boiss. and Heldr.

☆ *Gladiolus harmsianus* Vaupel

☆ *Gladiolus hirsutus* Jacq. (sect. Linearifolius) - small pink Afrikaner,lapmuis

☆ *Gladiolus hollandii* L. Bolus

☆ *Gladiolus horombensis* Goldblatt

☆ *Gladiolus huillensis* (Welw. ex Baker) Goldblatt

☆ *Gladiolus humilis* Stapf

☆ *Gladiolus huttonii* (N.E.Br.) Goldblatt and M.P.de Vos

☆ *Gladiolus hyalinus* Jacq.

☆ *Gladiolus illyricus* W.D.J.Koch - wild gladiolus

☆ *Gladiolus imbricatus* L.

☆ *Gladiolus inandensis* Baker

☆ *Gladiolus inflatus* Thunb.

☆ *Gladiolus inflexus* Goldblatt and J.C. Manning

☆ *Gladiolus insolens* Goldblatt and J.C. Manning

☆ *Gladiolus intonsus* Goldblatt

☆ *Gladiolus invenustus* G. J. Lewis

☆ *Gladiolus involutus* D.Delaroche (sect. Hebea)

☆ *Gladiolus iroensis* (A. Chev.) Marais

☆ *Gladiolus italicus* P. Mill. (sect. Gladiolus) - Italian gladiolus, cornflag

☆ *Gladiolus jonquilodorus* Eckl. ex G.J.Lewis

☆ *Gladiolus juncifolius* Goldblatt

☆ *Gladiolus kamiesbergensis* G. Lewis

☆ *Gladiolus karooicus* Goldblatt and J.C.Manning

☆ *Gladiolus kotschyanus* Boiss.

☆ *Gladiolus lapeirousioides* Goldblatt

☆ *Gladiolus laxiflorus* Baker

☆ *Gladiolus ledoctei* P.A. Duvign. and Van Bockstal

☆ *Gladiolus leonensis* Marais

Cultivation

In temperate zones, the corms of most species and hybrids should be lifted in autumn and stored over winter in a frost-free place, then replanted in spring. Some species from Europe and high altitudes in Africa, as well as the small 'Nanus' hybrids, are much hardier (to at least -15 °F/-26 °C) and can be left in the ground in regions with sufficiently dry winters. 'Nanus' is hardy to Zones 5-8. The large-flowered types require moisture during the growing season, and must be individually staked as soon as the sword-shaped flower heads appear. The leaves must be allowed to die down naturally before lifting and storing the corms. Plants are propagated either from small cormlets produced as offsets by the parent corms, or from seed. In either case, they take several years to get to flowering size. Clumps should be dug up and divided every few years to keep them vigorous.

In Culture

The Mancunian singer Morrissey is known to dance with gladioli, hanging from his back pocket or in his hands, especially during the era of The Smiths. This trait of his was made known in the music video for "This Charming Man", where he swung a bunch of yellow gladioli while singing.

The Australian comedian and personality Dame Edna Everage's signature flowers are gladioli, which she refers to as "glads". Gladioli are the flowers associated with a fortieth wedding anniversary. Gladiolus is the flower of August. "Gladiolus" was the word Frank Neuhauser correctly spelled to win the 1st National Spelling Bee in 1925.

Lilium

"Lily" and "Lilies" redirect here. For other uses, see Lily (disambiguation) and Lilies (disambiguation). Lilium (members of which are true lilies) is a genus of herbaceous flowering plants growing from bulbs, all with large prominent flowers. Lilies are a group of flowering plants which are important in culture and literature in much of the world. Most species are native to the temperate northern hemisphere, though their range extends into the northern subtropics. Many other plants have "lily" in their common name but are not related to "true" lilies.

Description

Lilies are tall perennials ranging in height from 2–6 ft (60–180 cm). They form naked or tunicless scaly underground bulbs which are their overwintering organs. In some North American species the base of the bulb develops intorhizomes, on which numerous small bulbs are found. Some species developstolons. Most bulbs are deeply buried, but a few species form bulbs near the soil surface. Many species form stem-roots. With these, the bulb grows naturally at some depth in the soil, and each year the new stem puts out adventitious roots above the bulb as it emerges from the soil. These roots are in addition to the basal roots that develop at the base of the bulb.

The flowers are large, often fragrant, and come in a range of colours including whites, yellows, oranges, pinks, reds and purples. Markings include spots and brush strokes. The plants are late spring- or summer-flowering. Flowers are borne in racemes or umbels at the tip of the stem, with six tepals spreading or reflexed, to give flowers varying from funnel shape to a "Turk's cap". The tepals are free from each other, and bear a nectary at the base of each flower. The ovary is 'superior', borne above the point of attachment of the anthers. The fruitis a three-celled capsule.

Seeds ripen in late summer. They exhibit varying and sometimes complexgermination patterns, many adapted to cool temperate climates. Naturally most cool temperate species are deciduous and dormant in winter in their native environment. But a few species which distribute in hot summer and mild winter area (*Lilium candidum, Lilium catesbaei, Lilium longiflorum*) lose leaves and remain relatively short dormant in Summer or Autumn, sprout from Autumn to winter, forming dwarf stem bearing a basal rosette of leaves until accept enough chilling requirement, the stem begins to elongate while warming.

Taxonomy

Taxonomical division in sections follows the classical division of Comber, species acceptance follows the World Checklist of Selected Plant Families, the taxonomy of section Pseudolirium is from the Flora of North America, the taxonomy of Section Liriotypus is given in consideration of Resetnik *et al.*, 2007, the taxonomy of Chinese species (various sections) follows the Flora of China and the taxonomy of Section Sinomartagon follows Nishikawa *et al.*, as does the taxonomy of Section Archelirion.

The World Checklist of Selected Plant Families, as of January 2014, considers Nomocharis a separate genus in its own right, however some authorities consider Nomocharis to be embedded within Lilium, rather than treat it as a separate genus.

There are seven sections:

☆ Martagon

☆ Pseudolirium

☆ Liriotypus

☆ Archelirion

☆ Sinomartagon

☆ Leucolirion

☆ Daurolirion

For a full list of accepted species with their native ranges, see List of *Lilium* species.

Some species formerly included within this genus have now been placed in other genera. These genera include Cardiocrinum, Notholirion, Nomocharis and Fritillar.

Etymology

The botanic name Lilium is the Latin form and is a Linnaean name. The Latin name is derived from the Greek ñéïí, leírion, generally assumed to refer to true, white lilies as exemplified by the Madonna lily. The word was borrowed from Coptic (dial. Fayyumic) hleri, from standard hreri, from Demoti chrry, from Egyptian hr?t "flower". Meillet maintain that both the Egyptian and the Greek word are possible loans from an extinct, substratum language of the Eastern Mediterranean. The Greeks also used the word ñí, krînon, albeit for non-white lilies.

The term "lily" has in the past been applied to numerous flowering plants, often with only superficial resemblance to the true lily, including water lily, fire lily, lily of the Nile, calla lily, trout lily, kaffir lily, cobra lily, lily of the valley, daylily, ginger lily, Amazon lily, leek lily, Peruvian lily, and others. All English translations of the Bible render the Hebrew shûshan, shôshan, shôshannâ as "lily", but the "lily among the thorns" of Song of Solomon, for instance, may be the honeysuckle.

For a list of other species described as lilies, see Lily (disambiguation).

Distribution and Habitat

The range of lilies in the Old World extends across much of Europe, across most of Asia to Japan, south to India, and east to Indochina and the Philippines. In the New World they extend from southern Canada through much of the United States. They are commonly adapted to either woodland habitats, often montane, or sometimes to grassland habitats. A few can survive in marshland and epiphytes are known in tropical southeast Asia. In general they prefer moderately acidic or lime-free soils.

Ecology

Lilies are used as food plants by the larvae of some Lepidoptera species including the Dun-bar.

Cultivation

Many species are widely grown in the garden in temperate and sub-tropical regions. They may also be grown as potted plants. Numerous ornamental hybrids have been developed. They can be used in herbaceous borders, woodland and shrub plantings, and as patio plants. Some lilies, especially Lilium longiflorum, form important cut flower crops. These may be forced for particular markets; for instance, Lilium longiflorum for the Easter trade, when it may be called the Easter lily.

Lilies are usually planted as bulbs in the dormant season. They are best planted in a south-facing (northern hemisphere), slightly sloping aspect, in sun or part shade, at a depth 2½ times the height of the bulb (except *Lilium candidum* which should be planted at the surface). Most prefer a porous, loamy soil, and good drainage is essential. Most species bloom in July or August (northern hemisphere). The flowering periods of certain lily species begin in late spring, while others bloom in late summer or early autumn.[20] They have contractile roots which pull the plant down to the correct depth, therefore it is better to plant them too shallowly than too deep. A soil pH of around 6.5 is generally safe. The soil should be well-drained, and plants must be kept watered during the growing season. Some plants have strong wiry stems, but those with heavy flower heads may need staking.

Awards

Below is a list of lily species and cultivars that have gained the Royal Horticultural Society's Award of Garden Merit:

- ☆ Lilium African Queen Group (VI-/a) 2002 H6 Reconfirmed 2013
- ☆ Lilium 'Casa Blanca' (VIIb/b-c) 1993 H6 Reconfirmed 2013
- ☆ Lilium 'Fata Morgana' (Ia/b) 2002 H6 Reconfirmed 2013

☆ Lilium 'Garden Party' (VIIb/b) 2002 H6 Reconfirmed 2013

☆ Lilium Golden Splendor Group (VIb-c/a) 2002 H6 Reconfirmed 2013

☆ Lilium henryi (IXc/d) 1993 H6 Reconfirmed 2013

☆ Lilium mackliniae (IXc/a) 2012 H5

☆ Lilium martagon (IXc/d) 2002 H7 Reconfirmed 2013

☆ Lilium pardalinum (IXc/d) 2002 H6

☆ Lilium Pink Perfection Group (VIb/a) 1993 H6 Reconfirmed 2013

☆ Lilium regale (IXb/a) 1993 H6 Reconfirmed 2013

Classification of Garden Forms

Numerous forms, mostly hybrids, are grown for the garden. They vary according to the species and interspecific hybrids that they derived from, and are classified in the following broad groups:

Asiatic Hybrids (Division I)

These are derived from hybrids between species in Lilium section Sinomartagon. They are derived from central and East Asian species and interspecific hybrids, including *Lilium amabile, Lilium bulbiferum, Lilium callosum, Lilium cernuum, Lilium concolor, Lilium dauricum, Lilium davidii, Lilium × hollandicum, Lilium lancifolium* (syn. *Lilium tigrinum*), *Lilium lankongense, Lilium leichtlinii, Lilium × maculatum, Lilium pumilum, Lilium × scottiae, Lilium wardii* and *Lilium wilsonii*.

These are plants with medium-sized, upright or outward facing flowers, mostly unscented. Dwarf (Patio, Border) varieties are much shorter, c.36–61 cm in height and were designed for containers. They often bear the cultivar name 'Tiny', such as the 'Lily Looks' series, *e.g.* 'Tiny Padhye', 'Tiny Dessert'.

Martagon Hybrids (Division II)

These are based on *Lilium dalhansonii, Lilium hansonii, Lilium martagon, Lilium medeoloides,* and *Lilium tsingtauense*.

The flowers are nodding, Turk's cap style (with the petals strongly recurved). This includes mostly European species: *Lilium candidum, Lilium chalcedonicum, Lilium kesselringianum, Lilium monadelphum, Lilium pomponium, Lilium pyrenaicum* and *Lilium × testaceum*.

American Hybrids (Division IV)

These are mostly taller growing forms, originally derived from *Lilium bolanderi, Lilium × burbankii, Lilium canadense,Lilium columbianum, Lilium grayi, Lilium humboldtii, Lilium kelleyanum, Lilium kelloggii, Lilium maritimum, Lilium michauxii,Lilium ichiganense, Lilium occidentale, Lilium × pardaboldtii, Lilium pardalinum, Lilium parryi, Lilium parvum, Lilium philadelphicum, Lilium pitkinense, Lilium superbum, Lilium ollmeri, Lilium washingtonianum,* and *Lilium wigginsii*.

Many are clump-forming perennials with rhizomatous rootstocks.

Longiflorum Hybrids (Division V)

These are cultivated forms of this species and its subspecies. They are most important as plants for cut flowers, and are less often grown in the garden than other hybrids.

Trumpet Lilies (Division VI), including Aurelian Hybrids (with L. henryi)

This group includes hybrids of many Asiatic species and their interspecific hybrids, including *Lilium × aurelianense, Lilium brownii, Lilium × centigale, Lilium henryi, Lilium × imperiale, Lilium × kewense, Lilium leucanthum, Lilium regale,Lilium rosthornii, Lilium sargentiae, Lilium sulphureum* and *Lilium × sulphurgale.*

The flowers are trumpet shaped, facing outward or somewhat downward, and tend to be strongly fragrant, often especially night-fragrant.

Oriental Hybrids (Division VII)

These are based on hybrids within *Lilium* section *Archelirion*, specifically *Lilium auratum* and *Lilium speciosum*, together with crossbreeds from several species native to Japan, including *Lilium nobilissimum, Lilium rubellum, Lilium alexandrae,* and *Lilium japonicum*. They are fragrant, and the flowers tend to be outward facing. Plants tend to be tall, and the flowers may be quite large. The whole group are sometimes referred to as "stargazers" because many of them appear to look upwards. (For the specific cultivar, see Lilium 'Stargazer'.)

Other Hybrids (Division VIII)

Includes all other garden hybrids.

Species (Division IX)

All natural species and naturally occurring forms are included in this group.

The flowers can be classified by flower aspect and form:

Flower Aspect

 a. Up-facing
 b. Out-facing
 c. Down-facing

Flower Form

 a. Trumpet-shaped
 b. Bowl-shaped
 c. Flat (or with tepal tips recurved)
 d. Tepals strongly recurved (with the Turk's cap form as the ultimate state)

Many newer commercial varieties are developed by using new technologies such as ovary culture and embryo rescue.

Pests and Diseases

Scarlet Lily Beetles, Oxfordshire, UK

Aphids may infest plants. Leatherjackets feed on the roots. Larvae of the Scarlet lily beetle can cause serious damage to the stems and leaves. The scarlet beetle lays its eggs and completes its life cycle only on true lilies (Lilium) and fritillaries (Fritillaria). Oriental, rubrum, tiger and trumpet lilies as well as Oriental trumpets (orienpets) and Turk's cap lilies and native North American Lilium species are all vulnerable, but the beetle prefers some types over others. The beetle could also be having an effect on native Canadian species and some rare and endangered species found in northeastern North America. Daylilies (Hemerocallis, not true lilies) are excluded from this category. Plants can suffer from damage caused by mice, deer and squirrels. Slugs, snails and millipedes attack seedlings, leaves and flowers. Brown spots on damp leaves may signal botrytis (also known as lily disease). Various fungal and viral diseases can cause mottling of leaves and stunting of growth.

Propagation and Growth

☆ Lilies can be propagated in several ways;

☆ by division of the bulbs by growing-on bulbils which are adventitious bulbs formed on the stem

☆ by scaling, for which whole scales are detached from the bulb and planted to form a new bulb

☆ by seed; there are many seed germination patterns, which can be complex

☆ by micropropagation techniques (which include tissue culture); commercial quantities of lilies are often propagated in vitro and then planted out to grow into plants large enough to sell.

According to a study done by Anna Pobudkiewicz and Jadwiga the use of flurprimidol foliar spray helps aid in the limitation of stem elongation in oriental lilies.

Toxicity

Some Lilium species are toxic to cats. This is known to be so especially for Lilium longiflorum though other Lilium and the unrelated Hemerocallis can also cause the same symptoms. The true mechanism of toxicity is undetermined, but it involves damage to the renal tubular epithelium (composing the substance of the kidney and secreting, collecting, and conducting urine), which can cause acute renal failure. Veterinary help should be sought, as a matter of urgency, for any cat that is suspected of eating any part of a lily – including licking pollen that may have brushed onto its coat.

Culinary and Herb Uses

China

Lilium bulbs are starchy and edible as root vegetables, although bulbs of some species may be very bitter. The non-bitter bulbs of *Lilium lancifolium, Lilium pumilum,* and especially *Lilium brownii* (Chinese; pinyin: bihé gân) and *Lilium davidii* var. *unicolor* are grown on a large scale in China as a luxury or health food, and are most often sold in dry form for herb, the fresh form often appears with other vegetables. The dried bulbs are commonly used in the south to flavor soup. Lily flowers are also said to be efficacious in pulmonary affections, and to have tonic properties. Lily flowers and bulbs are eaten especially in the summer, for their perceived ability to reduce internal heat. They may be reconstituted and stir-fried, grated and used to thicken soup, or processed to extract starch. Their texture and taste draw comparisons with thepotato, although the individual bulb scales are much smaller. There are also species which are meant to be suitable for culinary and/or herb uses. There are five traditional lily species whose bulbs are certified and classified as "vegetable and non-staple foodstuffs" on the National geographical indication product list of China.

Culinary Use

Lilium brownii, Lilium brownii var. *viridulum, Lilium concolor, Lilium dauricum, Lilium davidii, Lilium distichum, Lilium lancifolium, Lilium martagon* var. *pilosiusculum,* ŽR'O *Lilium pumilum, Lilium rosthornii, Lilium speciosum* var. *gloriosoides.*

Herb Use

Lilium brownii, Lilium brownii var. *viridulum, Lilium concolor, Lilium dauricum, Lilium lancifolium, Lilium pumilum, Lilium rosthornii, Lilium speciosum* var. *gloriosoides, Lilium sulphureum.* And there are researches about the selection of new varieties of edible lilies from the horticultural cultivars, such as 'Batistero' and 'California' among 15 lilies in Beijing, and 'Prato' and 'Small foreigners' among 13 lilies in Ningbo.

Japan

Culinary Use

Yuri-ne (lily-root) is also common in Japanese cuisine, especially as an ingredient of chawan-mushi (savoury egg custard). The major lilium species cultivated as vegetable are *Lilium leichtlinii* var. *maximowiczii, Lilium lancifolium,* and *Lilium auratum.*

Herb Use

Lilium lancifolium, Lilium brownii var. *viridulum, Lilium brownii* var. *colchesteri, Lilium pumilum.*

Taiwan

Culinary Use

The parts of lilium species which are officially listed as food material are the flower and bulbs of *Lilium lancifolium* Thunb., *Lilium brownii* var. *viridulum* Baker, *Lilium pumilum* DC., *Lilium candidum* Loureiro. Most edible lily bulbs which can be purchased in a market are mostly imported from mainland China (only in the scale form, and most marked as *Lilium davidii* var. *unicolor*) and Japan (whole bulbs, should mostly be *Lilium leichtlinii* var. *maximowiczii*). There are already commercially available organic growing and normal growing edible lily bulbs. The varieties are selected by the Taiwanese Department of Agriculture from the Asiatic lily cultivars that are imported from the Netherlands; the seedling bulbs must be imported from the Netherlands every year.

Herb Use

Lilium lancifolium Thunb., *Lilium brownii* var. *viridulum* Baker, *Lilium pumilum* DC.

South Korea

Herb Use

The lilium species which are officially listed as herbs are *Lilium lancifolium* Thunberg; *Lilium brownii* var. *viridulun* Baker.

Not Lilium

The "lily" flower buds known as jînzhçn ("golden needles") in Chinese cuisine are actually from *Hemerocallis citrina*.

Orchidaceae

The Orchidaceae are a diverse and widespread family of flowering plants, with blooms that are often colourful and often fragrant, commonly known as the orchid family. Along with the Asteraceae, they are one of the two largest families of flowering plants. The Orchidaceae have about 28,000 currently accepted species, distributed in ca 763 genera. The determination of which family is larger is still under debate, because verified data on the members of such enormous families are continually in flux. Regardless, the number of orchid species nearly equals the number of bony fishes and is more than twice the number of bird species, and about four times the number of mammal species. The family also encompasses about 6–11 per cent of all seed plants. The largest genera are *Bulbophyllum* (2,000 species), *Epidendrum* (1,500 species), *Dendrobium* (1,400 species) and *Pleurothallis* (1,000 species).

The family also includes Vanilla (the genus of the vanilla plant), Orchis (type genus), and many commonly cultivated plants such as Phalaenopsis and Cattleya. Moreover, since the introduction of tropical species into cultivation in the 19th century, horticulturists have produced more than 100,000 hybrids and cultivars.

Description

Orchids are easily distinguished from other plants, as they share some very evident, shared derived characteristics, or "apomorphies". Among these are: bilateral symmetry of the flower (zygomorphism), many resupinate flowers, a nearly always highly modified petal (labellum), fused stamens and carpels, and extremely small seeds.

Stem and Roots

Germinating seeds of the temperate orchid *Anacamptis coriophora*. The protocorm is the first organ that will develop into true roots and leaves. All orchids are perennial herbs that lack any permanent woody structure. They can grow according to two patterns:

> ☆ **Monopodial:** The stem grows from a single bud, leaves are added from the apex each year and the stem grows longer accordingly. The stem of

orchids with a monopodial growth can reach several metres in length, as in Vanda and Vanilla.

☆ **Sympodial:** Sympodial orchids have a front (the newest growth) and a back (the oldest growth). The plant produces a series of adjacent shoots which grow to a certain size, bloom and then stop growing and are replaced. Sympodial orchids grow laterally rather than vertically, following the surface of their support. The growth continues by development of new leads, with their own leaves and roots, sprouting from or next to those of the previous year, as in Cattleya. While a new lead is developing, the rhizome may start its growth again from a so-called 'eye', an undeveloped bud, thereby branching. Sympodial orchids may have visible pseudobulbs joined by a rhizome, which creeps along the top or just beneath the soil.

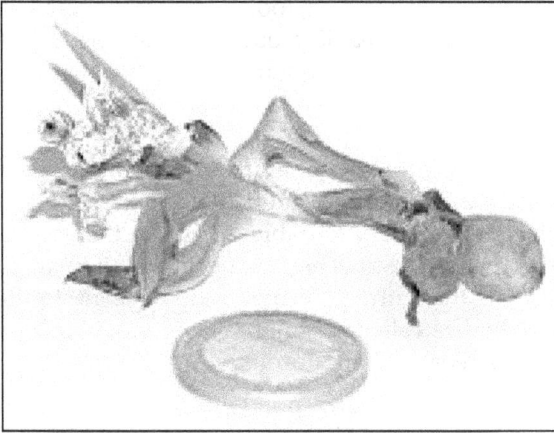

Anacamptis lactea showing the Two Tubers.

Terrestrial orchids may be rhizomatous or form corms or tubers. The root caps of terrestrial orchids are smooth and white. Some sympodial terrestrial orchids, such as Orchis and Ophrys, have two subterranean tuberous roots. One is used as a food reserve for wintry periods, and provides for the development of the other one, from which visible growth develops.

In warm and constantly humid climates, many terrestrial orchids do not need pseudobulbs. Epiphytic orchids, those that grow upon a support, have modified aerial roots that can sometimes be a few meters long. In the older parts of the roots, a modified spongy epidermis, called velamen, has the function to absorb humidity. It is made of dead cells and can have a silvery-grey, white or brown appearance. In some orchids, the velamen includes spongy and fibrous bodies near the passage cells, called tilosomes.

The cells of the root epidermis grow at a right angle to the axis of the root to allow them to get a firm grasp on their support. Nutrients mainly come from animal droppings and other organic detritus collecting among on their supporting surfaces.

The base of the stem of sympodial epiphytes, or in some species essentially the entire stem, may be thickened to form apseudobulb that contains nutrients and water for drier periods. The pseudobulb has a smooth surface with lengthwise grooves, and can have different shapes, often conical or oblong. Its size is very variable; in some small species of Bulbophyllum, it is no longer than two millimeters, while in the largest orchid in the world, Grammatophyllum speciosum (giant orchid), it can reach three meters. Some Dendrobium species have long, canelike pseudobulbs with short, rounded leaves over the whole length; some other orchids have hidden or extremely small pseudobulbs, completely included inside the leaves.

With ageing, the pseudobulb sheds its leaves and becomes dormant. At this stage, it is often called a backbulb. Backbulbs still hold nutrition for the plant, but then a pseudobulb usually takes over, exploiting the last reserves accumulated in the backbulb, which eventually dies off, too. A pseudobulb typically lives for about five years. Orchids without noticeable pseudobulbs are also said to have growths, an individual component of a sympodial plant.

Leaves

Like most monocots, orchids generally have simple leaves with parallel veins, although some Vanilloideae have reticulate venation. Leaves may be ovate, lanceolate, or orbiculate, and very variable in size on the individual plant. Their characteristics are often diagnostic. They are normally alternate on the stem, often folded lengthwise along the centre ("plicate"), and have no stipules. Orchid leaves often have siliceous bodies called stegmata in the vascular bundle sheaths (not present in the Orchidoideae) and are fibrous.

The structure of the leaves corresponds to the specific habitat of the plant. Species that typically bask in sunlight, or grow on sites which can be occasionally very dry, have thick, leathery leaves and the laminae are covered by a waxy cuticle to retain their necessary water supply. Shade-loving species, on the other hand, have long, thin leaves.

The leaves of most orchids are perennial, that is, they live for several years, while others, especially those with plicate leaves as in Catasetum, shed them annually and develop new leaves together with new pseudobulbs.

The leaves of some orchids are considered ornamental. The leaves of the *Macodes sanderiana*, a semiterrestrial or rock-hugging ("lithophyte") orchid, show a sparkling silver and gold veining on a light green background. The cordate leaves of *Psychopsis limminghei* are light brownish-green with maroon-puce markings, created by flower pigments. The attractive mottle of the leaves of lady's slippers from tropical and subtropical Asia (Paphiopedilum), is caused by uneven distribution of chlorophyll. Also, *Phalaenopsis schilleriana* is a pastel pink orchid with leaves spotted dark green and light green. The jewel orchid (*Ludisia discolor*) is grown more for its colorful leaves than its white flowers.

Some orchids, as *Dendrophylax lindenii* (ghost orchid), Aphyllorchis and Taeniophyllum depend on their green roots for photosynthesis and lack normally developed leaves, as do all of the heterotrophic species.

Orchids of the genus Corallorhiza (coralroot orchids) lack leaves altogether and instead wrap their roots around the roots of mature trees and use specialized fungi to harvest sugars.

Vanda Cultivar.

Flowers

The Orchidaceae are well known for the many structural variations in their flowers. Some orchids have single flowers, but most have a racemose inflorescence, sometimes with a large number of flowers. The flowering stem can be basal, that is, produced from the base of the tuber, like in Cymbidium, apical, meaning it grows from the apex of the main stem, like in Cattleya, or axillary, from the leaf axil, as in Vanda.

As an apomorphy of the clade, orchid flowers are primitively zygomorphic (bilaterally symmetrical), although in some genera like Mormodes, Ludisia, and Macodes, this kind of symmetry may be difficult to notice.

The orchid flower, like most flowers of monocots, has two whorls of sterile elements. The outer whorl has three sepals and the inner whorl has three petals. The sepals are usually very similar to the petals (thus called tepals, 1), but may be completely distinct.

The medial petal, called the labellum or lip (6), which is always modified and enlarged, is actually the upper medial petal; however, as the flower develops, the inferior ovary (7) or the pedicel usually rotates 180°, so that the labellum arrives at the lower part of the flower, thus becoming suitable to form a platform for pollinators. This characteristic, called resupination, occurs primitively in the family and is considered apomorphic, a derived characteristic all Orchidaceae share. The torsion of the ovary is very evident from the longitudinal section shown (below right). Some orchids have secondarily lost this resupination, *e.g.* Zygopetalum and *Epidendrum secundum*.

The normal form of the sepals can be found in Cattleya, where they form a triangle. In Paphiopedilum (Venus slippers), the lower two sepals are fused into

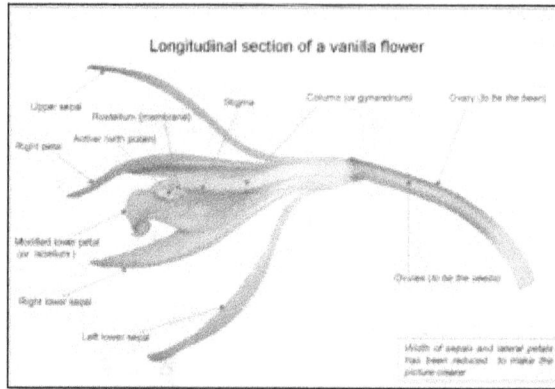

Longitudinal Section of a Flower of *Vanilla planifolia.*

a synsepal, while the lip has taken the form of a slipper. In Masdevallia, all the sepals are fused. Orchid flowers with abnormal numbers of petals or lips are called peloric. Peloria is a genetic trait, but its expression is environmentally influenced and may appear random.

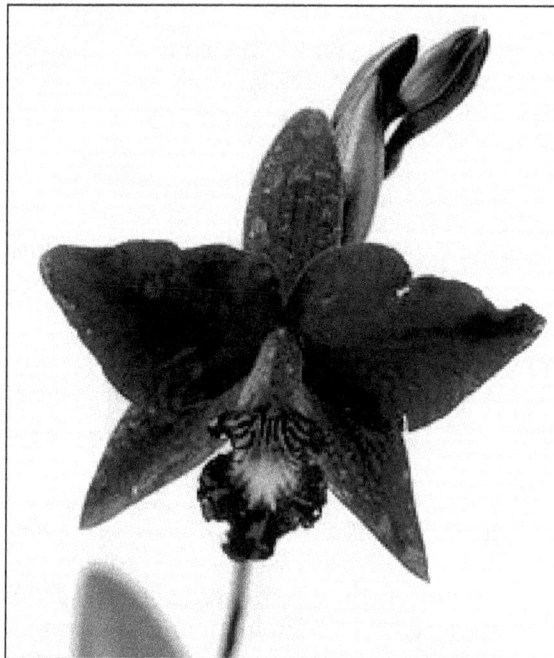

Laelia Cultivar Shows the Normal Form of Sepals.

Orchid flowers primitively had three stamens, but this situation is now limited to the genus Neuwiedia. Apostasia and the Cypripedioideae have two stamens, the central one being sterile and reduced to a staminode. All of the other orchids, the clade called Monandria, retain only the central stamen, the others being reduced to staminodes (4). The filaments of the stamens are always adnate (fused) to the style to form cylindrical structure called the gynostemium or column (2). In the primitive Apostasioideae, this fusion is only partial; in the Vanilloideae, it is more deep; in Orchidoideae and Epidendroideae, it is total. The stigma (9) is very asymmetrical, as all of its lobes are bent towards the centre of the flower and lie on the bottom of the column.

Pollen is released as single grains, like in most other plants, in the Apostasioideae, Cypripedioideae, and Vanilloideae. In the other subfamilies, which comprise the great majority of orchids, the anther (3), carries and two pollinia.

A pollinium is a waxy mass of pollen grains held together by the glue-like alkaloid viscin, containing both cellulosic strands and mucopolysaccharides. Each pollinium is connected to a filament which can take the form of a caudicle, as in Dactylorhizaor Habenaria, or a stipe, as in Vanda. Caudicles or stipes hold the pollinia to the viscidium, a sticky pad which sticks the pollinia to the body of pollinators.

At the upper edge of the stigma of single-anthered orchids, in front of the anther cap, is the rostellum (5), a slender extension involved in the complex pollination mechanism. As mentioned, the ovary is always inferior (located behind the flower). It is three-carpelate and one or, more rarely, three-partitioned, with parietal placentation (axile in the Apostasioideae). In 2011, *Bulbophyllum nocturnum* was discovered to flower nocturnally.

Fruits and Seeds

The ovary typically develops into a capsule that is dehiscent by three or six longitudinal slits, while remaining closed at both ends. The seeds are generally almost microscopic and very numerous, in some species over a million per capsule. After ripening, they blow off like dust particles or spores. They lack endosperm and must enter symbiotic relationships with various mycorrhizal basidiomyceteous fungi that provide them the necessary nutrients to germinate, so all orchid species aremycoheterotrophic during germination and reliant upon fungi to complete their lifecycles.

As the chance for a seed to meet a suitable fungus is very small, only a minute fraction of all the seeds released grow into adult plants. In cultivation, germination typically takes weeks.

Horticultural techniques have been devised for germinating orchid seeds on an artificial nutrient medium, eliminating the requirement of the fungus for germination and greatly aiding the propagation of ornamental orchids. The usual medium for the sowing of orchids in artificial conditions is agar agar gel combined with a carbohydrate energy source. The carbohydrate source can be combinations of discrete sugars or can be derived from other sources such as banana, pineapple, peach, or even tomato puree or coconut water. After the preparation of the agar agar medium, it is poured into test tubes or jars which are then autoclaved (or cooked in a pressure cooker) to sterilize the medium. After cooking, the medium begins to gel as it cools.

Pollination

The complex mechanisms which orchids have evolved to achieve cross-pollination were investigated by Charles Darwin and described in Fertilisation of Orchids (1862). Orchids have developed highly specialized pollination systems, thus the chances of being pollinated are often scarce, so orchid flowers usually remain receptive for very long periods, rendering unpollinated flowers long-lasting in cultivation. Most orchids deliver pollen in a single mass. Each time pollination succeeds, thousands of ovules can be fertilized.

Pollinators are often visually attracted by the shape and colours of the labellum. However, some Bulbophyllum species attract male fruit flies (Bactrocera spp.) solely via a floral chemical which simultaneously acts as a floral reward (*e.g.* methyl eugenol, raspberry ketone, or zingerone) to perform pollination.[8] The flowers may produce attractive odours. Although absent in most species, nectar may be produced in a spur of the labellum (8 in the illustration above), or on the point of the sepals, or in the septa of the ovary, the most typical position amongst the Asparagales.

In orchids that produce pollinia, pollination happens as some variant of the following sequence: when the pollinator enters into the flower, it touches a viscidium, which promptly sticks to its body, generally on the head or abdomen. While leaving the flower, it pulls the pollinium out of the anther, as it is connected to the viscidium by the caudicle or stipe. The caudicle then bends and the pollinium is moved forwards and downwards. When the pollinator enters another flower of the same species, the pollinium has taken such position that it will stick to the stigma of the second flower, just below the rostellum, pollinating it. The possessors of orchids may be able to reproduce the process with a pencil, small paintbrush, or other similar device.

Some orchids mainly or totally rely on self-pollination, especially in colder regions where pollinators are particularly rare. The caudicles may dry up if the flower has not been visited by any pollinator, and the pollinia then fall directly on the stigma. Otherwise, the anther may rotate and then enter the stigma cavity of the flower (as in *Holcoglossum amesianum*). The slipper orchid *Paphiopedilum parishii* reproduces by self-fertilization. This occurs when the anther changes from a solid to a liquid state and directly contacts the stigma surface without the aid of any pollinating agent or floral assembly.

The labellum of the Cypripedioideae is poke bonnet-shaped, and has the function of trapping visiting insects. The only exit leads to the anthers that deposit pollen on the visitor. In some extremely specialized orchids, such as the Eurasian genus Ophrys, the labellum is adapted to have a colour, shape, and odour which attracts male insects via mimicry of a receptive female. Pollination happens as the insect attempts to mate with flowers.

Many neotropical orchids are pollinated by male orchid bees, which visit the flowers to gather volatile chemicals they require to synthesize pheromonal attractants. Males of such species as Euglossa imperialis or *Eulaema meriana* have been observed to leave their territories periodically to forage for aromatic compounds, such as cineole, to synthesize pheromone for attracting and mating with females. Each type of orchid places the pollinia on a different body part of a different species of bee, so as to enforce proper cross-pollination. A rare achlorophyllous saprophytic orchid growing entirely underground in Australia, Rhizanthella slateri, is never exposed to light, and depends on ants and other terrestrial insects to pollinate it.

Catasetum, a genus discussed briefly by Darwin, actually launches its viscid pollinia with explosive force when an insect touches a seta, knocking the pollinator off the flower. After pollination, the sepals and petals fade and wilt, but they usually remain attached to the ovary.

Asexual Reproduction

Some species, such as Phalaenopsis, Dendrobium, and Vanda, produce offshoots or plantlets formed from one of the nodes along the stem, through the accumulation of growth hormones at that point. These shoots are known as keiki.

Taxonomy

The taxonomy of this family is in constant flux, as new studies continue to clarify the relationships between species and groups of species, allowing more taxa at several ranks to be recognized. The Orchidaceae is currently placed in the order Asparagales by the APG III system of 2009.

Five subfamilies are recognised. The cladogram below was made according to the APG system of 1998. It represents the view that most botanists had held up to that time. It was supported by morphological studies, but never received strong support in molecular phylogenetic studies.

Apostasioideae: 2 genera and 16 species, south-western Asia

Cypripedioideae: 5 genera and 130 species, from the temperate regions of the world, as well as tropical America and tropical Asia

Monandrae: Vanilloideae: 15 genera and 180 species, humid tropical and subtropical regions, eastern North America

Epidendroideae: more than 500 genera and more or less 20,000 species, cosmopolitan

Orchidoideae: 208 genera and 3,630 species, cosmopolitan

In 2015, a phylogenetic study showed strong statistical support for the following topology of the orchid tree, using 9 kb of plastid and nuclear DNA from 7 genes, a topology that was confirmed by a phylogenomic study in the same year.

Apostasioideae

> **Vanilloideae**

>> **Cypripedioideae**

>>> **Epidendroideae**

>>>> **Orchidoideae**

Evolution

A study in the scientific journal Nature has hypothesised that the origin of orchids goes back much longer than originally expected. An extinct species of stingless bee, Proplebeia dominicana, was found trapped in Miocene amber from about 15-20 million years ago. The bee was carrying pollen of a previously unknown orchid taxon, Meliorchis caribea, on its wings. This find is the first evidence of fossilised orchids to date and shows insects were active pollinators of orchids then. This extinct orchid, M. caribea, has been placed within the extant tribe Cranichideae, subtribe Goodyerinae (subfamily Orchidoideae).

Genetic sequencing indicates orchids may have arisen earlier, 76 to 84 million years ago during the Late Cretaceous. According to Mark W. Chase *et al.* (2001), the overall biogeography and phylogenetic patterns of Orchidaceae show they are even older and may go back roughly 100 million years.

Using the molecular clock method, it was possible to determine the age of the major branches of the orchid family. This also confirmed that the subfamily Vanilloideaeis a branch at the basal dichotomy of the monandrous orchids, and must have evolved very early in the evolution of the family. Since this subfamily occurs worldwide in tropical and subtropical regions, from tropical America to tropical Asia, New Guinea and West Africa, and the continents began to split about 100 million years ago, significant biotic exchange must have occurred after this split (since the age of Vanilla is estimated at 60 to 70 million years).

Genera

The following are amongst the most notable genera of the orchid family:

☆ *Aa*

☆ *Abdominea*

☆ *Acampe*

☆ *Acanthephippium*

☆ *Aceratorchis*

☆ *Acianthus*

☆ *Acineta*

☆ *Acrorchis*

☆ *Ada*

☆ *Aerangis*

☆ *Aeranthes*

☆ *Aerides*

☆ *Aganisia*

☆ *Agrostophyllum*

☆ *Amitostigma*

☆ *Anacamptis*

- ☆ *Ancistrochilus*
- ☆ *Angraecum*
- ☆ *Anguloa*
- ☆ *Ansellia*
- ☆ *Aorchis*
- ☆ *Aplectrum*
- ☆ *Arachnis*
- ☆ *Arethusa*
- ☆ *Armodorum*
- ☆ *Ascocenda*
- ☆ *Ascocentrum*
- ☆ *Ascoglossum*
- ☆ *Australorchis*
- ☆ *Auxopus*
- ☆ *Baptistonia*
- ☆ *Barkeria*
- ☆ *Barlia*
- ☆ *Bartholina*
- ☆ *Beloglottis*
- ☆ *Biermannia*
- ☆ *Bletilla*
- ☆ *Brassavola*
- ☆ *Brassia*
- ☆ *Bulbophyllum*
- ☆ *Calypso*
- ☆ *Catasetum*
- ☆ *Cattleya*
- ☆ *Cirrhopetalum*
- ☆ *Cleisostoma*
- ☆ *Clowesia*
- ☆ *Coelogyne*
- ☆ *Coryanthes*
- ☆ *Cymbidium*
- ☆ *Cyrtopodium*
- ☆ *Cypripedium*
- ☆ *Dactylorhiza*

- ☆ *Dendrobium*
- ☆ *Disa*
- ☆ *Dracula*
- ☆ *Encyclia*
- ☆ *Epidendrum*
- ☆ *Epipactis*
- ☆ *Eria*
- ☆ *Eulophia*
- ☆ *Gongora*
- ☆ *Goodyera*
- ☆ *Grammatophyllum*
- ☆ *Gymnadenia*
- ☆ *Habenaria*
- ☆ *Herschelia*
- ☆ *Ida*
- ☆ *Ionopsis*
- ☆ *Laelia*
- ☆ *Lepanthes*
- ☆ *Liparis*
- ☆ *Ludisia*
- ☆ *Lycaste*
- ☆ *Masdevallia*
- ☆ *Maxillaria*
- ☆ *Meliorchis*
- ☆ *Mexipedium*
- ☆ *Miltonia*
- ☆ *Mormodes*
- ☆ *Odontoglossum*
- ☆ *Oeceoclades*
- ☆ *Oncidium*
- ☆ *Ophrys*
- ☆ *Orchis*
- ☆ *Paphiopedilum*
- ☆ *Papilionanthe*
- ☆ *Paraphalaenopsis*
- ☆ *Peristeria*

☆ *Phaius*

☆ *Phalaenopsis*

☆ *Pholidota*

☆ *Phragmipedium*

☆ *Platanthera*

☆ *Pleione*

☆ *Pleurothallis*

☆ *Pomatocalpa*

☆ *Promenaea*

☆ *Pterostylis*

☆ *Renanthera*

☆ *Renantherella*

☆ *Restrepia*

☆ *Restrepiella*

☆ *Rhynchostylis*

☆ *Roezliella*

☆ *Saccolabium*

☆ *Sarcochilus*

☆ *Satyrium*

☆ *Selenipedium*

☆ *Serapias*

☆ *Sobralia*

☆ *Sophronitis*

☆ *Spiranthes*

☆ *Stanhopea*

☆ *Stelis*

☆ *Thrixspermum*

☆ *Tolumnia*

☆ *Trias*

☆ *Trichocentrum*

☆ *Trichoglottis*

☆ *Vanda*

☆ *Vanilla*

☆ *Yoania*

☆ *Zeuxine*

☆ *Zygopetalum*

Étymology

The type genus (*i.e.* the genus after which the family is named) is Orchis. The genus name comes from the Ancient Greek (órkhis), literally meaning "testicle", because of the shape of the twin tubers in some species of Orchis. The term "orchid" was introduced in 1845 by John Lindley in School Botany, as a shortened form of Orchidaceae.

Distribution

Orchidaceae are cosmopolitan, occurring in almost every habitat apart from glaciers. The world's richest diversity of orchid genera and species is found in thetropics, but they are also found above the Arctic Circle, in southern Patagonia, and two species of Nematoceras on Macquarie Island at 54° south.

The following list gives a rough overview of their distribution:

Oceania: 50 to 70 genera

North America: 20 to 26 genera

Tropical America: 212 to 250 genera

Tropical Asia: 260 to 300 genera

Tropical Africa: 230 to 270 genera

Europe and temperate Asia: 40 to 60 genera

Ecology

A majority of orchids are perennial epiphytes, which grow anchored to trees or shrubs in the tropics and subtropics. Species such as *Angraecum sororium* arelithophytes, growing on rocks or very rocky soil. Other orchids (including the majority of temperate Orchidaceae) are terrestrial and can be found in habitat areas such as grasslands or forest.

Some orchids, such as Neottia and Corallorhiza, lack chlorophyll, so are unable to photosynthesise. Instead, these species obtain energy and nutrients byparasitising soil fungi through the formation of orchid mycorrhizas. The fungi involved include those that form ectomycorrhizas with trees and other woody plants, parasites such as Armillaria, and saprotrophs. These orchids are known as myco-heterotrophs, but were formerly (incorrectly) described as saprophytes as it was believed they gained their nutrition by breaking down organic matter. While only a few species are achlorophyllous holoparasites, all orchids are myco-heterotrophic during germination and seedling growth, and even photosynthetic adult plants may continue to obtain carbon from their mycorrhizal fungi.

Uses

A flower of a Blc. Paradise Jewel 'Flame' hybrid orchid plant. Blooms of the Cattleyaalliance are often used in ladies' corsages.

Perfumery

The scent of orchids is frequently analysed by perfumers (using headspace technology and gas-liquid chromatography) to identify potential fragrance chemicals.

Horticulture

The other important use of orchids is their cultivation for the enjoyment of the flowers. Most cultivated orchids are tropical or subtropical, but quite a few which grow in colder climates can be found on the market. Temperate species available at nurseries include *Ophrys apifera* (bee orchid), *Gymnadenia conopsea* (fragrant orchid), *Anacamptis pyramidalis* (pyramidal orchid) and *Dactylorhiza fuchsii* (common spotted orchid).

Orchids of all types have also often been sought by collectors of both species and hybrids. Many hundreds of societies and clubs worldwide have been established. These can be small, local clubs such as the Sutherland Shire Orchid Society, or larger, national organisations such as the American Orchid Society. Both serve to encourage cultivation and collection of orchids, but some go further by concentrating on conservation or research.

The term "botanical orchid" loosely denotes those small-flowered, tropical orchids belonging to several genera that do not fit into the "florist" orchid category. A few of these genera contain enormous numbers of species. Some, such as Pleurothallis and Bulbophyllum, contain approximately 1700 and 2000 species, respectively, and are often extremely vegetatively diverse. The primary use of the term is among orchid hobbyists wishing to describe unusual species they grow, though it is also used to distinguish naturally occurring orchid species from horticulturally created hybrids.

Use as Food

Further Information: Vanilla

The dried seed pods of one orchid genus, Vanilla (especially *Vanilla planifolia*), are commercially important as a flavouring in baking, for perfume manufacture and aromatherapy. The underground tubers of terrestrial orchids [mainly Orchis mascula (early purple orchid)] are ground to a powder and used for cooking, such as in the hot beverage salep or in the Turkish frozen treat dondurma. The name salep has been claimed to come from the Arabic expression ?asyu al-tha'lab, "fox testicles", but it appears more likely the name comes directly from the Arabic name sa?lab. The similarity in appearance to testes naturally accounts for salep being considered an aphrodisiac.

The dried leaves of *Jumellea fragrans* are used to flavour rum on Reunion Island. Some saprophytic orchid species of the group Gastrodia produce potato-like tubers and were consumed as food by native peoples in Australia and can be successfully cultivated, notably *Gastrodia sesamoides*. Wild stands of these plants can still be found in the same areas as early aboriginal settlements, such as Ku-ring-gai Chase National Park in Australia. Aboriginal peoples located the plants in habitat by observing where bandicoots had scratched in search of the tubers after detecting the plants underground by scent.

Traditional Medicinal Uses

Orchids have been used in traditional medicine in an effort to treat many diseases and ailments. They have been used as a source of herbal remedies in China since 2800 BC. *Gastrodia elata* is one of the three orchids listed in the earliest known Chinese Materia Medica (Shennon bencaojing) (c. 100 AD). Theophrastus mentions orchids in his Enquiry into Plants (372–286 BC).

Cultural Symbolism

Orchids have many associations with symbolic values. For example, the orchid is the City Flower of Shaoxing, China. Cattleya mossiae is the national Venezuelan flower, while Cattleya trianae is the national flower of Colombia. Vanda 'Miss Joaquim' is the national flower of Singapore. Guarianthe skinneri is the national flower ofCosta Rica. Orchids native to the Mediterranean are depicted on the Ara Pacis in Rome, until now the only known instance of orchids in ancient art, and the earliest in European art.

Aster Flower History

Celebrate September birthdays with asters, the "herb of Venus." Named for the stars because of the shape of its blossoms, the aster symbolizes elegance and refinement. It is also considered a love charm and was thought to have mystical powers that could ward of evil serpents. It is said they came to be when the Greek goddess Asterea cried from the lack of stars on earth. Where her tears fell, asters grew.

Cuttings from these plants were laid on the graves of French soldiers to symbolize the longing for a revised ending to their battle. This flower's unique beauty continues to perpetuate present-day folklore and have even been used medicinally. These ancient wildflowers are closely related to mums, marigolds, daisies, and even artichokes. They are considered native plants in many parts of the U.S., but the China aster is the most common variety used by florists.

Asters have previously gone by other names, such as "Michaelmas Daisy" and "starwort." Commonly sporting purple or blue blossoms, these flowers are popular in garden borders and floral arrangements. There are both annual and perennial varieties. Often seen around a cornucopia announcing a plentiful harvest, these are fall-blooming plants of half a foot to three feet in height. Like most stars, asters enjoy the spotlight, so grow them in a sunny location in just about any hardiness zone. Soil should be kept moist, especially in warmer weather.

Aster is a genus of flowering plants in the family Asteraceae. Its circumscription has been narrowed, and it now encompasses around 180 species, all but one of which are restricted to Eurasia; many species formerly in Aster are now in other genera of the tribe Astereae.

Circumscription

The genus Aster once contained nearly 600 species in Eurasia and North America, but after morphologic and molecular research on the genus during the 1990s, it was decided that the North American species are better treated in a series of other related genera. After this split there are roughly 180 species within the genus, all but one being confined to Eurasia. The name Aster comes from the Ancient Greekword (astér), meaning "star", referring to the shape of the flower head. Many species and a variety of hybrids and varieties are popular as garden plants because of their attractive and colourful flowers. Aster species are used as food plants by thelarvae of a number of Lepidoptera species—see list of Lepidoptera that feed on Aster. Asters can grow in all hardiness zones.

The genus Aster is now generally restricted to the Old World species, with Aster amellus being the type species of the genus, as well as of the family Asteraceae. The New World species have now been reclassified in the genera Almutaster, Canadanthus, Doellingeria, Eucephalus, Eurybia, Ionactis, Oligoneuron,Oreostemma, Sericocarpus and Symphyotrichum, though all are treated within the tribe Astereae. Regardless of the taxonomic change, all are still widely referred to as "asters" (popularly "Michaelmas daisies" because of their typical blooming period) in the horticultural trades. See the List of Aster synonyms for more information.

Some common North American species that have now been moved are:

☆ *Aster breweri* (now *Eucephalus breweri*), Brewer's aster

☆ *Aster cordifolius* (now *Symphyotrichum cordifolium*), blue wood aster

☆ *Aster dumosus* (now *Symphyotrichum dumosum*), New York aster

☆ *Aster divaricatus* (now *Eurybia divaricata*), white wood aster

☆ *Aster ericoides* (now *Symphyotrichum ericoides*), heath aster

☆ *Aster laevis* (now *Symphyotrichum laeve*), smooth aster

☆ *Aster lateriflorus* (now *Symphyotrichum lateriflorum*), "Lady in Black", calico aster

☆ *Aster novae-angliae* (now *Symphyotrichum novae-angliae*), New England aster

☆ *Aster novi-belgii* (now *Symphyotrichum novi-belgii*), New York aster

☆ *Aster peirsonii* (now *Oreostemma peirsonii*), Peirson's aster

☆ *Aster protoflorian* (now *Symphyotrichum pilosum*), frost aster

☆ *Aster scopulorum* (now *Ionactis alpina*), lava aster

☆ *Aster sibiricus* (now *Eurybia sibirica*), Siberian aster

The "China aster" is in the related genus Callistephus.

Species

Aster alpinus is the only species of Aster (*Sensu stricto*) that grows natively in North America; it is found in mountains across the Northern Hemisphere.

In the United Kingdom, there are only two native members of the genus: goldilocks, which is very rare, and Aster tripolium, the sea aster. *Aster alpinus* spp. vierhapperiis the only species native to North America.

Some common species are:

☆ Aster alpinus, Alpine aster

☆ Aster amellus, European Michaelmas daisy or Italian aster

☆ Aster linosyris, goldilocks aster

☆ Aster scaber

☆ Aster tataricus, Tatarian aster

☆ Aster tongolensis

☆ Aster tripolium, sea aster

Hybrids and Cultivars

(those marked agm have gained the Royal Horticultural Society's Award of Garden Merit:

☆ Aster × frikartii (A. amellus × A. thomsonii) Frikart's aster

❑ Aster × frikartii 'Mönch'agm

❑ × frikartii 'Wunder von Stäfa'agm

☆ 'Kylie' (A. novae-angliae 'Andenken an Alma Pötschke' × A. ericoides 'White heather')

☆ 'Ochtendgloren'agm (A. pringlei hybrid)

☆ 'Photograph'agm

China aster is a popular winter annual grown for its showy flowers. Potted plants of china aster are used for various decorative/display purposes. Popular cultivars of china aster are Pot'n Patio Blue, Pot'n Patio Pink, Giant Massagno, Comet, Totem Pole, Benihanabi, White Kurenai, Ariake Pompon, American Beauty, Queen of the Market, Crego Azure, Crego, Ostrich Plume, Pompon, Invincible, Early Bird, Blue Wonder, Cactus Flowered, Giants of California, Bouquet Powderpuff, Bouquet Mid Blue, Bouquet White and Azure Blue. Major Indian Varieties of china aster include Poornima, Violet Cushion, Kamini, Phule Ganesh White Shashank, Phule Ganesh Violet, Phule Ganesh Pink, and Phule Ganesh Purple.

Propagation

China aster can be propagated by seeds. Their sowing time depends upon climate. It varies in different parts of India. In the northern plains, late-flowering types of China aster are sown during August–September. In south India, September–October is ideal time for sowing annuals. The seeds of annual flowers are sown in nursery beds, earthen pots, seed pans or wooden seed trays. The seed compost should consist of one part each of garden soil, coarse sand, farmyard manure and leaf-mould. For preparing the nursery beds, the soil should be dug up thoroughly and sufficient farmyard manure, should be mixed in soil. Raised nursery beds of convenient size (normally 60cm wide and 15cm high) should be prepared. If soil is heavy, some quantity of sand may be added. It is better if the soil of nursery bed or earthen pots is sterilized with 2 per cent formalin. For this, soil is drenched with formalin solution and is covered with polythene sheet for 45hr. Afterwards, the polythene is removed and soil is dried before sowing the seeds. Before sowing, the seeds should be treated with Cerason (0.2 per cent) and Captaf (0.2 per cent) to prevent the seedlings from damping off disease. The seeds should be sown thinly and evenly as thick sowing causes damping off of seedlings. In nursery beds, the seeds are sown in rows spaced 6cm apart. Then, they are covered with finely sieved leaf-mould. Watering is to be done with a watering can having a fine rose both in beds and pots. In beds, when germination is over, water is given for proper moisture. Thereafter, the beds should be kept weed-free. Cultivation Practices Planting The seedlings are transplanted 25 days after sowing at 4-leaf stage. Before transplanting, seedlings are hardened off by withholding water for 1 or 2 days or by exposing them gradually to sunlight. Transplanting is, generally, done either on a cloudy day or in the evening. Transplanting in evening is good as the night cool temperature is beneficial for the establishment. Light watering every day in early morning or late in the afternoon is required for about a week for proper establishment of the seedlings. Dwarf annuals like china aster should be planted at a distance of 30cm × 30cm.

Manuring and Fertilization

The farmyard manure or compost @ 3kg/m2 is mixed in the soil. Chemical fertilizers—20g urea, 60–120g superphosphate and 30–60g muriate of potash/m^2 should also be added. Half quantity of urea and full of superphosphate and muriate of potash should be applied at the time of bed preparation. The remaining quantity of urea must be applied one month after transplantation. Spraying plants with 2 per cent urea twice or thrice is beneficial for good growth and flowering. Fertilizers should never come in the direct contact of the foliage since they cause scorching. Fertilizers should never be applied in the pot-grown annual flowers. However, some readymade pot-mixtures can be used. The pot-mixture should consist of 2 parts of garden soil and one part each of coarse sand and farmyard manure. Instead of fertilizers, it is better if pot-grown plants are given liquid feeding. The liquid manure is prepared by fermenting 1–2kg each of fresh cowdung and oil cake in 10 litres of water in a drum for one week. It is diluted to tea colour and sieved with the help of a muslin cloth. It is applied @ 500–1000ml/pot at 7–10 days intervals.

Growth and Flowering

Environmental factors and various cultural conditions affect growth and flowering of many annuals. China aster (*Callistephus chinensis*) is a typical short-day plant. Growth ratarding or promoting substances play a major role in getting dwarf plants or higher flower yield. GA3 (100–400ppm) gives beneficial results in respect of growth and flowering in China aster. Application of these growth substances is more effective at vegetative stage. Aftercare After transplanting, beds are weeded, hoed and watered regularly. As soon as seedlings are established in beds, pinching is done for making the plants bushy. Sometimes, seedlings of China aster produce flower buds at an early stage. These buds should be removed as soon as they appear. The number of buds/stem is reduced by disbudding the axillary buds, if large blooms are desired. Irrigation Little water is needed everyday up to 7–10 days after transplantation. When the seedlings start new growth, profuse watering once or twice a week is required in beds. Later, frequency and quantity of watering depend upon soil and season. In lighter soils, more frequent irrigation is needed than that in heavy soils. The season of planting also determines the frequency of irrigation. During summer season, irrigation should be done at weekly intervals in beds, while at 10–12 days intervals in winters. Irrigation during rainy season depends upon prevailing weather conditions. Potted plants need daily watering during summer, whereas on alternate days in winter.

Harvesting and Postharvest Management

Most of the annual flowers are grown for garden display purpose in various ways. However, China aster is grown commercially for cut flower or loose flower purpose. Their flowers are harvested when they are fully open and are sold in the local markets. China aster flowers are cut along with their stems when they develop their original colour. The flowers, in general, are cut either late in the afternoon or very early in the morning. After harvesting, cut flowers should be put in a bucket of water filled up to one-fourth of the volume as it helps in their recovery from the shock of being cut away from the plant. As far as possible, the freshly opened flowers should be cut as freshness enhances their shelf-life. Since china asters are used as cut flowers, proper postharvest management is necessary for prolonging their vase-life. The flowers are graded according to stem length, flower size, flower shape, flower colour and freshness. The cut flowers/loose flowers are marketed in local markets. If flowers are not sold the day they are harvested, to store them in a cold storage is imperative. Various formulations are used to prolong the vase-life of cut flowers. In China aster, a preservative containing 60g/litre sucrose + 250mg/litre 8-HQS + 70mg/litre CCC + 50mg/litre $AgNO_3$ extends their vase-life.

Gerbera Daisies

If you have the gift of courage to send gerbera daisies are sure your selection. Distinguished by large flowering heads that closely resemble those of sunflowers, gerbera daisies come in a vibrant rainbow of colors. Bright pink, snow white, sunny yellow and ruby ??red are just a few of the beautiful colors that these happy flowers boast.

The gerbera daisy was discovered in 1884 near Barberton, South Africa, by Scotsman Robert Jameson. While the flower scientific name *Gerbera jamesonii*, remembers the name of its founder, the significance of its common name from the German naturalist Traugott Gerber. Breeding programs that started in England in 1890 improved the flower quality and color variations. The gerbera daisy's popularity soon traveled to growers in the Netherlands, together with Colombia is the main distributor of the cut of the flower version today. The gerbera is currently the fifth most popular flower in the world behind the rose, carnations, chrysanthemums and tulips.

The meaning of gerbera daisies stem from those attributed to the general daisy family. These meanings are innocence and purity, and daisies are also a classic symbol of beauty. However, the gerbera variety holds an extra sense of cheerfulness, which stems from the assortment of colors available. A varied bouquet of gerbera daisies can lift the spirit and sending one is an ideal way to brighten someone's day too. The vast multitude of available varieties has helped the gerbera daisy become a favorite choice for many different occasions - birthdays, get well, congratulations, thank you, and much more.

There are many types of flowers that can help to our thoughts and feelings for our loved ones, but the gerbera daisy and its meaning stand out as one of the most pronounced bright and cheerful. With its bold and striking appearance, the gerbera daisy become the most highly prized daisy variety. When you send gerbera daisies, know that these fresh flowers a deep meaningful message across and make a vivid impression.

Gerbera L. is a genus of plants in the Asteraceae (daisy family). It was named in honour of German botanist and medical doctor Traugott Gerber | (1710-1743) who travelled extensively in Russia and was a friend of Carl Linnaeus. Gerbera is native to tropical regions of South America, Africa and Asia. The first scientific description of a Gerbera was made by J.D. Hooker in Curtis's Botanical Magazine in 1889 when he described *Gerbera jamesonii*, a South African species also known as Transvaal daisy or Barberton Daisy. Gerbera is also commonly known as the African Daisy.

Gerbera species bear a large capitulum with striking, two-lipped ray florets in yellow, orange, white, pink or red colours. The capitulum, which has the appearance of a single flower, is actually composed of hundreds of individual flowers. Themorphology of the flowers varies depending on their position in the capitulum. The flower heads can be as small as 7 cm (Gerbera mini 'Harley') in diameter or up to 12 cm (Gerbera 'Golden Serena').

Gerbera is very popular and widely used as a decorative garden plant or as cut flowers. The domesticated cultivars are mostly a result of a cross between *Gerbera jamesonii* and another South African species *Gerbera viridifolia*. The cross is known as Gerbera hybrida. Thousands of cultivars exist. They vary greatly in shape and size. Colours include white, yellow, orange, red, and pink. The centre of the flower is sometimes black. Often the same flower can have petals of several different colours.

Gerbera is also important commercially. It is the fifth most used cut flower in the world (after rose, carnation, chrysanthemum, and tulip). It is also used as a model organism in studying flower formation. Gerbera contains naturally occurring coumarin derivatives. Gerbera is a tenderperennial plant. It is attractive to bees, butterflies and/or birds, but resistant to deer. Their soil should be kept moist but not soaked.

Species

- ☆ *Gerbera ambigua*
- ☆ *Gerbera aurantiaca : Hilton Daisy*
- ☆ *Gerbera bojeri*
- ☆ *Gerbera bonatiana*
- ☆ *Gerbera connata*
- ☆ *Gerbera cordata*
- ☆ *Gerbera crocea*
- ☆ *Gerbera curvisquama*
- ☆ *Gerbera delavayi*
- ☆ *Gerbera diversifolia*
- ☆ *Gerbera elliptica*
- ☆ *Gerbera emirnensis*
- ☆ *Gerbera galpinii*
- ☆ *Gerbera gossypina*
- ☆ *Gerbera hypochaeridoides*
- ☆ *Gerbera jamesonii : Barberton Daisy, Gerbera Daisy, Transvaal Daisy*
- ☆ *Gerbera kunzeana*
- ☆ *Gerbera latiligulata*
- ☆ *Gerbera leandrii*
- ☆ *Gerbera leiocarpa*
- ☆ *Gerbera leucothrix*
- ☆ *Gerbera lijiangensis*
- ☆ *Gerbera linnaei*
- ☆ *Gerbera macrocephala*
- ☆ *Gerbera maxima*
- ☆ *Gerbera nepalensis*
- ☆ *Gerbera nivea*
- ☆ *Gerbera parva*
- ☆ *Gerbera perrieri*
- ☆ *Gerbera petasitifolia*
- ☆ *Gerbera piloselloides*
- ☆ *Gerbera pterodonta*
- ☆ *Gerbera raphanifolia*
- ☆ *Gerbera ruficoma*

☆ *Gerbera saxatilis*

☆ *Gerbera serotina*

☆ *Gerbera serrata*

☆ *Gerbera tomentosa*

☆ *Gerbera viridifolia*

☆ *Gerbera wrightii*

Primary Significance

The fifth most popular flower in the world, *Gerbera daisies* can mean innocence, purity, and cheerfulness. These large daisy variations come in a number of vibrant colors, and sending them is the perfect way to brighten someone's day.

When you want to send the gift of cheer, Gerbera daisies are your sure choice. Distinguished by large flowering heads that closely resemble those of sunflowers, Gerbera daisies come in a vibrant rainbow of colors. Bright pink, snow white, sunny yellow, and ruby red are just a few of the gorgeous colors that these happy flowers boast. The Gerbera daisy was discovered in 1884 near Barberton, South Africa, by Scotsman Robert Jameson. While the flower's scientific name, Gerbera jamesonii, recollects the name of its founder, the meaning of its common name draws from German naturalist Traugott Gerber. Breeding programs that began in England in 1890 enhanced the flower's quality and color variations.

The Gerbera daisy's popularity soon traveled to growers in the Netherlands which, along with Columbia, is the primary distributor of the flower's cut version today. The Gerbera currently ranks as the fifth most popular flower in the world behind the rose, carnations, chrysanthemum, and tulip. Their meanings stem from those attributed to the general daisy family and include innocence and purity, as well as being a classic symbol of beauty. However, the Gerbera variety holds an added meaning of cheerfulness, which is attributed to their perky variety of colors. An assorted bouquet of Gerbera daisies can quickly lift the spirit and are an ideal way to brighten someone's day.

The sheer multitude of available colors has helped the Gerbera daisy become a favorite choice for many different occasions from birthdays, get well, congratulations, thank you and many more. There are many types of flowers which can help to express our thoughts and feelings for our loved ones, but the Gerbera daisy and its meaning stand out as one of the most distinctly bright and merry. With its bold and striking appearance, the Gerbera daisy has become the most highly-prized daisy variety. When you send Gerbera daisies, know that these fresh flowers convey a deeply meaningful message and make a lively lasting impression.

Carnation Flower

The carnation is also known as *Dianthus caryophyllus* or Chinese pink. The carnation traditionally celebrates fascination and admiration, although different colors of the carnation flower convey different meanings. White carnations, for example, wish "good luck," while red carnations denote passion, and pink carnations portray reminiscence or a mother's love. Striped carnations, on the other hand, are said to signal refusal. These single flowers have a long stem of 18-24 inches that supports many petals also seen in green, yellow, frosted, and multi-colored.

Legend has it that the first carnation was the earthly manifestation of Mother Mary's undying love for her son, Jesus, resultant from a tear splashed upon the ground under the cross. Carnation flowers are in season year-round to help commemorate a wide array of emotions. These flowers, after all, are the stuff boutonnieres are made of. Carnations are also not difficult to grow at home.

There are over 300 varieties of Carnations *Dianthus* because hybrids occur naturally due to cross-pollination in the wild. Carnations can be grown from seed planted indoors or in the garden and will also fill out a container beautifully. Carnations like to be in well-drained soil and full sun. Annual varieties are most common while Perennial varieties should be mulched like crazy but are typically hardy. Neither flowers until the second year, but it's easy to forge their multi-blossomed stems into a single long-stem carnation. As with many things of beauty and intrigue, Carnations are poisonous if ingested.

Carnation flowers celebrate a rich history indeed. They can be traced all the way back to Greek ad Roman times. More recently, the British National Carnation Society was established in 1949 to celebrate this wonderful flower. The BNCS is the offspring of the National Society for Exhibition ofCarnations and Picotees, which was founded in 1850!

Primary Significance

The carnation, scientifically known as Dianthus caryophyllus, is a historically rich and meaningful flower choice. With its scientific name dianthus roughly translating to "flower of love" or "flower of the gods", depending on the source, this flower is one that has been revered for centuries. One of the world's oldest cultivated flowers, the carnation is appreciated for its ruffled appearance, clove-like scent, and extended blooming period.

The carnation's history dates back to ancient Greek and Roman times, when it was used in art and d?cor. Christians believe that the first carnation bloomed on earth when Mary wept for Jesus as he carried his cross. Carnations in these early times were predominantly found in shades of pale pink and peach, but over the years the palette of available colors has grown to include red, yellow, white, purple, and even green. Throughout so many centuries of change, the popularity of the carnation has remained undiminished. The fact that the carnation continues to endure is a testament to its vast appeal.

The meanings of carnations include fascination, distinction, and love. Like many other flowers, different messages can also be expressed with the flower's different color varieties. Light red carnations, for example, are often used to convey admiration, whereas the dark red version expresses deeper sentiments of love and affection. White carnations are associated with purity and luck, and pink carnations are often given as a sign of gratitude. In the early part of the 20th century, carnations became the official flower of Mother's Day in addition finding particular significance in many other cultures worldwide.

To this day, carnations remain a favorite flower choice for many different occasions. They are immediately recognizable flowers, and they possess a charm and allure that continues to captivate people around the globe. In fact, in many parts of the world, the popularity of carnations surpasses that of any other flower including roses. The powerful sentiments these flowers can express are a perfect complement to their classic beauty and long-lasting freshness. By retaining its status as a floral mainstay for such a long time, the carnation has proven itself to be a lasting flower in more ways than one.

Carnations

According to a Christian legend, Carnations first appeared on earth as Jesus carried the Cross. Carnations sprang up from where the Virgin Mary's tears fell as she cried over her son's plight.

Miss. Anna Jarvis (The Founder of Mother's Day) used Carnations at the first Mother's Day **celebration because Carnations were her mother's favorite flower.**

Of the several kinds of Carnations, the three most common are the annual carnations, border carnationsand perpetual-flowering carnations.

Carnations are also commonly referred to by their scientific name, "Dianthus", the name given by the Greek botanist Theopharastus. Carnations got the name Dianthus from two Greek Words - "dios", referring to the god Zeus, and "anthos", meaning flower. Carnations are thus known as the "The Flowers of God".

Kingdom

Plantae

Division

Magnoliophyta

Class

Magnoliopsida

Order

Caryophyllales

Family

Caryophyllaceae

Genus

Dianthus

Carnations - Meanings

Another reason why carnations have become popular is because they come in numerous colors and each color of carnation has a different meaning. Some of these meanings are listed below.

Carnations	What they Mean
Carnations in general	Fascination, Woman's Love
Pink Carnations	Mother's Love
Light red Carnations	Admiration
Dark red Carnations	Deep Love and a Woman's Affection
White Carnations	Pure Love and Good Luck
Striped Carnations	Regret, Refusal
Green Carnations	St. Patrick's Day
Purple Carnations	Capriciousness
Yellow Carnation	Disappointment, Dejection

It is a good idea to check the meaning of the particular color or type of carnation before you gift them to someone.

Some Interesting Facts about Carnations

Carnations Wxpress Love, Fascination and Distinction

☆ Carnations are native to Eurasia.

☆ Historically, Carnations are known to have been used for the first time by Greeks and Romans in garlands.

☆ Carnations are exotic to Australia but have been grown commercially as a flower crop since 1954.

☆ Carnation blooms last a long time even after they are cut.

☆ Carnation flowers have become symbolic of mother's love and also of Mother's Day. Learn why you should select carnations as **Mother's day flowers.**

About the Carnation Flower and Plant

The single flowers of the Carnations species, Dianthus caryophyllus have 5 petals and vary from white to pink to purple in color. Border Carnation cultivars may have double flowers with as many as 40 petals.

When grown in gardens, Carnations grow to between 6 and 8.5 cm in diameter. Petals on Carnations are generally clawed or serrated.

Carnations are bisexual flowers and bloom simply or in a branched or forked cluster. The stamens on Carnations can occur in one or two whorls, in equal number or twice the number of the petals.

The Carnation leaves are narrow and stalk less and their color varies from green to grey-blue or purple. Carnations grow big, full blooms on strong, straight stems.

Types of Carnations

☆ **Carnation cultivars are mainly of three types:**

☆ **Large floweredCarnations** - one large flower per stem.

☆ **Spray Carnations (Mini Carnations** - with lots of smaller flower.

☆ **Dwarf flowered Carnations** - several small flowers on one stem.

Growing Carnations

☆ Carnations grow readily from cuttings made from the suckers that form around the base of the stem, the side shoots of the flowering stem, or the main shoots before they show flower-buds.

☆ The cuttings from the base make the best plants in most cases.

☆ These cuttings may be taken from a plant at any time through fall or winter, rooted in sand and potted up.

☆ They may be put in pots until the planting out time in spring, which is usually in April or in any time when the ground is ready to be handled.

☆ The soil should be deep, friable and sandy loam.

Carnation Plant Care

☆ Carnations need some hours of full sun each day and should be kept moist.

☆ Avoid over-watering as this may tend to turn the foliage yellow.

☆ Spent flowers should be removed promptly to promote continued blooming.

☆ The quality of the bloom depends on the soil and irrigation aspects for growing carnations.

☆ Those who grow carnations should know the importance of pinching, stopping and disbudding.

☆ At the time of plucking carnations, leave three to four nodes at the base and remove the stem.

☆ The plant foliage should not be exposed to the direct heat of a stove or the sun.

Dahlias

Dahlia is a genus of bushy, tuberous, herbaceousperennial plants native to Mexico. A member of the Asteraceae (or Compositae, dicotyledonous plants, related species include the sunflower, daisy, chrysanthemum, and zinnia. There are 42 species of dahlia, with hybrids commonly grown as garden plants. Flower forms are variable, with one head per stem; these can be as small as 5 cm (2 in) diameter or up to 30 cm (1 ft) ("dinner plate"). This great variety results from dahlias being octoploids—that is, they have eight sets of homologouschromosomes, whereas most plants have only two. In addition, dahlias also contain many transposons—genetic pieces that move from place to place upon an allele—which contributes to their manifesting such great diversity.

The stems are leafy, ranging in height from as low as 30 cm (12 in) to more than 1.8–2.4 m (6–8 ft). The majority of species do not produce scented flowers orcultivars. Like most plants that do not attract pollinating insects through scent, they are brightly colored, displaying most hues, with the exception of blue.

The dahlia was declared the national flower of Mexico in 1963. The tubers were grown as a food crop by the Aztecs, but this use largely died out after the Spanish Conquest. Attempts to introduce the tubers as a food crop in Europe were unsuccessful.

Description

Perennial plants, with mostly tuberous roots. While some have herbaceous stems, others have stems which lignify in the absence of secondary tissue and resprout following winter dormancy, allowing further seasons of growth. As a member of the Asteraceae the flower head is actually a composite (hence the older name Compositae) with both central disc florets and surrounding ray florets. Each floret is a flower in its own right, but is often incorrectly described as a petal, particularly by horticulturalists. The modern name Asteraceae refers to the appearance of a star with surrounding rays.

Taxonomy

History

Early History

Spaniards reported finding the plants growing in Mexico in 1525, but the earliest known description is by Francisco Hernández, physician to Philip II, who was ordered to visit Mexico in 1570 to study the "natural products of that country". They were used as a source of food by the indigenous peoples, and were both gathered in the wild and cultivated. The Aztecs used them to treat epilepsy, and employed the long hollow stem of the (Dahlia imperalis) for water pipes. The indigenous peoples variously identified the plants as "Chichipatl" (Toltecs) and "Acocotle" or "Cocoxochitl" (Aztecs). From Hernandez' perception of Aztec, to Spanish, through various other translations, the word is "water cane", "water pipe", "water pipe flower", "hollow stem flower" and "cane flower". All these refer to the hollowness of the plants' stem.

Hernandez described two varieties of dahlias (the pinwheel-like Dahlia pinnata and the huge *Dahlia imperialis*) as well as other medicinal plants of New Spain. Francisco Dominguez, a Hidalgo gentleman who accompanied Hernandez on part of his seven-year study, made a series of drawings to supplement the four volume report. Three of his drawings showed plants with flowers: two resembled the modern bedding dahlia, and one resembled the species Dahlia merki; all displayed a high degree of doubleness. In 1578 the manuscript, entitled Nova Plantarum, Animalium et Mineralium Mexicanorum Historia, was sent back to the Escorial in Madrid; they were not translated into Latin by Francisco Ximenes until 1615. In 1640, Francisco Cesi, President of the Academia Linei of Rome, bought the Ximenes translation, and after annotating it, published it in 1649-1651 in two volumes as

Rerum Medicarum Novae Hispaniae Thesaurus Seu Nova Plantarium, Animalium et Mineraliuím Mexicanorum Historia. The original manuscripts were destroyed in a fire in the mid-1600s.

European Introduction

In 1787, the French botanist Nicolas-Joseph Thiéry de Menonville, sent to Mexico to steal the cochineal insect valued for its scarlet dye, reported the strangely beautiful flowers he had seen growing in a garden in Oaxaca. In 1789, Vicente Cervantes, Director of the Botanical Garden at Mexico City, sent "plant parts" to Abbe Antonio José Cavanilles, Director of the Royal Gardens of Madrid. Cavanilles flowered one plant that same year, then the second one a year later. In 1791 he called the new growths "Dahlia" for Anders Dahl. The first plant was called Dahlia pinnata after its pinnate foliage; the second, Dahlia rosea for its rose-purple color. In 1796 Cavanilles flowered a third plant from the parts sent by Cervantes, which he named Dahlia coccinea for its scarlet color.

In 1798, Cavanilles sent D. Pinnata seeds to Parma, Italy. That year, the Marchioness of Bute, wife of The Earl of Bute, the English Ambassador to Spain, obtained a few seeds from Cavanilles and sent them to Kew Gardens, where they flowered but were lost after two to three years.

In the following years Madrid sent seeds to Berlin and Dresden in Germany, and to Turin and Thiene in Italy. In 1802, Cavanilles sent tubers of "these three" (D. pinnata, D. rosea, D. coccinea) to Swiss botanistAugustin Pyramus de Candolle at University of Montpelier in France, Andre Thouin at theJardin des Plantes in Paris and Scottish botanist William Aiton at Kew Gardens. That same year, John Fraser, English nurseryman and later botanical collector to the Czar of Russia, brought D. coccineaseeds from Paris to the Apothecaries Gardens in England, where they flowered in his greenhouse a year later, providingBotanical Magazine with an illustration.

In 1804, a new species, Dahlia sambucifolia, was successfully grown at Holland House, Kensington. Whilst in Madrid in 1804, Lady Holland was given either dahlia seeds or tubers by Cavanilles. She sent them back to England, to Lord Holland's librarian Mr Buonaiuti at Holland House, who successfully raised the plants. A year later, Buonaiuti produced two double flowers. The plants raised in 1804 did not survive; new stock was brought from France in 1815. In 1824, Lord Holland sent his wife a note containing the following verse:

Classification

Since 1789 when Cavanilles first flowered the dahlia in Europe, there has been an ongoing effort by many growers, botanists and taxonomists, to determine the development of the dahlia to modern times. At least 85 species have been reported: approximately 25 of these were first reported from the wild, the remainder appeared in gardens in Europe. They were considered hybrids, the results of crossing between previously reported species, or developed from the seeds sent by Humboldt from Mexico in 1805, or perhaps from some other undocumented seeds that had found their way to Europe. Several of these were soon discovered to be identical with earlier reported species, but the greatest number are new varieties. Morphological variation is highly pronounced in the dahlia. William John Cooper Lawrence, who hybridized hundreds of families of dahlias in the 1920s, stated: "I have not yet seen any two plants in the families I have raised which were not to be distinguished one from the other. Constant reclassification of the 85 reported species has resulted in a considerably smaller number of distinct species, as there is a great deal of disagreement today between systematists over classification.

In 1829, all species growing in Europe were reclassified under an all-encompassing name of *D. variabilis, Desf.*, though this is not an accepted name. Through the interspecies cross of the Humboldt seeds and the *Cavanilles* species, 22 new species were reported by that year, all of which had been classified in different ways by several different taxonomists, creating considerable confusion as to which species was which.

In 1830 William Smith suggested that all dahlia species could be divided into two groups for color, red-tinged and purple-tinged. In investigating this idea Lawrence determined that with the exception of *D. variabilis*, all dahlia species may be assigned to one of two groups for flower-colour: Group I (ivory-magenta) or Group II (yellow-orange-scarlet).

Cultivation

Dahlias grow naturally in climates which do not experience frost (the tubers are hardy to USDA Zone 8),consequently they are not adapted to withstand sub-zero temperatures. However, their tuberous nature enables them to survive periods ofdormancy, and this characteristic means that gardeners in temperate climates with frosts can grow dahlias successfully, provided the tubers are lifted from the ground and stored in cool yet frost-free conditions during the winter. Planting the tubers quite deep (10 – 15 cm) also provides some protection. When in active growth, modern dahlia hybrids perform most successfully in well-watered yet free-draining soils, in situations receiving plenty of sunlight. Taller cultivars usually require some form of staking as they grow, and all garden dahlias need deadheading regularly, once flowering commences.

Horticultural Classification

History

The inappropriate term *D. variabilis* is often used to describe the cultivars of *Dahlia* since the correct parentage remains obscure, but probably involves *Dahlia coccinea*. In 1846 the Caledonia Horticultural Society of Edinburgh offered a prize of 2,000 pounds to the first person succeeding in producing a blue dahlia. This has to date not been accomplished. While dahlias produce anthocyanin, an element necessary for the production of the blue, to achieve a true blue color in a plant, the anthocyanin delphinidin needs six hydroxyl groups. To date dahlias have only developed five, so the closest that breeders have come to achieving a "blue" specimen are variations of mauve, purples and lilac hues.

By the beginning of the twentieth century a number of different types were recognised. These terms were based on shape or colour, and the National Dahlia Society included cactus, pompon, single, show and fancy in its 1904 guide. Many national societies developed their own classification systems until 1962 when the International Horticultural Congress agreed to develop an internationally recognised system at it Brussels meeting that year, and subsequently in Maryland in 1966. This culminated in the 1969 publication of *The International Register of Dahlia Names* by the Royal Horticultural Society which became the central registering authority.

This system depended primarily on the visibility of the central disc, whether it was open centred or whether only ray florets were apparent centrally (double bloom). The double bloom cultivars were then subdivided according to the way in which they were folded along their longitudinal axis, flat, involute (curled inwards) or revolute (curling backwards). If the end of the ray floret was split, they were considered fimbriated. Based on these characteristics, nine groups were defined plua a tenth miscellaneous group for any cultivars not fitting the above characteristics. Fimbriated dahlias were added in 2004, and two further groups (Single and Double orchid) in 2007. The last group to be added, Peony, first appeared in 2012. In many cases the bloom diametre was then used to further label certain groups from miniature through to giant. This practice was abandoned in 2012.

Modern System (RHS)

There are now more than 57,000 registered cultivars, which are officially registered through the Royal Horticultural Society (RHS). The official register is The International Register of Dahlia Names 1969 (1995 reprint) which is updated by annual supplements. The original 1969 registry published about 14,000 cultivars adding a further 1700 by 1986 and in 2003 there were 18,000. Since then about a hundred new cultivars are added annually.

Flower Type

The official RHS classification lists fourteen groups, grouped by flower type, together with the abbreviations used by the RHS;

☆ **Group 1** – Single-flowered dahlias (Sin) — Flower has a central disc with a single outer ring of florets (which may overlap) encircling it, and which may be rounded or pointed.

☆ **Group 2** – Anemone-flowered dahlias (Anem) — The centre of the flower consists of dense elongated tubular florets, longer than the disc florets of Single dahlias, while the outer parts have one or more rings of flatter ray florets. Disc absent.

☆ **Group 3** – Collerette dahlias (Col) — Large flat florets forming a single outer ring around a central disc and which may overlap a smaller circle of florets closer to the centre, which have the appearance of a collar.

☆ **Group 4** – Waterlily dahlias (WL) — Double blooms, broad sparse curved, slightly curved or flat florets and very shallow in depth compared with other dahlias. Depth less than half the diameter of the bloom.

☆ **Group 5** – Decorative dahlias (D) — Double blooms, ray florets broad, flat, involute no more than seventy five per cent of the longitudinal axis, slightly twisted and usually bluntly pointed. No visible central disc.

☆ **Group 6** – Ball dahlias (Ba)— Double blooms that are ball shaped or slightly flattened. Ray florets blunt or rounded at the tips, margins arranged spirally, involute for at least seventy five per cent of the length of the florets. Larger than Pompons.

☆ **Group 7** – Pompon dahlias (Pom) — Double spherical miniature flowers made up entirely from florets that are curved inwards (involute) for their entire length (longitudinal axis), resembling a pompon.

☆ **Group 8** – Cactus dahlias (C) — Double blooms, ray florets pointed, with majority revolute (rolled) over more than fifty per cent of their longitudinal axis, and straight or incurved. Narrower than Semi cactus.

☆ **Group 9** – Semi cactus dahlias (S–c)— Double blooms, very pointed ray florets, revolute for greater than twenty five per cent and less than fifty per cent of their longitudinal axis. Broad at the base and straight or incurved, almost spiky in appearance.

☆ **Group 10** – Miscellaneous dahlias (Misc) — not described in any other group.

☆ **Group 11** – Fimbriated dahlias (Fim) — ray florets evenly split or notched into two or more divisions, uniformly throughout the bloom, creating a fimbriated (fringed) effect. The petals may be flat, involute, revolute, straight, incurving or twisted.

☆ **Group 12** – Single Orchid (Star) dahlias (SinO) — single outer ring of florets surround a central disc. The ray florets are either involute or revolute.

☆ **Group 13** – Double Orchid dahlias (DblO) — Double blooms with triangular centres. The ray florets are narrowly lanceolate and are either involute or revolute. The central disc is absent.

☆ **Group 14** – Peony-flowered dahlias (P) — Large flowers with three or four rows of rays that are flattened and expanded and arranged irregularly. The rays surround a golden disc similar to that of Single dahlias.

Flower Size

Earlier versions of the registry subdivided some groups by flower size. Groups 4, 5, 8 and 9 were divided into five subgroups (A to E) from Giant to Miniature, and Group 6 into two subgroups, Small and Miniature. Dahlias were then described by Group and Subgroup, *e.g.* 5(d) 'Ace Summer Sunset'. Some Dahlia Societies have continued this practice, but this is neither official nor standardised. As of 2013 The RHS uses two size descriptors.

☆ Dwarf Bedder (Dw.B.) — not usually exceeding 600 mm (24 in) in height, *e.g.* 'Preston Park' (Sin/DwB)

☆ Lilliput dahlias (Lil) — not usually exceeding 300 mm (12 in) in height, with single, semi-double or double florets up to 26 mm (1.0 in) in diameter. ("baby" or "top-mix" dahlias), *e.g.* 'Harvest Tiny Tot' (Misc/Lil)

Sizes can range from tiny micro dahlias with flowers less than 50mm to giants that are over 250mm in diameter. The groupings listed here are from the New Zealand Society.

☆ Giant flowered cultivars have blooms with a diameter of over 250mm.

☆ Large flowered cultivars have blooms with a diameter between 200mm-250mm.

☆ Medium flowered cultivars have blooms with a diameter between 155mm-200mm.

☆ Small flowered cultivars have blooms with a diameter between 115mm-155mm.

☆ Miniature flowered cultivars have blooms with a diameter between 50mm-115mm.

☆ Pompom flowered cultivars have blooms with a diameter less than 50mm.

In addition to the official classification and the terminology used by various dahlia societies, individual horticulturalists use a wide range of other descriptions, such as 'Incurved' and abbreviations in their catalogues, such as CO for Collarette.

Double Dahlias

"A double dahlia delights the eye" and he was right. This corpulent flower grows so tall that it shows its bright purple and white, yellow, deep red and orange ruffles against the blue skies and is very regal. It has two names thanks to two men, who fell in love with it! It is called dahlia after Swedish botanist Andreas Dahl. In eastern Europe and Russia, it is called Georgina named for a botany professor, Johann Georgi, who was so thrilled when he saw it for the first time on a holiday in Paris, that he took several flowers home for his friends.

Europeans discovered the dahlias during the Spanish conquest of Mexico early in the 16th century. The Aztecs grew them for their tuberous roots, which were used in medicine. It was also used in cooking till people thankfully realised that it wasn't very tasty and stopped eating it. Georginas attract butterflies because of their stunning colours. They often get mildew in cold weather. If you give them good air circulation by pulling off the bottom row of leaves, with 8 to 12 inches of clear space, under the plant, you will solve the mildew problem.

Dahlias are water guzzlers and regularly watered produce more flowers. If these get pests, don't use garden chemicals. A study has shown that children of non-chemical gardeners were six times less likely to contract leukaemia. Dahlias discourage nematodes when you grow them around a flower or vegetable affected by them. They also discourage moles. When we use pesticides, 60 to 90 per cent (by volume) of what we spray misses the intended target. It goes into the air or water table. To get rid of slugs affecting your dahlias and other flowers, keep a couple of boards in the garden. If you flip them open and scoop up the slugs and drown them in boiling water, your problem is solved.

In 1805, several new species were reported with red, purple, lilac, and pale yellow coloring, and the first true double flower was produced in Belgium. One of the more popular concepts of dahlia history, and the basis for many different interpretations and confusion, is that all the original discoveries were single flowered types, which, through hybridization and selective breeding, produced double forms. [79] Many of the species of dahlias then, and now, have single flowered blooms. *coccinea*, the third dahlia to bloom in Europe, was a single. But two of the three drawings of dahlias by Dominguez, made in Mexico between 1570–77, showed definite characteristics of doubling. In the early days of the dahlia in Europe, the word "double" simply designated flowers with more than one row of petals. The greatest effort was now directed to developing improved types of double dahlias.

During the years 1805 to 1810 several people claimed to have produced a double dahlia. In 1805 Henry C. Andrews made a drawing of such a plant in the collection of Lady Holland, grown from seedlings sent that year from Madrid. Like other doubles of the time it did not resemble the doubles of today. The first modern double, or full double, appeared in Belgium; M. Donckelaar, Director of the Botanic Garden at Louvain, selected plants for that characteristic, and within a few years secured three fully double forms. By 1826 double varieties were being grown almost exclusively, and there was very little interest in the single forms. Up to this time all the so-called double dahlias had been purple, or tinged with purple, and it was doubted if a variety untinged with that color was obtainable.

In 1843, scented single forms of dahlias were first reported in Neu Verbass, Austria. *D. crocea*, a fragrant variety grown from one of the Humboldt seeds, was probably interbred with the single *D. coccinea*. A new scented species would not be introduced until the next century when the *D. coronata* was brought from Mexico to Germany in 1907.

The exact date the dahlia was introduced in the United States is uncertain. One of the first Dahlias in the USA may be the D. coccinea speciosissima grown by Mr William Leathe, of Cambridgeport, near Boston, around 1929. According to Edward Sayers "it attracted much admiration, and at that time was considered a very elegant flower, it was however soon eclipsed by that splendid scarlet, the Countess of Liverpool". However 9 cultivars were already listed in the catalog from Thornburn, 1825. And even earlier reference can be found in a catalogue from the Linnaean Botanical Garden, New York, 1820, that includes one scarlet, one purple, and two double orange Dahlias for sale.

Sayers stated that "No person has done more for the introduction and advancement of the culture of the Dahlia than George C. Thorburn, of New York, who yearly flowers many thousand plants at his place at Hallet's Cove, near Harlaem. The show there in the flowering season is a rich treat for the lovers of floriculture : for almost every variety can be seen growing in two large blocks or masses which lead from the road to the dwelling-house, and form a complete field of the Dahlia as a foreground to the house. Mr T. Hogg, Mr William Read, and many other well known florists, have also contributed much in the vicinity of New York, to the introduction of the Dahlia. Indeed so general has become the taste that almost every garden has its show of the Dahlia in the season." In Boston too there were many collections, a collection from the Messrs Hovey of Cambridgeport was also mentioned.

In 1835 Thomas Bridgeman, published a list of 160 double dahlias in his Florist's Guide. 60 of the choicest were supplied by Mr. G. C. Thornburn of Astoria, N.Y. who got most of them from contacts in the UK. Not a few of them had taken prices "at the English and American exhibitions".

Recommended Varieties

Picking a favorite dahlia is like going through a button box. There is a great spectrum of color, size, and shape. Here are some popular choices:

☆ 'Bishop of Llandaff': small, scarlet, intense flowers with handsome, dark-burgundy foliage

☆ 'Miss Rose Fletcher': an elegant, spiky, pink cactus plant with 6-inch globes of long, quilled, shell-pink petals

☆ 'Bonne Esperance', aka 'Good Hope': a foot-tall dwarf that bears 1-½-inch, rosy-pink daisies all summer that are reminiscent of Victoria bedding dahlias (though it debuted in 1948)

☆ 'Kidd's Climax': the ultimate in irrational beauty with 10-inch "dinnerplate" flowers with hundreds of pink pentals suffused with gold

☆ 'Jersey's Beauty': a 7-foot tall pink plant with hand-size flowers that brings great energy to the fall garden.

Uses

Floriculture

The asterid eudicots contain two economically important geophyte genera, *Dahlia* and *Liatris*. Horticulturally the garden dahlia is usually treated as the cultigen *D. variabilis* Hort., which while being responsible for thousands of cultivars has an obscure taxonomic status (see also Cultivation).

Characteristics of different Dahlia Hybrids

There are literally thousands of cultivated varieties of Dahlias that have been hybridized throughout the years.Dahlia plants range in height from as low as 12 inches to as tall as 6-8 feet.

The flowers can be as small as 2 inches or up to a foot in diameter. You should therefore consider the ultimate goal of your endeavor, as well as your available space in choosing the varieties you wish to grow. Some specimens may provide an abundance of cut flowers for the home, while others give you the opportunity to make a bold statement in your landscape by pruning, disbudding and ultimately forcing the plant to create a few single, gigantic blooms.

Novice Dahlia growers may want to start by selecting a few plants of varying colors, sizes and types. Most Dahlia gardeners will be happy to share their thoughts and experiences with you regarding their successes, failures and favorites. They may

even be willing to share a few tubers with you. Once you've grown your first crop of these beauties, you will have a much better idea of which types of Dahlias to grow in subsequent years.

Other

Today the dahlia is still considered one of the native ingredients in Oaxacan cuisine; several cultivars are still grown especially for their large, sweet potato-like tubers. Dacopa, an intense mocha-tasting extract from the roasted tubers, is used to flavor beverages throughout Central America.

In Europe and America, prior to the discovery of insulin in 1923, diabetics—as well as consumptives—were often given a substance called *Atlantic starch* or *diabetic sugar*, derived from inulin, a naturally occurring form of fruit sugar, extracted from dahlia tubers. Inulin is still used in clinical tests for kidney functionality.

Orchids

Orchids comprise of unique group of plants and are considered the most beautiful flower in God's creation. In the original Indian Vedic scriptures these are called "Vanda". This has been adopted as a generic name in one of the most beautiful group of orchids. The Orchids comprise the largest family with 35,000 species of flowering plants. They have always fetched fabulous prices. In 1904, Odontoglossum crispum, an orchid, was sold for £1500 in Sussex (England).

Most of the orchids are perennial herbs with simple leaves.Although, the specialized flower structure conforms to a standard plan, the vegetative parts show great variation. Exotic varieties of orchids cultivated on a large scale in South-East-Asian countries and a few outstanding hybrids imported from Thailand, Singapore and Malaysia to run bulk of collections with the growers themselves. Indian dendrobiums, cymbidiums and vandas have also played a major role in the development of major orchid industry in the world.

Orchids form 90 per cent of our flora and are the longest botanical family of higher plants in India. It is estimated that about 1300 species (140 genera) or orchids are found in this country with Himalayas as their main home and others scattered in Eastern and Western Ghats. North-Eastern India owing to its peculiar gradiant and varied climatic conditions contains largest group of temperate, sub-tropical and tropical orchids.

Variety of Orchids

India has a very large variety of orchids and hilly regions have one or the other orchid flowering almost throughout the year. The diversity is so large that there are large-flowered, small-flowered, terrestrial, epiphytic and also saprophytic orchids. In general, terrestrial orchids are more common in North-Western India, epiphytic orchids in North-Eastern India and small-flowered orchids in Western Ghats. The largest terrestrial genus is Habenaria (ca. 100 spp.) and the largest epiphytic genus is Dendrobium (ca. 70 spp.). Most of the Paphiopedilum (Lady's Slipper) species are restricted to N.E. Himalayas except for P. druryi which has been reported from Kerala but now is almost extinct from its original habitat.

Some orchids are endemic to India and are not found any where else in the world. They are Cryptochilus, Anthogonium, Risleya, Sirhookera and Cleisocentron.

Asking me what types of orchids there are is like asking me what types of people there are. Well, there are tall and short people, smart people and not-so-smart people, introverts and extroverts, people with different skin colors, and people from different parts of the world. Likewise, you will find many different types of orchids—different colors, species and hybrids, miniature and standard-sized, all different genera from different parts of the world. So I will narrow the question a bit, and look at it in terms of the orchids you will likely to buy and grow.

Moth Orchids

The most common type of orchid is probably the moth orchid, or the Phalaenopsis orchids. These are the plants that you can buy from a standard grocery store. Or if you happen to live close by an Asian supermarket, you will find truck-loads of the white or purplish-pink variety. Some fashion/interior designers even dubbed this purplish-pink color as "orchid." (This is ultra confusing for an orchid grower, as orchids come in a million different shades and colors, but well, these are the same people who invented colors like sour lemon and spiced mustard. It's just all marketing!)

Phalaenopsis Orchid.

In the wild,Phalaenopsisare epiphytes and grow on trees in a constantly moist environment. They usually have long flower spikes and therefore look very graceful. These types of flowers have the power to add elegance to the home and brighten up a dull atmosphere, even if your boyfriend leaves his socks on the floor. Phalaenopsis are now widely grown as houseplants. Even though they usually bloom only once or twice a year, their flowers can often last for two to three months. Learn more about Phalaenopsis orchids.

Dendrobium Orchids

Another type of orchid you can grow at home is what I call the "Thai restaurant orchids." These types of orchids areDendrobium hybrids and can be found at Thai restaurants that decorate their tables with freshly-cut flowers. They come in many colors (white, green, purple, pink, yellow and more) and require a fairly warm environment. In fact, they could loose their leaves in the winter when the ambient temperature drops below 60°F (15°C) or so. But the "Thai restaurant orchid" is really just one small group of hybrids within the Dendrobium genus. Dendrobium is actually one of the largest orchid genera.

Antelope Type Dendropbium Orchid.

Dendrobiumspecies live as epiphytes and lithophytes in New Guinea, Southern China, Thailand, Japan, Australia, New Zealand, Tahiti and more. As a result, it's hard to generalize how to care for these types of orchids. Some of them require cool nights of 50 to 59°F (10 to 15°C) while some of them require warm temperatures in the 80s°F (27°C). That's why it's so important to understand their natural environment so that you can make them happy. Learn more about Dendrobium orchids.

Slipper Orchids

My favorite type of orchid is the lady slipper orchidbecause of their strange appearance. They are like no other types of flowers and have pouch-shaped lips. The mostly terrestrial and lithophytic slipper orchids include four genera— Paphiopedilum, Phragmipedium, Cypripedium and Selenipedium. But most Cypripedium and Selenipedium are not plants for the beginner because they can be quite difficult to grow under cultivation unless you live in temperate regions with Cypripedium growing wildly in your back yard!

You can readily buyPaphiopedilum at fancy grocery stores, and if you can't find them, find a fancier store where women shop with little dogs in their bag! While these types of flowers come in girlie soft pink, eye-catching yellow, innocent white and other soft colors, many of them are dark red, brown and green with hairy and warty petals. The infamous Paphiopedilum sanderianum from Borneo has lateral sepals (the side petals) that can hang down 3 feet (1 meter) long! This highly sought-after plant can cost hundreds of dollars. But there are plenty of wonderfulPaphiopedilum plants out there that don't cost an arm and a leg.

Paphiopedilum Kobold's Doll.

Boat Orchids

Here in Southern California, one orchid that grows extremely well is the boat orchid, Cymbidium. These types of orchids have been grown and depicted in drawings and poetry for more than two thousand years since the time of Confucius. They are still popular plants today because of the big, showy and long-lasting flowers. The pink, yellow, green, red, brown, peach or combination colored flowers also last superbly as cut flowers—if you change water daily and cut back the bottom of the spike, they can look pristine for a month or two in the vase.

Pink Cymbidium Orchid.

Speaking of cut Cymbidium flowers, my friend has several out door Cymbidium and the once-a-year flowering always brought her lots of joy and pride. It's like an annual EXPLOSION of flowers that fills her whole back yard! Right before the prom, their teenage neighbor forgot to get his date some flowers, and for some reason, he thought it was okay to make a bouquet out of these Cymbidium flowers without any permission. She was so upset about the loss that the thoughtless boy had to make up the mistake by working in her garden for the rest of the year. The moral of the story? If you decide to steal someone's orchids, you'd better not get caught! Learn more about Cymbidium orchids.

I can go on and on about other types of orchids, but I would probably go a little crazy by the time I am done going through the 700 or so genera, more than 25,000 orchid species and more than 100,000 hybrids. However, there are a lot of fascinating orchids you really should not miss, so visit my other pages to learn more about those orchids.

Some of the Indian species are so ornamental and in demand that their natural populations have been over exploited. Some species in the genera like Arundina, Cymbidium, Coelogyne, Dendrobium, Paphiopedilum, Renanthera, and Vanda are almost extinct. The provisional list of 150 endangered plants of India includes many orchids like *Acanthephippium sylhetense, Anoectochilus sikkimensis, Aphyllorchis montana, Arachnanthe clarkei, Arundina graminifolia, Cymbidium macrorhizon, Dendrobium densiflorum, Didiciea cunninghamii, Eria crassicaulis, Galeola lindleyana, Gastrodia exillis, Paphiopedilum faireanum, P. cordigerum, P.druryi, Pleione humilis, Renanthera imschootiana, Vanda coerulea, V.pumila* and *V. roxburghii.*

In order to check further extinction of this precious wealth from the country, it is essential to arrange vigorous propagation. Orchids, like other horticultural crops, maybe propagated either sexually or asexually. Since most of the commercial orchids are highly heterozygous, they are not raised through seed and are propagated through vegetative means to get true-to type plants. Conventional methods like cuttings, division of shoots or keikis, are followed along with mericloning through tissue-culture techniques.

Trade in Orchid

Though India has a vast wealth of orchids, the orchid industry in the country is not well organized. There are only a few orchid exporters located in Kalimpong and Darjeeling and their exports do not exceed ₹5 lakh per annum. The industry urgently needs restructuring. Due to varied topography and climatic conditions as well as rainfall pattern in the different parts of the country, it is possible to propagate orchid species in different countries for commercialization.

Future of Orchid

It has been observed that much of the country's orchid treasure has been depleted due to deforestation through burning of forests and felling of trees in the hills and mountains for timber. Effective steps need to be taken to assist further depletion of this wealth. However, the orchid lovers who are the ultimate buyers of the plant and who always go for novelty, uniqueness and rarity are helping immensely in commercializing the orchid industry. Further, the growing of orchid seed with the help of culture media has revolutionized the commercial orchid growing and hybridization, as every viable seed can be turned into a new plant. The ICAR-Indian Institute of Horticultural Research, Bangalore, CSIR-National Botanical Research Institute, Lucknow, Bhabha Atomic Research Centre, Mumbai are doing commendable work in this regard. There is, however, a strong need to start a few more units for the propagation of orchid through meristem culture. Further, a selected group of orchid growers may be given training for the same at different research centers. It may be reiterated that this God gifted treasure of the country should not be allowed further depletion and the necessary impetus should be given to the orchid enthusiasts for preservation and development of this wealth.

The Corpse Flower

New Yorkers can breathe easy again

The "Corpse flower" has passed its smelly peak. But the plant dubbed "Baby" may soon have some of its own. It was the first time in 67 years that the odoriferous Amorphophallus titanium bloomed in New York City. The plant began blooming around 2 pm on Thursday (3rd August, 2006) at the Brooklyn Botanic Garden. By that evening, it was giving off its signature pungent odour, a smell akin to rotten eggs or garbage left out overnight, which emanate well into Friday morning. But by Friday afternoon, visitors had to get right next to the unusual flower to take the smell. Alessandro Chiari, the BBG's plant propagator, used camel hair brushes to apply fine yellow pollen to the female part of the plant. The pollen came from Virginia Tech University, which has its own Amorphophallus titanium. If the pollen takes, the plant will sprout small red fruits in the next few weeks from which the museum will take seeds and grow more flowers.

"Its really about sex," observed Leeann Lavin, the museum's chief spokeswoman. The plant's male section should have pollen available for the taking by Sunday, Mr. Chiari said. Workers plan to store the pollen and send it to other groups who have similar plants so that they can keep the breeding going. The plant does not self-pollinate.

The exotic flower, native only to Sumatra and named "Baby" by its handlers, burst into bloom after 10 years of nurturing from seeds. It can grow as much as seven inches a day and up to 9 feet tall. While obnoxious to humans, its smell is ambrosia to the insects – sweat bees and carrion beetles – that pollinate the flower in the wild. In captivity, the plant needs to be pollinated artificially.

> Hundreds of visitors filed through a Virginial Tech greenhouse to get a glimpse, and whiff, of a powerfully malodorous "crorpse flower" as it bloomed. The large Indonesian plant, whose botanical name is Amorphophallus titanium, began opening up about 6pm on Friday (4th August) and was in full bloom by early Saturday morning, curator Debbie Wiley-Vawter said on Monday. The plant emits a stench to attract decaying flesh-eating beetles, and sweat bees for pollination. Once it blooms, the odour lingers for about eight hours, then it takes several more years before the plant has enough energy to bloom again.

The plant is expected to start wilting and closing in sometime Sunday, Mr. Chiari said. Its lime-coloured, central protuberance will likely collapse in the next few days. But it will not die, and the botanic garden will keep it. Thousands have stopped by to witness the plant's rapid growth and smell its offensive stink. But even those who missed the worst spell of smell said they were amazed. "This is something that everyone here is going to remember", said a 20 year old, who watched from the sidelines on Friday. "I think it's beautiful, said an other visitor.

The last time one of the plants, formally known as an inflorescence, blossomed locally was at the New york Botanical Garden in the Bronx in 1939. It actually was the Bronx's "official flower" until then-Borough President Fernando Ferrer changed it six years ago to the more mundane day lily. On the net, the garden is at www.bbg. org "Good thing it only lasts a day", said a visitor, alluding to the intense odour.

A Few Garden Flowers

Neat and dwarf like, cyclamen Primulaceae are charming tuberous plants. Originally found in the Mediterranean region and also known by its common name sowbread, cyclamen grow quite well in fairly rich soil containing plenty of leaf-mould and is particularly happy in a shady pocket of the rockery, under the trees or in pots. Different species bloom for many weeks and most of the species lose their foliage for a brief period each year, generally during summers.

Cyclamen Persicumlaceae, the Florists' or Persian cyclamen is the large flowering species which is now offered in many strains: colours of white, pink, salmon, crimson to mauve or cherry-red flowers and plain or marbled foliage, it grows to about a foot tall. Puck – a hybrid, blooms almost continuously bearing about two inch flowers on six to nine plants and is highly recommended.

Most small flowered cyclamens are suited for gardens at medium elevations and do best in light shaded soil which is enriched with compost and leaf-mould, provided by an annual spring mulch of about half an inch.

However, the Florists' cyclamen and Puck can be planted in autumn for late winter or spring blossoms. As houseplants, they do best in bright, indirect or curtain-filtered sunlight where night temperature is around 13-15.5 degrees Celsius and day temperature is no more than 20-22 degree Celsius.

Corms should be planted a couple of inches below the surface with roots uppermost, four inches apart in rockeries or ground in well-drained soil enriched with bonemeal and peat moss or leaf mould. Seeds, however, should be sown around August-September in boxes of open soil placing them in a cool corner. Seedlings should not be exposed to too much hot sun. As the leaves begin to appear, the seedlings may be lifted gently and potted into small pots.

Cyclamen likes an open soil at all stages of growth and absolutely loves lime in soil. A mixture of equal parts of loam, sand and sifted leaf-mould with a little complete fertilizer is good for cyclamen. Do not press the soil firmly and always water at the side of the pot to avoid wetting central growth. Water them with care as too much water rots the corms and too little injures them beyond recovery. Let the crown of the tuber be on level with the top of the soil. As plants grow and are about to flower feed with weak liquid manure once a fortnight. Keep the foliage clean by syringing with clean water.

When the blossoms are over, and the foliage shows signs of discoloration, decrease watering gradually and after the foliage dies down completely, withhold water altogether, take them out of soil and dry till potting time in next season.

Corms give the largest flowers in first year of blooming, thereafter there may be more flowers, but smaller. Cyclamens from seeds usually take about 18 months to flowers but Puck blossoms in about five months.

Hibiscus

Hibiscus is a genus of flowering plants in the mallow family, Malvaceae. The genus is quite large, containing several hundred species that are native to warm-temperate, subtropical and tropical regions throughout the world. Member species are often noted for their showy flowers and are commonly known simply as hibiscus, or less widely known as rose mallow. The genus includes both annual and perennial herbaceous plants, as well as woody shrubs and small trees. The generic name is derived from the Greek word (hibískos), which was the name Pedanius Dioscorides (ca. 40–90) gave to *Althaea officinalis*.

Orgin

The hibiscus is thought to have originated in South China. It has been cultivated in China, Japan and the Pacific Islands. There are two species thought to be native to Hawaii, including the *Hibiscus arnottianus* and *Hibiscus waimeae*, and there are approximately 300 related species of hibiscus in the world.

History of Hibiscus

There was a time when the forerunners of the present day hibiscus hybrids were scattered all over the globe, especially in places having warm and tropical climatic conditions. Botanists believe that there are eight original hibiscus species that can be considered as ancestors of the striking hibiscus hybrids found today. These ancestors are believed to be native to India, China, Mauritius, Hawaii, Fiji, or Madagascar.

The forerunners of the present day hibiscus hybrids are similar to their successors in several ways. These ancestors could be differentiated from other plants as they were tall growing, mostly supple trees that flowered freely and formed seeds through self-pollination, which gave rise to new plants, genetically akin to their parents. However, the blooms of these ancestors were smaller compared to the present day hibiscus hybrids. Nevertheless, the plants produced copious single-hued flowers. These plants as well as their flowers were introduced in Europe way back in the 1700s and about 100 years later in the United States.

Among the annually growing hibiscus species, Manihot and the African are the most familiar varieties seen cultivated ingardens. They are very attractive plants having a somewhat distinctive appearance. They are considered to be different from the other hibiscus plants.

Hibiscus plants bearing white and pink flowers are among the perennially growing plants that are most often cultivated in gardens. Both these varieties are outstanding plants that produce large attractive and bold flowers measuring approximately 5 inches across.

Generally, hibiscus flowers remain on the tree just for a day after blooming. They open in the morning and the flowers begin to wilt by the time it is late afternoon. As if the plants compensate for this poor show by producing copious buds that grow very fast and all keep opening every time. Propagation of hibiscus is usually done by its seeds. Alternatively, they can also be propagated by means of root divisions and should preferably be planted along the borders in your garden.

Provided you sow the seeds at the onset of spring, the plants will flower frequently during the first season itself, and the size as well as the beauty of the flowers will keep enhancing for the initial five years of their existence. At the same time, the plants will maintain their excellent looks.

As hibiscus plants grow quite tall, ideally they should be planted close to the borders and once the planting is over, they require very little care. All that you need to do is cut the flowering stems once they stop blooming for the season. While most hibiscus flowers usually do not have any aroma, some flowers are slightly fragrant.

Archaeologists have unearthed Chinese porcelain from the era of the Ming Dynasty (1368 - 1644) that is ornamented with drawings of hibiscus flowers.

Hibiscus plants from the tropical regions are members of the Malvaceae (mallow) family. Other plants include the hardy hibiscus that is cultivated in the north, the Confederate Rose, the rose-of-sharon (also known as shrubby althea), hollyhock and a number of other species.

Hibiscus rosa-sinesis, which is native to Asia as well as the Pacific islands, is Malaysia's national flower. This species has a close association with Hawaii, but the national flower of Hawaii is different - a native hibiscus species called H. brackenridgei. Flowers of this species occur in several thousand hues and color combinations (baring pure blue or black), while some plants produce blooms whose diameter range from 2 inches to as big as 10 inches - 12 inches. A number of varieties even grow into bushes, which have an extremely sluggish growth - for instance, growing just about a foot in many years. On the other hand, some hibiscus varieties grow up to a height of 15 feet if they are allowed to grow unrestricted. The range of flowers of plants belonging to the tropical hibiscus family is amazing - some blooming in singles and doubles, while there are others that bloom nearly daily.

Till the turn of the 19th century, people knew little about the hibiscus native to Hawaii and interest regarding these species developed by the end of the century. While a number of hibiscus varieties were brought from China and hybridized with the species native to Hawaii. Gradually, the interest regarding these plants spread to mainland United States and Florida turned out to be the hub of this interest, giving rise to the Reasoner family, one of the earliest pioneers. In 1950, the American Hibiscus Society was set up and Norman Reasoner became the first president of the society.

Apart from Hawaii, people in Australia too have a great interest in hibiscus. It is believed that hibiscus plants were introduced in this island continent as early as the beginning of the 19th century. However, the actual interest in the genus started much later when the Brisbane city council imported 30 hibiscus plants from India for landscaping purpose around the city. Gradually, people in the northern region of New Zealand also got involved in this hibiscus culture.

If you are growing hibiscus in areas having frosts, it is advisable that you should grow your favourite grafted plants in containers or pots and bring them indoors during the winter months. In fact, several gardeners grow their entire hibiscus plants in pots.

People inhabiting the regions lying in the edge of the Indian Ocean and Pacific Ocean have cultivated hibiscus for several centuries now. To some extent, the expansion of colonialism during the 18th as well as the 19th centuries contributed in developing the present day romantic image of the flowers. It is believed that the name Hibiscus has its origin in the Greek term "hibiskos". Ancient Greek physician Dioscorides gave this name to a marshmallow plant having close association with the hibiscus way back in the 1st century A.D. People in China have been growing a very old species called Hibiscus rosa-sinensis for its ornamental flowers. It is believed that this species have been grown for several hundred, if not several thousand years, as so far botanists have not yet been able to find any record of these plants growing in the wild. While it has been found that plants of this species were in cultivation in many regions of Asia, the earliest available documents reveal that H. rosa-sinensis was grown in the areas around temples in China. This not only indicates that this species is native to China, but also gives it its name "sinensis".

It is interesting to note that hybridists in Hawaii have used just three native species, counting the Hibiscus arnottianus, which bears pearl-white aromatic flowers, along with over 33 species brought in from several other countries, especially H. schizopetalus and H. cameronii, which are native to East African nations, to hybridize and re-hybridize them extensively to produce over 5,000 horticultural hibiscus varieties that we see today. When we talk about "horticultural" here, it denotes that botanists recorded the genetic parentage as well as identified the offsprings, but not the entire 5,000 varieties merited further propagation and promotion. Conversely, it was found during hybridizing trials that roughly only one variety out of 100 or 200 was capable of producing the desirable attributes, such as excellent flowers, foliage, form and an overall good performance.

Of these successful varieties, hybridists again developed novel tropical hibiscus cross breeds having large and luscious blooms, fascinating hues and a remarkable range. At this stage, the hybrid hibiscus varieties had already overshadowed the original species and the Hawaiians adopted them so eagerly that they not only established the first ever hibiscus society anywhere in the world there in 1911, but also passed a law in 1923 which made the hibiscus flower the symbol of Hawaiian territory.

Botanists and hybridists in Florida had accomplished intensive hybridizing of hibiscus by the middle of the 20th century and now the focus of work moved from Hawaii to the United States' south-eastern regions. Much later, horticulturists in Australia started developing further new cultivars with unparallel success. By the time we entered the 1980s, already people were cultivating more than 4,000 recognized hybrids of the tropical variety.

As of now, over 10,000 hibiscus hybrids are being cultivated in different places across the globe. All through this breeding endeavours, the species H. rosa-sinensis continued to be the most significant parent genetically, but there were several other species that were used simultaneously. There has been so extensive hybridization that now it is extremely difficult to exactly establish the tropical hybrids that are actually offsprings of Hibiscus rosa-sinensis and which were developed from other original species. There is, however, no doubt that it is high time that there ought to be a clarification regarding the common use of the expression "rosa-sinensis hybrids". In fact, there are several Hawaiian hybrids as well as Fijian hybrids that have not been derived from the original species Hibiscus rosa-sinensis. However, there are some varieties that do have their ancestry related to H. rosa-sinensis, but the breeding happened so long back that they hardly possess any attribute of the species today. Thus, there is no recorded parentage of these Hawaiian and Fijian hybrids along with several others.

Description

The leaves are alternate, ovate to lanceolate, often with a toothed or lobed margin. The flowers are large, conspicuous, trumpet-shaped, with five or more petals, color from white to pink, red, orange, peach, yellow or purple, and from

4–18 cm broad. Flower color in certain species, such as H. mutabilis and H. tiliaceus, changes with age. The fruit is a dry five-lobed capsule, containing several seeds in each lobe, which are released when the capsule dehisces (splits open) at maturity. It is of red and white colours. It is an example of complete flowers.

Uses

Symbolism and Culture

The hibiscus is the national flower of Haiti and is used in their national tourism slogan of Haïti: Experience It the Hibiscus species also represents several other nations. The Hibiscus syriacus is the national flower of South Korea, and Hibiscus rosa-sinensis is the national flower of Malaysia. The red hibiscus is the flower of the Hindu goddess Kali, and appears frequently in depictions of her in the art of Bengal, India, often with the goddess and the flower merging in form. The hibiscus is used as an offering to goddess Kali and Lord Ganesha in Hindu worship.

In the Philippines, the gumamela (local name for hibiscus) is used by children as part of a bubble-making pastime. The flowers and leaves are crushed until the sticky juices come out. Hollow papaya stalks are then dipped into this and used as straws for blowing bubbles. Together with soap, hibiscus juices produce more bubbles.

The hibiscus flower is traditionally worn by Tahitian and Hawaiian girls. If the flower is worn behind the left ear, the woman is married or in a relationship. If the flower is worn on the right, she is single or openly available for a relationship. The hibiscus is Hawaii's state flower.

Nigerian author Chimamanda Ngozi Adichie named her first novel Purple Hibiscus after the delicate flower. The bark of the hibiscus contains strong bast fibres that can be obtained by letting the stripped bark set in the sea to let the organic material rot away.

Flowers are intrinsic part of hindu worship and rituals. the fragrance of the flower pleases the deity. they create a pleasant and aesthetic value to the pooja ritual. they are symbol of happiness, prosperity and completion of purpose of life.

Its good to follow the rituals correctly like offering right flower to the deity. but its also important to have a deep devotion and trust in the deity. have a child like innocence and love for your chosen deity, whom you worship regularly. any prayer done with sincerity is heard by the almighty.

Possibly Effective for

High blood pressure. Some early research shows that drinking hibiscus tea for 2-6 weeks decreases blood pressure in people with mildly high blood pressure. Other early research shows that taking a hibiscus extract by mouth for 4 weeks may be as effective as the prescription drug captopril for reducing blood pressure in people with mild to moderate high blood pressure. However, an analysis of results from various clinical studies suggests that there is not enough evidence to draw strong conclusions about the effects of hibiscus in reducing high blood pressure.

Insufficient Evidence for

High cholesterol. Some early research suggests that taking hibiscus extract bymouth or consuming hibiscus tea might lower cholesterol levels in people withmetabolic syndrome or diabetes. However other early research shows that taking a specific extract of hibiscus leaves (Green Chem, Bangalore, India) for 90 days does not improve cholesterol levels in people with high cholesterol. Also, taking hibiscus extract by mouth for 12 weeks does not appear to reduce cholesterol compared to the drug pravastatin and may actually increasecholesterol levels in people with high cholesterol.

☆ Loss of appetite.

☆ Colds.

☆ Constipation.

☆ Irritated stomach.

☆ Fluid retention.

☆ Heart disease.

☆ Nerve disease.

☆ Other conditions.

Species

There are many types (species) of Hibiscus. The most popular ones in gardens are:

☆ *Hibiscus brackenridgei* (Hawaiian hibiscus)

☆ *Hibiscus rosa-sinensis* (Chinese hibiscus)

☆ *Hibiscus syriacus* (garden hibiscus)

Hibiscus trionum is a common weed in gardens and farms.

Hibiscus (Hibiscus spp.), in all its varieties and forms, grows in full sun and produces large, funnel-shaped flowers with soft petals and attractive large stamens. The flowers come in a range of colors, some with veins of different colors toward the center. If your climate doesn't allow you to grow hibiscus outdoors, you can still enjoy them in large containers that spend summers outdoors and winters indoors.

Huge Flowers

Hibiscus has some of the largest flowers of any plant. Rose mallow (Hibiscus moscheutos) produces the largest flowers of all hibiscus from late spring until the first frost, with some reaching 1 foot across. Flowers on the Chinese hibiscus (Hibiscus rosa-sinensis) come in single or double forms and can be 4 to 8 inches wide. Rose mallow thrives in USDA plant hardiness zones 5b through 11 and Chinese hibiscus in USDA zones 8a through 11.

A National Flower

Hibiscus rosa-sinensis has been cultivated for centuries in tropical Asia and is honored as the national flower of Malaysia. It grows up to 30 feet tall in warm tropical climates like Hawaii, but up to 15 feet tall in the U.S. mainland. This tropical hibiscus features single or double flowers in a full range of colors.

A State Flower

Two hibiscus are native to Hawaii, but only one has been named the official state flower: Hibiscus brackenridgei. This hibiscus grows as either a shrub or a tree and produces pure yellow flowers with red veining near the center and a prominent yellow stamen. It grows to 15 feet tall and 8 feet wide in USDA plant hardiness zones 10a through 11.

Additional Facts

Resembling a bouquet of hollyhocks, rose of Sharon (*Hibiscus syriacus*) grows as a deciduous shrub up to 12 feet tall and 6 feet wide in USDA plant hardiness zones 5b through 11 and can also be trained into a single trunk with a treelike top or as an espalier. An evergreen tree hibiscus (*Hibiscus tiliaceus*) is native to tropical Asia and Polynesia. It grows 30 feet tall and wide in USDA zones 10b and 11 with 4-inch flowers that open yellow in the morning and deepen to orange by the end of each day.

Hibiscus Overview Information

Hibiscus is a bushy annual plant. Parts of the flower are used to make a popular drink in Egypt called Karkade. Various parts of the plant are also used to make jams, spices, soups, and sauces. The flowers are used to make medicine.

Hibiscus is used for treating loss of appetite, colds, heart and nerve diseases, upper respiratory tract pain and swelling (inflammation), fluid retention, stomach irritation, and disorders of circulation; for dissolving phlegm; as a gentle laxative; and as a diuretic to increase urine output.

In foods and beverages, hibiscus is used as a flavoring. It is also used to improve the odor, flavor, or appearance of tea mixtures.

Hibiscus Side Effects and Safety

Hibiscus is Likely Safe for most people in when consumed in food amounts. It is possibly Safe when taken by mouth appropriately in medicinal amounts. The possible side effects of hibiscus are not known.

Special Precautions and Warnings

Pregnancy and breast-feeding: Hibiscus is Possibly Unsafe when taken by mouth during pregnancy. There is some evidence that hibiscus might startmenstruation, and this could cause a miscarriage. There is not enough reliable information about the safety of taking hibiscus if you are breast-feeding. Stay on the safe side, and avoid use.

Diabetes: Hibiscus might decrease blood sugar levels. The dose of your diabetes medications might need to be adjusted by your healthcare provider.

Low blood pressure: Hibiscus might lower blood pressure. In theory, taking hibiscus might make blood pressure become too low in people with low blood pressure.

Surgery: Hibiscus might affect blood sugar levels, making blood sugar control difficult during and after surgery. Stop using hibiscus at least 2 weeks before a scheduled surgery.

7

Export and Import of Floriculture in India

Introduction

Floriculture is an age old farming activity in India having immense potential for generating gainful self-employment among small and marginal farmers. In the recent years it has emerged as a profitable agri-business in India and worldwide as improved standards of living and growing consciousness among the citizens across the globe to live in environment friendly atmosphere has led to an increase in the demand of floriculture products in the developed as well as in the developing countries worldwide. The production and trade of floriculture has increased consistently over the last 10 years.

Floricultural products consist of a wide variety of different types of plants and plant materials. They are broadly classified into four category, namely, (i) bulbs, tubers, corms, *etc.*, chicory plant (non-food) (ii) other live plants, roots, cuttings, mushroom spawn, (iii) cut flowers, dried flowers for bouquets, etc and (iv) foliage etc except flowers for ornamental purposes.

During the ten year period 1996-2005 production and export of floricultural products from India has increased manifold. In 1996, India's export of this commodity was around 18 million US $ which has increased to almost four times in 2005. In a country like India floriculture as an industry has tremendous potential for generation of gainful employment in rural as well as urban areas. Since the global demand of floricultural products in increasing day by day, India can effectively capitalize this opportunity to solve the problem of unemployment and poverty through achieving a consistent growth in production and export of this commodity as well earn valuable foreign exchange. In comparison to the developed countries and also to some developing countries floriculture in India is still in a nascent stage.

We first make a review of the global scenario of floriculture and compare the export growth of Indian floriculture with other countries across the globe considering data for the period 1996-2005. Finally, we try to identify some emerging markets of floriculture and make a review of India's export performance in these markets during the last five years that is from 2001 to 2005. The data used for the purpose of making comparisons and study of export performance has been collected from the website of the United Nations Commodity Trade Statistics Database.

The climate in is an important resource for agrarian production, namely for the flowers productions, because, the territory of some countries, as the Mediterranean countries, has sun in the majority of the months, during the year. Nevertheless, these productions are not totally explored in some countries landscape, because a lot of reasons.

The most important reason, in the European Union countries, is because the common agricultural policy. This European policy is totally oriented to the agricultural sector of the north countries and few oriented to the south and Mediterranean. In this way the Mediterranean farmers are induced to produce the agricultural products typical of the European north countries and not their products, like the flowers and others products from the horticulture and fruit production. The farmers are induced, because they receive more supports if they choose some productions in detriment of others. As, the agricultural productions have a lot of risks and uncertainties, the farmers opt for productions with at least one guarantee that is the public support, in form of subsidies.

In another way, it is needed some adjustments in the international and European policies for the flowers productions and commercialization and, of course, for the international trade of these productions. Because, it is needed to demonstrate to the farmers that despite the flowers productions can have few national and European supports, this sector well organized could be a profitable sector, with benefits to the rural populations and the countries.

The landscape of some countries origin small farms, but there are some works that demonstrate that is needed small area for be profitable the flowers production.

So, in this work, it is tried to analyze the data about the international trade of flowers with Portugal. This, because the international trade is the first step to become profitable the production of a sector with tradable products, like said the Keynesian theorist. From here it is analyzed, with the convergence theory and some test of stationary, the stability of the data.

This analyze is an approach to conclude about the evolution of the international flowers sector, the perspective for the future and some adjustments of policy need for become more profitable this sector with great possibility of grow in some countries. The data used are from 2006 to 2010 and were obtained from the INE (Statistics Portugal), gently given by the AICEP (Trade and Investment Agency).

The works available about the flower sector, analyze, this question, in different perspective. For example, Reinten *et al.* (2011) analyzed the cut flowers activity in the southern African and it potential for the international trade. The influence of the United Kingdom retailers in the international trade of flowers was analyzed by (Hughes, 2000). The relationship between the international trade of flowers and some plant pest dissemination was studied by (Areal *et al.*, 2007). The labor conditions in the cut flowers activities were considered by Riisgaard (2008). The African flowers growers potential in the European markets were researched by Cunden and Van Heck (2004). In another way, Vringer and Blok (2000) analyzed the environment implications of the decorative cut flowers.

Agri Export Zone (AEZ) Scheme

The Agri Export Zones (AEZ) scheme has been introduced by the Ministry of Commerce for promoting export of specific produce/products grown in a contiguous area, with the objective of providing remunerative returns to farmers on a sustained basis by improved access to exports. This objective is aimed to be achieved by encouraging the state governments to identify product specific Agri Export Zones for 'end to end' development for export of products from a geographically contiguous area.

The Export Import (EXIM) Policy 2002-07 suggests that the corporate sector with proven credentials may be encouraged to sponsor Agri Export Zone for boosting agro exports. They may provide for pre/postharvest treatment and operations, plant protection, processing, packaging, storage and related Research and Development.

In keeping with the Government's new policy, APEDA has proposed setting up AEZs in areas contiguous to the existing floriculture units (including those under rehabilitation). The value chain for floriculture exports can be summarized as follows:

☆ Supply of planting material

☆ Supply of inputs

☆ Technical know-how

☆ Production infrastructure

☆ Post – harvest management (grading/sorting/packing/storage)

☆ Refrigerated transport

☆ Marketing

❏ Exports (including logistics)

❏ Strengthening of domestic market

☆ Marketing tie-ups foreign markets

Most of the floriculturists find it difficult to address all the above issues on their own as individually each is generally a small unit. As a result the units run into problems of lack of volume, not being able to maintain quality, or unable to market at a remunerative price. Hence, it is imperative to adopt an integrated approach, under the AEZ model, to ensure that all aspects of production to marketing are organized in a way so as to benefit even the small producers leading to increased exports.

Keeping the above in view, APEDA has selected 5 states namely Tamil Nadu, Karnataka, Maharashtra, Uttarakhand and Sikkim for setting up AEZ for floriculture. The districts/areas to be covered in these states under AEZ are as follows.

States for AEZ

Sl.No.	State	District/Area	Date of Approval	Date of signing MOU
1.	Tamil Nadu	Dharmapuri	27-11-2001	7-1-2002
		Nilgiri	28-8-2002	6-2-2003
2.	Karnataka	Bangalore (Urban)		
		Bangalore (Rural)	13-6-2002	1-7-2002
		Kolar		
		Tumkur		
		Uodagu		
		Belgaum		
3.	Maharashtra	Pune		
		Nasik	5-3-2002	10-6-2002
		Kohlapur		
		Sangli		
4.	Uttarakhand	Dehradun		
		Paulnagar	23-4-2002	30-5-2002
		Udhamsingh Nagar		
		Nainital		
		Uttarkashi		
5.	Sikkim	East Sikkim	13-6-2002	26-8-2002

When the AEZs became functional, in the second phase a few more districts/areas in each state may also be included. Needless to say, these selected districts have the maximum concentration of existing floriculture units and also have suitable climate and soil, *etc.* for the industry.

It may be reiterated that the main aim is to generate cost effective production units with focus on market requirements. Once compact area for the production of different cut flowers with common facilities is available, the growers will not have to invest huge amount on postharvest infrastructure, logistics and marketing individually. Once these facilities are available, the grower has only to concentrate on managing his farm/greenhouses. This concept may induce a sense of healthy competition, which may ultimately lead to qualitative and quantitative production.

Exports

Floriculture has been identified as a thrust export sector in India in the post-liberalization era. The global markets offer a vast potential and advantages for India.

However, India's share in the international markets for floricultural products is still negligible at less than one per cent. India's exports of floricultural products in the year 2007-08 decreased by 48 per cent to US$ 84.5 million (₹ 340 crores), from US$ 144 million (₹ 653 crores) in 2006-07, which further decreased by 5.18 per cent in the year 2008-09 to US$ 80.31 million. However, in rupee terms there was a marginal increase of 8.4 per cent from ₹ 340 crores in 2007-08 to ₹ 368.8 crores in 2008-09.

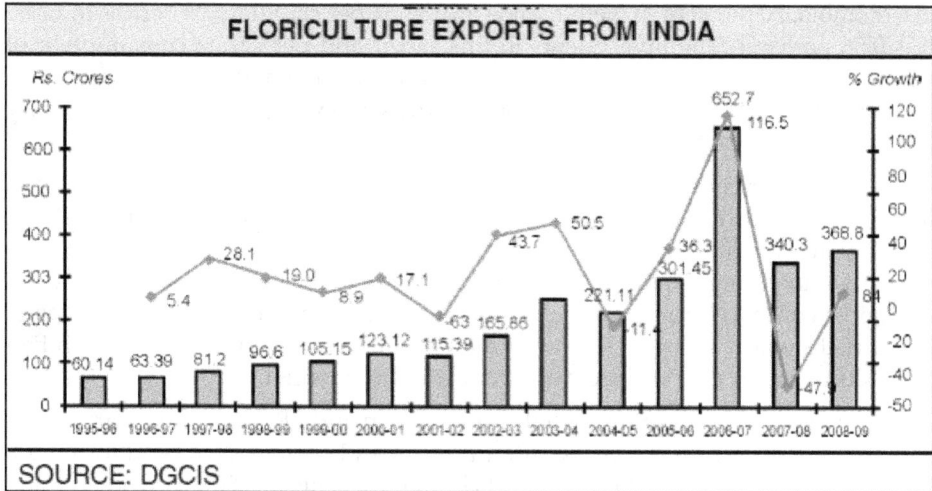

Figure 7.1: Floriculture Exports from India.

Commercial floriculture in India is going through a paradigm shift, where traditional flower cultivation is giving way to modern hi-tech flower cultivation, which is evident from India's rising production and exports. Exports of floricultural products have been growing at a CAGR of 15 per cent over the past decade. However, the growth of the industry has been significantly affected by the recent global recession largely due to decline in demand in all major markets. India's exports of floricultural products in the year 2007-08 decreased by 41 per cent to US$ 84.5 million (₹ 340 crores), from US$ 144 million (₹ 653 crores) in 2006-07, which further decreased by 5.18 per cent in the year 2008-09 to US$ 80.19 million. However, in 2008-09, in rupee terms, export of floriculture from India increased marginally. In the recent years, dried flowers and foliage have been forming a large part of floricultural product exports from India. During 2008-09, dried flowers constituted over 60 per cent of cut flowers exports, and dried foliage constituted over 95 per cent of total foliage exports from India. Fresh cut flowers are mainly exported from Tamil Nadu, Karnataka and Maharashtra. Dried flowers are exported mainly from Tamil Nadu and West Bengal, with the later accounting for around 70 per cent of the dried flower exports from the country. Europe continues to be the largest destination for Indian floriculture exports. However, in the recent years Indian exports of floriculture products have also been to the Japanese and Australian markets. India's distinctive advantages for development of the floriculture sector.

Though floriculture is flourishing both in India as well as in the State, it has not made any remarkable breakthrough in the domestic and international floriculture markets due to various constraints. The country's share in the world trade of fresh flowers is 0.40 per cent to 0.50 per cent as compared to Netherlands 65 per cent, Columbia 12 per cent, Italy 6 per cent, Israel 4 per cent, Kenya 1 per cent and other countries 20 per cent. The area under floriculture although high compared to many countries, the area under protected cultivation is low compared to these countries. The proportion of area under protected to total area floricultural area is 99 per cent in Colombia, 70 per cent in Netherlands and 57.51 per cent Italy. Where as in India it is 0.56 per cent. The investments in this sector and per capita consumption of flowers are also considerably low when compared to other developed countries like Western Europe, Japan and USA. In other words, the vast potential in the country does not seem to be fully tapped.

Over 95 per cent of Indian cut flower exports are different varieties of Rose. India is, thus at present, a negligible player in the international trade in fresh cut international destinations. Although Indian export of fresh cut flowers has been increasing in recent years, the volume as well as share in international trade is negligible compared to its competitors. India faces a major challenge in terms of infrastructure and awareness among small producers. The main feature of Indian floriculture is that the producers are small and fragmented.

Important cut flowers exported from India include Roses, Lilies, Carnations, and Orchids. In the recent years, dried flowers and foliage have been forming a large part of floricultural product exports from India. During 2007-08, dried flowers constituted around 65 per cent of cut flowers exports, and dried foliage constituted around 97 per cent of total foliage exports from India, which were 64 per cent and 99 per cent respectively in the year 2008-09 (upto February 2009).

The Government of India has identified floriculture as high export potential. It was proposed that export of floriculture should be increased ₹100 crore per annum in eighth plan period. For this purpose, National Commission on Agriculture set a target to bring five lakh hectares of land should be under floriculture up to 2000 A.D. due to planning with target and facilities given by supporting agencies, export of floriculture has been increasing year by year. At the beginning of current decade, India's flower export to world market was of about $ 50 billion crore per annum which less than 0.1 per cent, it was definitely negligible.

World floriculture trade in 1995 was US $6946161 thousands. Recently the global market for cut flowers is growing at the rate of 15 per cent per annum. India's growth rate was 13 per cent in year 2005-06.

Table 7.1 shows that the export of floriculture product has been raising tremendously during the last 12 years. The export of flowers from India in 2006-07 fetched a foreign exchange of ₹649.83 crores, which is more than 9 times over the year 1995-96. If we observed carefully, it indicates that, rapid growth has seen from the year 2002-03 onwards. However, between 2005-06 and 2006-07 net growth was 16 per cent. Export units are mainly concentrated around Pune, Nasik, Bangalore, Delhi, Gurgaon, Coimbatore, Faridabad, Chandigarh, Lucknow, Chennai, Calcutta, Vadodara, Jalpaiguri and Amritsar. The major importers for Indian cut flowers are Europe and Japan.

Table 7.1: Year-wise Export of Floriculture (₹ Crore)

Sl.No.	Years	Value (₹ Crore)
1	1995-96	60.14 (2.70)
2	1996-97	63.39 (2.84)
3	1997-98	81.20 (3.64)
4	1998-99	96.60 (4.33)
5	1999-00	105.15 (4.71)
6	2000-01	123.12 (5.52)
7	2001-02	115.39 (5.17)
8	2002-03	165.86 (7.44)
9	2003-04	249.55 (11.19)
10	2004-05	221.11 (9.91)
11	2005-06	299.41 (13.42)
12	2006-07	649.83 (29.13)
	Total	**2230.75 (100)**

Figures in the bracket indicate percentage to respective total.

Source: APEDA database (2008), Ministry of Agriculture Government of India, New Delhi.

Export Opportunity for India: A Report of a Trade Delegation

A trade delegation for floriculture produces visited USSR, Holland, West Germany, USA during 1980. The delegation had highlighted the following points. The import to these countries will be mainly during winter season *i.e.* between November-March when agro-climatic conditions are not suitable for the plants. Import of flowers grown under Glass-house conditions will be preferred for their uniformity in quality.

The ornamental flowers which have been highlighted for export from India to these countries include Gladioli, Roses of specific varieties, Chrysanthemum, Carnation orchids *etc.*

With the varied agro-climatic conditions of the country, no doubt, we have got good scope for the development of ornamental flowers like Rose, Gladioli, Tube rose *etc.* But, for all these we have to develop package of practices and postharvest technologies so that their quick dispatch to foreign markets will be ensured. The foreign markets will however depend much on the quality of the produces.

Consumption and demand of flowers is raising world over. There are 140 countries (Bhattacharji) growing flowers. However, European countries have developed their flora business very early but they could not produce enough quantity. However, new production centres have been developing in Asian countries. Even though, Latin America and African countries have increasing their production of flowers. Recently India and other Asian countries have emerged as development centres of floriculture. The details regarding development of floriculture given as below.

Markets for Exports of Floriculture Products

Over the years, Europe has been the largest destination for Indian floriculture exports, and EU auctions have been the preferred channel for the Indian flower exporters. However, in the recent years, Indian exporting units have been developing and increasing their concentration in the Japanese and Australian markets, which are reflected in the increase in the export realizations from Japan, Australia and New Zealand, over the past 5 years. Though the focus is slowly shifting to different markets, EU yet remains an important destination for India's cut flowers, which is evident in the fact that Europe, accounted for over 50 per cent of the total cut flower export realizations during 2008-09.

In the two key export markets, *viz.* Europe and Japan, India faces intense competition from East African countries like Kenya, Ethiopia and Tanzania (in the EU market), and from South Korea, Thailand, and Australia and New Zealand (in the Japanese market). These countries have immense production capacities, and provide the varieties with quality of flowers, which match the international standards.

Majority of the floricultural units are based in Southern part of the country mainly in Karnataka, Andhra Pradesh and Tamil Nadu. About 125 hi-tech units are concentrated in Karnataka covering 200 hectares; Maharashtra has 39 units covering 150 hectares, Tamil Nadu has 17 units covering 152 hectares and 153. Delhi has 12 units covering 50 hectares. There are around 70 high tech floriculture units in the country. These units are mostly concentrated in Bangalore and Pune belt.

The country has exported 22,947.23 MT of floriculture products to the world for the worth of ₹ 460.75 crores in 2014-15.

Major Export Destinations (2014-15): United States, United Kingdom, Germany, Netherland and United Arab Emirates were major importing countries of Indian floriculture during the same period.

Individual Sub – Products

Bulbs, Tubers, Tuberous Roots	Plant For Tissue Culture
Bulbs Horticultural	Flowering Plants
Chicory plants	Other Live Plants
Other Bulb/Tubers	Live Mushrooms Spawn
Unrooted Cuttings	Cut Flowers For Bouquet's/Fresh
Edible Fruit Trees Grafted or Not	Other Cut Flowers For Bouquet's
Cactus	Moosses and Lichens For Bouquet Fresh
Rhododendrons (Grafted Or Not)	Other Foliages/Buds For Bouquet Fresh
Roses Grafted Or Not	Foliages/Branch/Buds Not Fresh

Current Global Scenario of Floriculture Export

Globally, the export of floricultural products has increased phenomenally from 8 billion US $ in 1996 to 13 billion US $ in 2005. The number of countries reported to have exported floricultural products in 1996 was a mere 50 which has increased to 118 in the year 2005. In fact, this number gradually increased from 50 in 1996 to its peak at 137 in 2002 and 2003. However, the contribution of export of floriculture products to the global export has hovered in the same level during 1996-05. In fact, it was the highest (0.19 per cent) in 1996 and thereafter, it has mostly remained in the range of 0.14 per cent to 0.16 per cent. Figure 7.2 below shows the trend in the number of countries exporting floriculture products during 1996-2005.

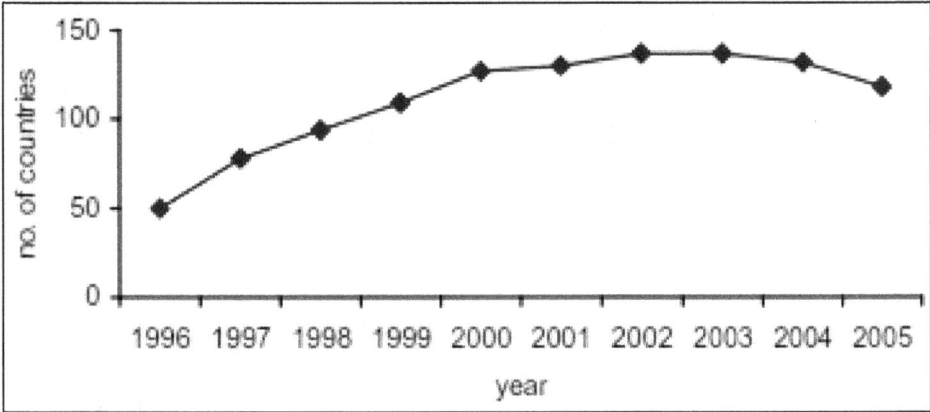

Figure 7.2: Distribution of the Number of Countries Reported to have Exported Floriculture Products during 1996-05.

The Netherlands has, traditionally dominated the world export market of floriculture products with its value of export growing from 4.6 billion US $ in 1996 to 6.7 billion US $ in 2005. It alone contributes more than 50 per cent of the total export of floricultural products worldwide. But its contribution to the global export has come down marginally from 57 per cent in 1996 to 52 per cent in the year 2005. This may be due to the fact that several other countries like Ecuador, Costa-Rica, India, China, Uganda, Austria and Kenya have emerged in the market of floriculture export after the opening up of the worldwide market in the WTO regime.

Columbia, the next highest floriculture product exporting country though has remained in the second position throughout the period 1996-2005, in terms of value of export of floricultural products it is far away from the Netherlands. Italy is the third largest exporter of floriculture products worldwide. During the year 2005, it has exported floriculture products to the tune of $ 696.90 million. Apart from these three countries, Belgium, Denmark, Germany, Ecuador, USA and Costa-Rica has made significant contribution in the global export of floricultural products during the period under reference. Figure 7.3 shows the global trend in the export of floriculture products during the period 1996-2005.

India's Contribution to Global Export

India is in the 18th position, just behind China in the export of floricultural products of the world. But India's contribution to the global export has increased notably during the period 1996-2005. In 1996, India exported floriculture products of worth 18 million US $ which has increased to 68 million US $ in 2005. In percentage

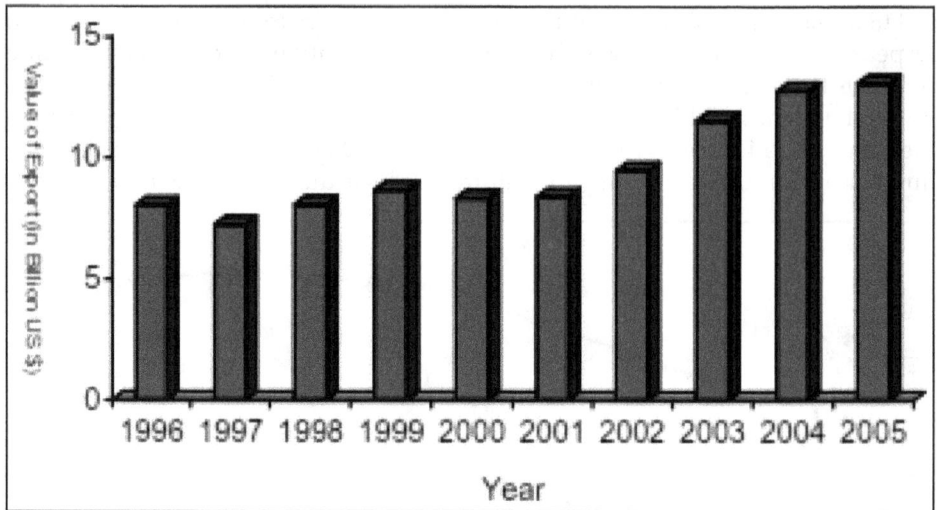

Figure 7.3: Growth in Global Export of Floricultural Products during 1996-2005.

terms, India's contribution to the total volume of export of floriculture products of the world has increased from 0.22 in 1996 to 0.52 in 2005. But contribution of this commodity to India's total volume of export has remained more or less at the same level during 1996-2005. This is because the volume of export of some other commodities likes engineering goods, textile and readymade garments *etc.* have been far more than that of floriculture products. On an average, floriculture contributed about 0.07 per cent of the country's total export volume during the period 1996-2005. The chart below shows the trend in India's export of floricultural products during 1996-2005.

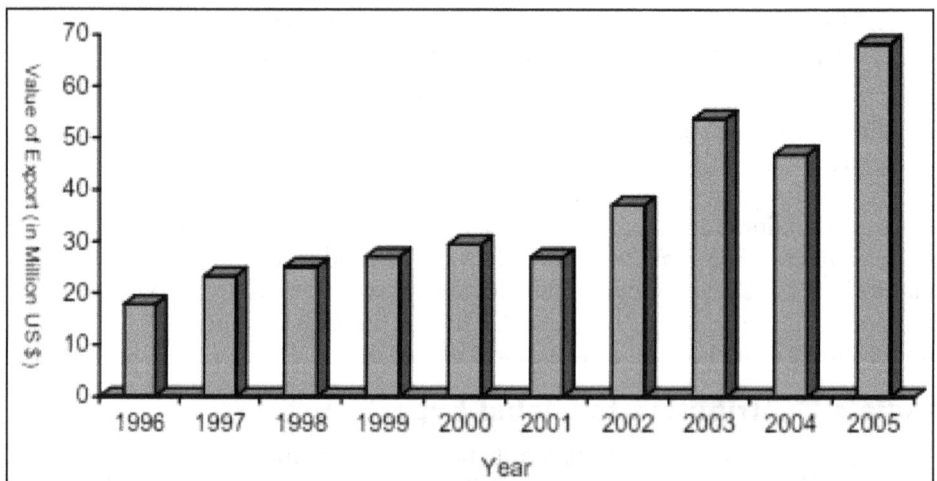

Figure 7.4: Growth in India's Export of Floricultural Products during 1996-2005.

The compound annual growth rate (CAGR) of India's floricultural product export during 1996-2005 is 14 per cent which is far ahead than that of the CAGR in global export of floriculture products (5 per cent). The CAGR of export of floricultural products for some selected countries is presented in the Table 7.2.

Table 7.2: CAGR of Export of Floricultural Products for some Selected Countries

Country/Period	CAGR (in per cent)
Denmark (1999-05)	1.09
Netherlands (1996-05)	3.82
Costa-Rica (1997-05)	4.23
Italy (1996-05)	4.32
Colombia (1996-05)	5.85
Belgium (1999-05)	7.23
China (1996-05)	9.91
Ecuador (1998-05)	9.55
South Africa (2000-05)	10.54
India (1996-05)	14.36
Kenya (1997-04)	16.00
Uganda (1996-05)	28.13
World (1996-05)	5.00

The growth in India's floriculture export in 1996-2005 has been quite noticeable in comparison to several other developed nations like the Netherlands, Denmark Italy, Belgium *etc*. During the period under reference, Uganda registered the highest CAGR in floriculture product export and India was the third country to register a CAGR more than 10.0. Such a consistent and robust increase in the export growth is quite encouraging.

In the year 2005, around 46 per cent of the total value of floriculture product export from India has been cut flowers, dried flowers for bouquets *etc*. and another 43 per cent was in the category of foliage, branches and other parts of plants without flowers or flower buds *etc*. In fact, export of cut flowers, dried flowers for bouquets *etc*. was the highest (83 per cent) in the year 1997 and then it gradually come down to 46 per cent of the total value of export of floriculture produce in the year 2005. On the other hand, export of foliage *etc*. has consistently increased from a mere 4 per cent in 1996 to around 43 per cent of the total floriculture product export in the year 2005. However, export of live plants, roots, cuttings, mushrooms, spawn *etc*. during 1996-05 has remained more or less in the same level and the contribution of bulbs, tubers, corms *etc*. has been 1 to 2 per cent of the total volume of export during this period. The trend in the percentage contribution of different categories of floricultural products to total floriculture export during 1996-2005 is presented in Figure 7.5.

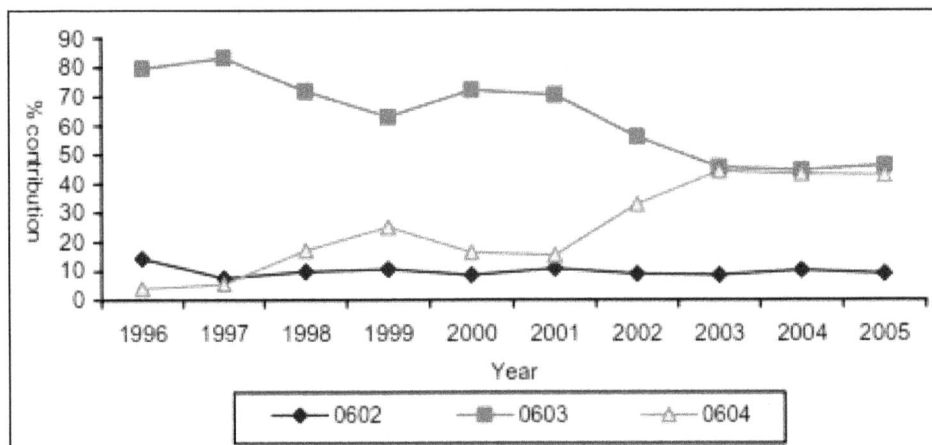

Figure 7.5: Trend in Percentage Contribution of different Categories of Floricultural Product to India's Total Floriculture Export during 1996-05

Major Export Destinations of India's Floriculture Products

From this sub-section onwards we will restrict our discussion to the period 2001-05, *i.e.*, the second half of the reference period 1996-2005 as discussed earlier. During the five year period 2001-05, USA has been the major importer of Indian floriculture products. In 2005, it has imported around one fourth of India's total floriculture export. Next to USA is Japan. In 2001 Japan imported floricultural product worth of 2.7 million US $ which has increased almost 5 times to 12.8 million US $ in the year 2005. The third major export destination for India's floriculture product is the UK and export to this country in 2005 has risen to 7.0 million US $ from 2.3 million US $ in the year 1996. Apart from these three destinations, the Netherlands, Germany, Italy, UAE and France also import a significant amount of our floriculture produce.

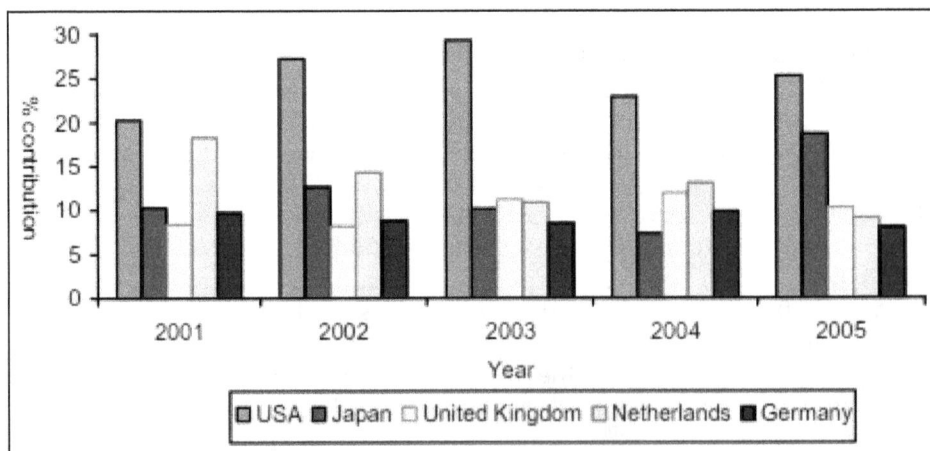

Figure 7.6: Trend in Percentage Contribution of India's Export of Floricultural Product to Top Five Countries 2001-05.

It is also heartening to note that Indian exporters have also been able to increase their share of floricultural product exports to some emerging markets like Russia, China, Thailand, South Africa and Austria. In case of China and Russia the growth has been quite substantial. Floriculture products of worth three thousand million US $ was exported to China in 2001 which has grown to more than one million US $ in 2005 while for Russia it has increased to 0.3 million US $ in 2005 from almost nil in 2001. On the other hand, export of this commodity to countries like France, Singapore, Belgium and Spain has either remained in the same level or registered a negative growth in 2001-05.

Global Importers of Floricultural Products and Identification of Potential Markets for India

Germany is the highest importer of floriculture products. The share of Germany in global import has increased from 1.7 billion US $ in 2001 to 2.5 billion 2005. It is followed by the USA with an import value of 1.6 billion US $ in 2005. Its share in global import of flower and flower products has come down to 12 per cent in 2005 from 15 per cent in 2001. UK is the third largest importer of floriculture products from across the globe and its import has more or less stagnated at the same level during the period 2001-05.

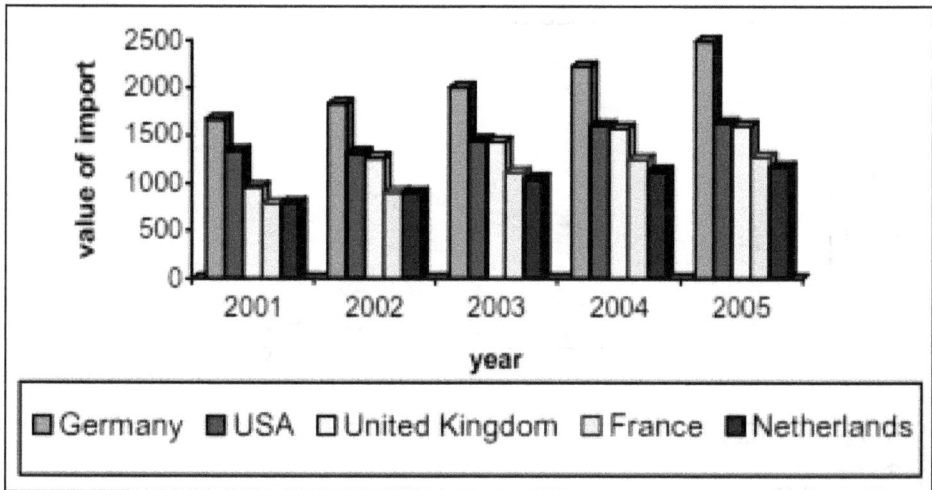

Figure 7.7: Trend in Import of Floriculture Products (in million US $) of Top Five Countries.

In the recent years import of floricultural products from countries all over the world has been quite impressive for the Russian Federation, Poland, Norway, Czech Republic, Hungary, China, Ukraine, Romania, Thailand and South Africa. The CAGR of import of flower during 2001-05 has been phenomenal for the countries mentioned Table 7.3.

Table 7.3: CAGR of Floriculture Product Import during
2001-05 for some Selected Countries

Country	CAGR	Country	CAGR
Thailand	39.81	Hungary	23.80
Romania	37.20	Poland	18.60
Ukraine	34.90	Czech Rep	16.51
Russian Federation	26.89	South Africa	13.21
China	25.41	Norway	11.82
		World	8.78

From the above table it is observed that while the global CAGR for import of flower from the countries all over the world has been 8.78 per cent, it has been far more for countries like Thailand, Romania, Ukraine, Russian Federation and China. In fact, the CAGR for import of flowers from across the world for the countries mentioned above has exceeded that of the developed countries like USA, UK, Germany, the Netherlands, Italy and Japan by a significant margin. The following Figure 7.8 shows the trend in import of floriculture products for some selected emerging flower importing countries.

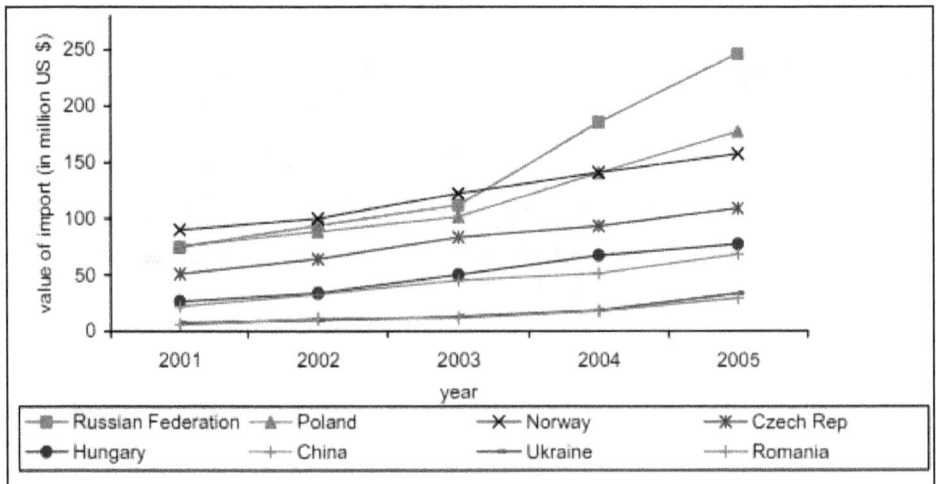

Figure 7.8: Trend in Import of Floriculture Products (in million US $)
of some Emerging Countries.

With a view to compare India's export performance vis-à-vis import of floricultural products world wide we compute the double relative measure (DRij) of trade intensity by using the formula given below:

$$DR_{ij} = \frac{X_{ij}X_w}{X_iM_j} = \frac{X_{ij}}{X_i} \Big/ \frac{M_j}{X_w} \quad \ldots\ldots (1)$$

where X_{ij} = exports of country i to trading partner j,

X_w = total world exports, X_i = total exports of country i,

M_j = total imports of country j.

As per the definition of this measure, the importance of the particular country as a destination is higher for India than for the world if $DR_{ij} > 1$. The following table shows the values of the trade intensity indices of India with respect to developed countries and to some emerging nations during the period 2001-05.

Table 7.3: Double Relative Measure of Trade Intensity Index for some Selected Countries

Country	Year				
	2001	2002	2003	2004	2005
Germany	0.50	0.46	0.50	0.57	0.42
United Kingdom	0.75	0.61	0.90	0.98	0.84
France	0.44	0.41	0.34	0.25	0.21
Switzerland	0.56	0.26	0.19	0.19	0.15
Austria	0.06	0.10	0.26	0.82	0.52
Canada	0.25	0.28	0.43	0.49	0.19
Denmark	0.16	0.15	0.19	0.22	0.15
Hungary	0.59	0.64	1.03	0.16	0.27
Romania	0.00	0.05	0.06	0.87	0.35
Russian Federation	0.00	0.05	0.39	0.41	0.23
Czech Rep.	0.03	0.31	0.30	0.25	0.15
USA	1.28	1.97	2.32	1.83	2.03
Netherlands	1.93	1.51	1.18	1.47	1.02
Italy	1.06	0.69	1.18	0.87	0.63
Japan	2.31	3.22	2.88	2.05	5.30
Poland	0.69	1.68	3.08	1.47	0.96
New Zealand	11.45	11.44	15.46	12.93	11.08
Spain	1.78	0.66	0.69	0.76	0.79
Sri Lanka	79.87	107.70	93.04	74.78	51.87
South Africa	6.68	3.75	3.04	15.44	13.55
Malaysia	7.55	7.43	9.47	7.80	10.23
Mauritius	7.68	10.24	6.63	14.41	9.20
Saudi Arabia	5.34	5.28	3.36	4.91	4.99
Thailand	0.31	1.20	0.88	9.69	4.36
China	0.56	0.94	0.69	4.69	2.90
Israel	2.72	4.47	7.73	3.27	2.78
Philippines	6.80	0.89	4.13	7.34	1.94
Singapore	3.65	3.34	2.76	2.70	1.94

From Table 7.3, it is easy to note that during the period 2001-05 the countries like the UK, Germany, France, Switzerland, Australia, Austria, Canada, Denmark, Czech Republic and Hungary were less important as markets for India's floriculture products than they were for the world, while countries such as USA, the Netherlands, Italy, Japan, Poland, South Africa, Singapore, Sri Lanka, China *etc.* appeared to be more important markets to India for export of floriculture products than the rest of the world. A comparison of Tables 7.2 and 7.3 also reveals that although the growth in import of floriculture products have been phenomenal for countries like the Russian Federation, Romania, Czech Republic and Hungary, Indian exporters were unable to make their presence felt in these markets even though in the other emerging economies of China, Thailand and South Africa Indian exporters have a significant presence.

Table 7.4: Trend in Import of Floriculture Products (in million US $) vis-à-vis per Capita GDP (in US $) during 2001-05

Country	Year				
	2001	2002	2003	2004	2005
(1)	(2)	(3)	(4)	(5)	(6)
Austria	230.61	254.43	313.94	303.85	329.47
	29927	30376	30994	32276	33662
Canada	225.25	228.44	248.91	276.60	298.45
	28223	29255	30014	31263	32886
China	22.14	32.94	45.27	51.38	68.67
	4335	4756	5265	5896	6572
Czech Republic	50.85	64.22	83.45	93.44	109.18
	16429	17211	18064	19448	21317
Denmark	181.96	194.59	239.75	257.36	264.91
	29507	30106	30558	31914	33722
France	790.47	909.10	1119.17	1254.30	1274.42
	26772	27608	28087	29300	30120
Germany	1670.33	1840.45	2008.80	2230.59	2494.77
	26406	26858	27196	28303	29309
Hungary	26.65	33.93	50.15	67.50	77.49
	13891	14711	15450	16813	18086
Israel	6.74	7.65	8.41	10.00	11.03
	23329	22880	23157	24382	25670
Italy	378.15	418.17	454.74	614.28	631.96
	26017	26577	27150	28180	29019
Japan	375.76	376.14	408.73	460.94	463.78
	26746	27051	27851	29251	30821
Malaysia	3.37	6.59	4.42	5.59	4.30
	8903	9183	9545	10276	10843

Country	Year				
	2001	2002	2003	2004	2005
(1)	(2)	(3)	(4)	(5)	(6)
Mauritius	0.23	0.30	0.47	0.48	0.48
	10234	10667	11052	11794	12456
Netherlands	801.05	905.06	1053.08	1135.02	1176.04
	29832	30520	30672	31790	32927
New Zealand	5.05	5.39	5.72	6.67	7.53
	20472	21381	22133	23932	22511
Philippines	0.86	0.87	0.87	0.91	0.91
	4020	4202	4368	4664	4920
Poland	75.69	88.45	101.63	140.94	177.59
	10850	11216	11959	13316	13980
Romania	6.11	11.25	11.99	18.22	29.70
	6460	7027	7648	8700	9208
Russian Federation	74.92	93.74	112.01	185.53	246.41
	7562	8130	9036	9899	10897
Saudi Arabia	9.58	8.55	9.92	13.35	14.80
	12460	12318	13210	13790	14729
Singapore	42.83	41.71	41.11	40.18	45.02
	22937	24843	25788	28860	29921
South Africa	4.00	4.49	5.84	6.81	7.44
	9878	10222	10634	11394	12347
Spain	145.52	171.88	222.80	234.48	264.03
	22757	23418	24105	24992	26125
Sri Lanka	0.64	0.41	0.48	0.39	0.41
	3591	3735	3939	4390	4569
Switzerland	327.49	365.93	426.29	450.59	450.37
	31835	32821	31747	33039	34355
Thailand	1.93	2.88	3.44	3.60	10.31
	6515	6909	7483	8090	8551
United Kingdom	962.66	1268.68	1435.08	1577.38	1603.80
	27515	28257	29265	30821	32005
USA	1342.65	1320.97	1454.32	1607.21	1630.96
	34802	35986	37501	39677	41854

We present below in Table 7.4, the trend of import in floriculture vis-à-vis per capita gross domestic product (GDP) (in US $) for some selected countries during 2001- 05. It may be noted from this table that for almost all the countries mentioned here, the consumption of floricultural products increases as the per capita GDP moves up though the rate of increase in consumption varies widely across the countries. This increase may be due to the fact that with an increase in the GDP the standard of living improves which might have a positive effect towards the consumption of flower products in the developed as well developing nations. What is encouraging to note from Table 7.4 that even in developing countries such as China, Thailand, Malaysia, Mauritius, Philippines, Sri Lanka and South Africa also consumption of flower and products has been growing up with the increase of per capita GDP and these countries are emerging as potential future markets for floricultural products. In Table 7.4, the import figures (in million US $) for a particular country has been shown in the first row while the second row represents the per capita GDP figures (in US $).

We now present below the CAGR of floricultural product import vis-à-vis CAGR of India's export of floricultural products to some selected countries during 2001- 2005. It is observed from Table 7.5 that even though the growth in import of floricultural products for countries like Belgium, Spain, France, Hungary, Italy, Switzerland and Oman has been quite substantial during 2001-05, India's contribution to the import of these countries was unsatisfactory. In fact, India's share to the import of Switzerland, Singapore and Oman has decreased during the period of reference.

Table 7.5: CAGR of Import vis-à-vis that of India's Export during 2001-05

Country	CAGR of Import of Floricultural Products during 2001-05 (in per cent)	CAGR of India's Export of Floricultural Products during 2001-05 (in per cent)
Belgium	9.72	8.61
Spain	12.65	5.40
France	10.02	4.53
Hungary	23.80	16.84
Italy	10.82	10.28
China, Hong Kong SAR	2.35	1.10
Switzerland	6.58	-9.42
Singapore	1.00	-1.96
Oman	10.40	-22.07

Data Analysis

In the Tables 7.6 and 7.7, with absolute values, it is presented the countries with more than ninety per cent of import and export, together, of flowers, in different forms, with Portugal.

The majority of import comes from the Europe, namely from Netherlands, Spain, France, Italy and Belgium (Table 7.6).

Portugal export flowers, also, namely, to the Europe. Netherlands, Spain, France, United Kingdom, Italy and Germany are the most important destinations of the Portuguese flowers (Table 7.7).

From South Africa Portugal import particularly other live plants (including roots), cuttings and slips and mushroom spawn, and flowers and buds cut for

Table 7.6: Flowers, in different Forms, Import Values (Euros)

Category	Year	South Africa	Angola	Cape Verde	Brazil	USA	Swiss	China	Israel	Thailand
Bulbs, tubers, roots, vegetation or flowers, etc., roots	2006	2.157	NA	NA	1.274	372	NA	1.647	7.986	138.327
	2007	NA	NA	NA	NA	NA	NA	NA	18.029	14.793
	2008	NA	NA	NA	NA	NA	NA	NA	18.437	5.881
	2009	2.044	NA	NA	NA	NA	NA	NA	18.118	15.538
	2010	799	NA	NA	NA	NA	NA	NA	7.329	26.048
Other live plants (including roots), cuttings and slips, mushroom spawn	2006	31.472	NA	NA	8.702	210.723	NA	33.211	46.549	2.643
	2007	17.763	NA	NA	15.291	6.626	340	123.611	23.472	6.706
	2008	33.954	NA	NA	2.232	38.996	NA	84.095	24.151	4.504
	2009	45.153	NA	NA	2.401	98.204	NA	43.796	25.149	NA
	2010	40.043	NA	NA	NA	6.831	NA	42.926	11.871	7.831
Flowers and buds of p/ branches/ ornamental purposes, fresh, dried, etc.	2006	403.975	NA	NA	621.467	NA	NA	4.768	747	NA
	2007	411.523	NA	NA	600.341	841	NA	11.067	2.263	9.908
	2008	351.874	NA	NA	394.230	NA	NA	377	3.461	9.243
	2009	94.730	NA	NA	269.030	NA	NA	10.439	NA	54.422
	2010	32.594	NA	NA	120.340	NA	NA	7.348	7.350	56.163
Foliage, branches and other parts of plants/ flowers/ buttons, etc.	2006	115.553	NA	NA	4.229	4.290	2.100	10.791	1.362	NA
	2007	100.953	NA	NA	13.542	5.583	NA	38.982	NA	NA
	2008	86.019	NA	NA	33.422	2.562	NA	56.646	3.864	165
	2009	23.300	NA	NA	66.852	482	NA	11.543	NA	168
	2010	21.388	NA	NA	50.288	556	NA	6.239	NA	550

	Year	India	Germany	Belgium	Denmark	Spain	France	Netherlands	Italy	UK
Bulbs, tubers, roots, vegetation or flowers, etc., roots	2006	NA	184.655	323.358	2.081	1.719.302	76.858	7.314.035	22.233	109
	2007	NA	218.944	420.957	13.790	1.414.294	24.020	8.108.684	83.434	NA
	2008	NA	172.713	263.807	5.610	979.793	121.148	7.380.338	208.148	5.262
	2009	3.095	45.533	20.095	5.599	311.182	44.927	18.140.983	1.762	NA
	2010	NA	158.095	44.527	NA	568.419	32.663	8.265.723	212	NA
Other live plants (including roots), cuttings and slips, mushroom spawn	2006	NA	812.604	1.935.345	155.717	19.648.803	1.275.583	23.872.143	4.873.415	235.806
	2007	NA	958.877	1.200.847	86.945	20.719.930	1.557.055	24.984.372	4.956.950	296.460
	2008	631	879.314	1.577.302	121.919	33.488.342	2.320.759	28.173.080	4.876.879	163.012
	2009	386	925.327	853.432	52.771	20.032.454	3.174.921	21.341.104	3.691.985	4.073
	2010	35.339	947.652	725.835	16.206	14.275.770	3.008.624	25.605.736	5.801.943	181.842
Flowers and buds of p/branches/ornamental purposes, fresh, dried, etc.	2006	3.420	18.119	742.311	NA	3.184.189	19.959	18.564.755	23.234	307.593
	2007	4.552	58.322	703.287	NA	4.340.428	233.570	17.597.788	571.456	47.485
	2008	13.745	63.523	296.022	NA	4.511.597	183.583	17.554.637	341.210	7.330
	2009	11.952	53.344	NA	NA	5.881.775	471.971	11.289.615	18.563	NA
	2010	23.714	83.297	NA	NA	3.699.776	1.096.432	14.337.539	38.959	1.309
Foliage, branches and other parts of plants/flowers/buttons, etc.	2006	72.208	44.484	111.946	9.682	694.007	233.900	2.701.125	39.742	104.062
	2007	66.024	64.743	24.685	NA	885.546	38.878	2.786.677	49.211	17.746
	2008	25.814	11.323	98.458	26	2.183.000	5.744	2.990.013	8.428	22.888
	2009	NA	35.927	231.130	NA	1.109.228	4.997	2.125.187	22.937	NA
	2010	73.237	92.800	85.399	NA	1.460.915	4.768	2.134.191	61.688	140

Table 7.7. Flowers, in different Forms, Export Values (Euros)

Product	Year	South Africa	Angola	Cape Verde	Brazil	USA	Swiss	China	Israel	Thailand
Bulbs, tubers, roots, vegetation or flowers, etc., roots	2006	NA	NA	NA	NA	NA	NA	NA	NA	NA
	2007	NA	NA	NA	NA	NA	NA	NA	NA	NA
	2008	NA	NA	NA	NA	NA	NA	NA	NA	NA
	2009	NA	222	NA	NA	NA	NA	NA	NA	NA
	2010	NA	804	NA	NA	NA	NA	NA	NA	NA
Other live plants (including roots), cuttings and slips, mushroom spawn	2006	2.009	24.638	3.586	6.014	152.809	201.578	11.466	NA	NA
	2007	14.629	20.104	5.367	3.085	188.855	78.588	18.110	1.216	NA
	2008	90	68.185	5.808	5.031	69.720	73.130	NA	1.265	NA
	2009	NA	71.048	3.838	NA	77.862	44.287	NA	NA	NA
	2010	NA	57.885	3.484	NA	161.294	29.904	NA	NA	NA
Flowers and buds of p/ branches/ ornamental purposes, fresh, dried, etc.	2006	NA	17.738	7.169	NA	NA	1.572	NA	NA	NA
	2007	NA	22.538	3.228	NA	NA	1.370	NA	NA	NA
	2008	NA	10.987	1.260	NA	NA	877	NA	NA	NA
	2009	NA	11.726	3.822	NA	1.368	840	NA	NA	NA
	2010	NA	50.352	11.272	NA	NA	1.152	NA	NA	NA
Foliage, branches and other parts of plants/ flowers/ buttons, etc.	2006	NA	1.360	602	NA	NA	NA	NA	NA	NA
	2007	NA	2.631	NA	NA	NA	20	NA	NA	NA
	2008	NA	4.785	NA	NA	NA	NA	NA	NA	NA
	2009	NA	4.344	31	NA	NA	NA	NA	NA	NA
	2010	NA	5.439	1.696	NA	NA	NA	NA	NA	NA

Category	Year	India	Germany	Belgium	Denmark	Spain	France	Netherlands	Italy
Bulbs, tubers, roots, vegetation or flowers, etc., roots	2006	NA	NA	NA	NA	12.708	NA	1.079.522	NA
	2007	585	NA	NA	NA	67.904	14.871	1.375.967	NA
	2008	620	NA	39	NA	63.563	NA	930.373	NA
	2009	NA	NA	NA	NA	3.230.676	NA	780.195	NA
	2010	NA	1.250	NA	NA	NA	10.280	1.090.755	NA
Other live plants (including roots), cuttings and slips, mushroom spawn	2006	NA	2.520.958	524.812	95	5.228.290	5.094.387	7.853.049	339.464
	2007	NA	2.404.217	566.594	NA	8.195.069	5.416.216	9.087.990	509.067
	2008	NA	1.168.196	556.362	13.743	6.155.503	5.717.727	9.803.493	787.499
	2009	NA	1.131.320	1.030.193	91.816	4.954.271	6.197.957	9.432.277	1.400.801
	2010	NA	456.796	716.690	102.716	5.046.533	6.361.183	11.731.354	2.948.997
Flowers and buds of p/branches/ornamental purposes, fresh, dried, etc.	2006	NA	314.610	163.381	NA	162.320	175.185	4.724.390	130.188
	2007	NA	243.326	21.188	NA	289.375	210.492	4.751.432	13.529
	2008	NA	253.031	277.834	NA	318.687	6.004	5.256.735	NA
	2009	NA	63.184	331.742	NA	618.525	35.044	7.508.691	NA
	2010	NA	NA	123.427	NA	330.097	3.537	3.213.584	NA
Foliage, branches and other parts of plants/flowers/buttons, etc.	2006	NA	504.076	202.770	NA	12.496.917	603.765	2.196.748	2.099.778
	2007	NA	539.568	228.525	NA	15.937.919	58.797	1.981.932	4.505.754
	2008	NA	576.795	262.566	NA	8.864.672	38.611	2.467.610	1.549.003
	2009	NA	387.664	289.549	NA	9.353.731	44.199	1.806.112	2.468.492
	2010	NA	636.750	296.178	NA	16.586.061	30.447	1.294.269	5.235.820

Table 7.8. Flowers, in different Forms, Import Percentage Relatively to the Total of each Country

	Year	South Africa	Angola	Cape Verde	Brazil	USA	Swiss	China	Israel	Thailand
Bulbs, tubers, roots, vegetation or flowers, etc., roots	2006	NA	NA	NA	NA	NA	NA	NA	41,20	47,10
	2007	NA	NA	NA	NA	NA	NA	NA	36,94	29,71
	2008	1,24	NA	NA	NA	NA	NA	NA	41,87	22,16
	2009	0,84	NA	NA	NA	NA	NA	NA	27,60	28,75
	2010	NA	NA	NA	NA	NA	NA	NA	4,95	NA
Other live plants (including roots), cuttings and slips, mushroom spawn	2006	3,35	NA	NA	2,43	50,77	100,00	71,18	53,63	21,35
	2007	7,20	NA	NA	0,52	93,84	NA	59,59	48,39	22,76
	2008	27,33	NA	NA	0,71	99,51	NA	66,58	58,13	NA
	2009	42,23	NA	NA	NA	92,47	NA	75,96	44,71	8,64
	2010	85,74	NA	NA	12,32	99,47	NA	27,16	95,05	3,73
Flowers and buds of p/branches/ ornamental purposes, fresh, dried, etc.	2006	77,61	NA	NA	95,42	6,44	NA	6,37	5,17	31,55
	2007	74,57	NA	NA	91,71	NA	NA	0,27	6,93	46,70
	2008	57,33	NA	NA	79,53	NA	NA	15,87	NA	77,60
	2009	34,37	NA	NA	70,53	NA	NA	13,00	27,68	62,00
	2010	8,06	NA	NA	86,52	NA	NA	0,12	NA	96,23
Foliage, branches and other parts of plants/flowers/ buttons, etc.	2006	19,04	NA	NA	2,15	42,78	NA	22,45	NA	NA
	2007	18,23	NA	NA	7,77	6,16	NA	40,14	7,74	0,83
	2008	14,10	NA	NA	19,76	0,49	NA	17,55	NA	0,24
	2009	22,56	NA	NA	29,47	7,53	NA	11,04	NA	0,61
	2010	6,20	NA	NA	1,16	0,53	NA	72,73	NA	0,03

	Year	India	Germany	Belgium	Denmark	Spain	France	Netherlands	Italy	UK
Bulbs, tubers, roots, vegetation or flowers, etc., roots	2006	NA	16,83	17,91	13,69	5,17	1,30	15,16	1,47	NA
	2007	NA	15,33	11,80	4,40	2,38	4,60	13,16	3,83	2,65
	2008	20,05	4,30	1,82	9,59	1,14	1,22	34,29	0,05	NA
	2009	NA	12,33	5,20	NA	2,84	0,79	16,42	0,00	NA
	2010	NA	9,50	24,19	NA	3,79	2,07	18,35	0,01	NA
Other live plants (including roots), cuttings and slips, mushroom spawn	2006	NA	73,71	51,10	86,31	75,73	84,01	46,72	87,56	81,96
	2007	1,57	78,03	70,55	95,58	81,36	88,20	50,22	89,74	82,13
	2008	2,50	87,28	77,26	90,41	73,29	85,88	40,34	98,84	100,00
	2009	26,71	73,93	84,82	100,00	71,36	72,63	50,86	98,29	99,21
	2010	47,86	80,52	55,27	97,77	69,26	84,48	44,94	92,16	95,54
Flowers and buds of p/ branches/ ornamental purposes, fresh, dried, etc.	2006	6,45	4,48	29,93	NA	15,86	12,60	32,91	10,09	13,13
	2007	34,20	5,64	13,24	NA	10,96	6,98	31,29	6,28	3,69
	2008	77,44	5,03	NA	NA	21,52	12,77	21,34	0,50	NA
	2009	17,93	6,50	NA	NA	18,49	26,47	28,48	0,66	0,71
	2010	NA	2,86	19,31	NA	17,43	13,28	30,10	2,02	2,02
Foliage, branches and other parts of plants/flowers/ buttons, etc.	2006	93,55	4,98	1,05	NA	3,24	2,10	5,21	0,87	4,91
	2007	64,23	1,00	4,40	0,02	5,30	0,22	5,33	0,16	11,53
	2008	NA	3,39	20,92	NA	4,06	0,14	4,02	0,61	NA
	2009	55,36	7,24	9,98	NA	7,30	0,12	4,24	1,05	0,08
	2010	52,14	7,12	1,23	2,23	9,52	0,17	6,61	5,81	2,44

branches and ornamental purposes, fresh, dried, *etc.* (Table 7.8). Do not import any flowers from Angola and Cape Verde. From Brazil import in particular flowers and buds cut for branches and ornamental purposes, fresh, dried, *etc.* The United States of America send to Portugal namely other live plants (including roots), cuttings and slips and mushroom spawn. The Swiss few importance has in the Portuguese flowers importations. The majority of imports from China are other live plants (including roots), cuttings and slips and mushroom spawn. From Israel are bulbs, tubers, roots, vegetation or flowers, *etc.* and roots and other live plants (including roots), cuttings and slips and mushroom spawn. From Thailand the majority are flowers and buds cut for branches and ornamental purposes, fresh, dried, *etc.* From India Portugal import namely flowers and buds cut for branches and ornamental purposes, fresh, dried, *etc.* and foliage, branches and other parts of plants, without flowers and buttons, *etc.* From the European countries Portugal import in particular other live plants (including roots), cuttings and slips and mushroom spawn. The differences in the import of flowers from different countries are normal, because their availability depend from their natural and climate conditions.

Portugal exports namely other live plants (including roots), cuttings and slips and mushroom spawn. However the Portuguese export has little importance to South Africa, Brazil, United States of America, China, Israel, India, Thailand and Denmark (Table 7.9).

Indian Exports

Indian Scenario

The Indian floriculture industry is facing a lot of production and export marketing problems. These *Inter alia* include exorbitant airfreight costs, high interest rates and lack of R and D efforts in the area. A number of government institutions are playing an important role in promoting exports of floricultural products. These include Agricultural and Processed Food Products Development Authority (APEDA). National Horticultural Board and National Bank for Agriculture and Rural Development (NABARD). Cultivation of floriculture at present is concentrated in the States of Karnataka, Tamil Nadu, Andhra Pradesh, West Bengal, Maharashtra, Rajasthan, Uttar Pradesh, Delhi and Haryana.

The APEDA over the years has been playing an important role in export promotion and development of floriculture and floriculture products. Growers/ exporters are provided assistance under its various schemes for making improvements in packaging, quality upgradation, creation of infrastructure, *etc.*, The liberalized policies of the government both at the centre and states level have resulted in mushrooming of floricultural units in the country over the years. Some of the recent policy initiatives taken include setting up of an exclusive agri-export zone for floricultural units in Bangalore and according approval for setting up 100 per cent export-oriented units (EoUs) by the government of India to about 191 units. Of this 70 EoUs are operational. As a result, a large number of reputed companies have set up 100 per cent EoUs in floriculture.

Table 7.9: Flowers, in different Forms, Export Percentage Relatively to the Total of each Country

	Year	South Africa	Angola	Cape Verde	Brazil	USA	Swiss	China	Israel	Thailand	UK
Bulbs, tubers, roots, vegetation or flowers, etc., roots	2006	NA	NA	NA	NA	NA	NA	NA	NA	NA	NA
	2007	NA	NA	NA	NA	NA	NA	NA	NA	NA	NA
	2008	NA	0,25	NA	NA	NA	NA	NA	NA	NA	NA
	2009	NA	0,70	NA	NA	NA	NA	NA	NA	NA	NA
	2010	NA	1,02	NA	NA	NA	NA	NA	NA	NA	NA
Other live plants (including roots), cuttings and slips, mushroom spawn	2006	100,00	44,41	62,44	100,00	100,00	98,26	100,00	100,00	NA	96,35
	2007	100,00	81,21	82,17	100,00	100,00	98,81	NA	100,00	NA	91,47
	2008	NA	81,35	49,90	NA	98,27	98,14	NA	NA	NA	87,74
	2009	NA	50,56	21,18	NA	100,00	96,29	NA	NA	NA	93,13
	2010	NA	59,94	44,58	NA	100,00	98,91	NA	NA	NA	98,17
Flowers and buds of p/ branches/ ornamental purposes, fresh, dried, etc.	2006	NA	49,78	37,56	NA	NA	1,71	NA	NA	NA	NA
	2007	NA	13,09	17,83	NA	NA	1,19	NA	NA	NA	0,98
	2008	NA	13,43	49,69	NA	1,73	1,86	NA	NA	NA	6,05
	2009	NA	43,98	68,51	NA	NA	3,71	NA	NA	NA	2,86
	2010	NA	24,26	54,08	NA	NA	1,09	NA	NA	NA	0,48
Foliage, branches and other parts of plants/flowers/ buttons, etc.	2006	NA	5,81	NA	NA	NA	0,03	NA	NA	NA	3,65
	2007	NA	5,70	NA	NA	NA	NA	NA	NA	NA	7,56
	2008	NA	4,97	0,40	NA	NA	NA	NA	NA	NA	6,20
	2009	NA	4,75	10,31	NA	NA	NA	NA	NA	NA	4,00
	2010	NA	14,77	1,34	NA	NA	NA	NA	NA	NA	1,35

	Year	India	Germany	Belgium	Denmark	Spain	France	Netherlands	Italy	UK
Bulbs, tubers, roots, vegetation or flowers, etc., roots	2006	100,00	NA	NA	NA	0,28	0,26	8,00	NA	NA
	2007	100,00	NA	0,00	NA	0,41	NA	5,04	NA	NA
	2008	NA	NA	NA	NA	17,79	NA	4,00	NA	NA
	2009	NA	0,11	NA	NA	NA	0,16	6,29	NA	NA
	2010	NA	0,03	NA	NA	0,00	NA	3,80	NA	NA
Other live plants (including roots), cuttings and slips, mushroom spawn	2006	NA	75,44	69,41	NA	33,46	95,02	52,85	10,12	96,35
	2007	NA	58,47	50,73	100,00	39,96	99,23	53,11	33,70	91,47
	2008	NA	71,50	62,38	100,00	27,29	98,74	48,30	36,20	87,74
	2009	NA	41,72	63,07	100,00	22,98	99,31	67,69	36,03	93,13
	2010	100,00	35,66	62,07	100,00	22,42	99,51	74,20	24,13	98,17
Flowers and buds of p/branches/ornamental purposes, fresh, dried, etc.	2006	NA	7,63	2,60	NA	1,18	3,69	27,63	0,27	NA
	2007	NA	12,66	25,33	NA	2,07	0,10	28,48	NA	0,98
	2008	NA	3,99	20,09	NA	3,41	0,56	38,45	NA	6,05
	2009	NA	NA	10,86	NA	1,50	0,06	18,54	NA	2,86
	2010	NA	NA	19,22	NA	5,15	NA	13,06	NA	0,48
Foliage, branches and other parts of plants/flowers/buttons, etc.	2006	NA	16,93	27,99	NA	65,08	1,03	11,52	89,61	3,65
	2007	NA	28,87	23,94	NA	57,55	0,67	13,37	66,30	7,56
	2008	NA	24,50	17,53	NA	51,52	0,70	9,25	63,80	6,20
	2009	NA	58,16	26,07	NA	75,52	0,48	7,47	63,97	4,00
	2010	NA	64,32	18,71	NA	72,43	0,49	8,94	75,87	1,35

Besides there is a full-fledged Department of Floriculture in the Ministry of Agriculture which has been making efforts to increase the area and production of floriculture from the country. About 34,000 hectares are under cultivation of floriculture in the country. The bulk of the area lies in the Southern states like Tamil Nadu, Karnataka and Andhra Pradesh.

In this context, it may be stated that West Bengal has made a significant stride in the export of floriculture. The state is doubling its supply of tube roses to some of the world's biggest flower auction centers at Naaldwijk and Aalsmeer near Amsterdam. Some 9000 stems are being dispatched to the centre in Holland twice a week on flights from the state. Further, talks are also on with a new auction centre in Dubai for regular supply of flowers. Once the process of registration with the centre is through, West Bengal will have another market for flowers in West Asia.

There is also overseas demand for flowers from the state during the Indian religious festive season. The West Bengal state Food Processing and Horticulture Development corporation get order during the festival season from private parties based in London and New Jersey for lotus and marigold – flowers traditionally required for worship.

Though India's place in the world trade in floriculture and its products is insignificant being only about 1 per cent, the country has made a significant stride during last one decade, in terms of export earnings. The total exports from this group increased from ₹ 14.4 crore during 1991 to ₹123 crore by 2000-01 and ₹ 250 crore in 2003-04. Interesting is the shift in composition of floriculture products. The concerted efforts at export oriented production of cut flowers since the last one decade could be attributed to this shift. During 1989-90, export of live plants and dried plants were the important components of exports from floriculture group. Though dried flowers still hold promise, the export earnings from cut flowers increased substantially during the last one decade. The cut flower export value has shown a tremendous increase during the period from ₹4 million to ₹253.0 million. This has come about with the establishment of a large number of export oriented cut flower units around Bangalore, Pune, Delhi and Hyderabad during the last five years. The major products has been rose being grown by more than 90 per cent of commercial units followed by tropical orchids (Dendrobium). Limited exports are also taking place in carnation, geranium, *etc.*

However, Indian can mint millions of dollars in foreign exchange by tapping the huge demand abroad for flowers. But the country will have to dole out sops like reduction in freight charges. India comes a poor third in floriculture export. Africa and Latin America occupy the first two slots. Our annual exports are in the range of ₹200 to 250 crores as against Kenya's $1billion and Ethiopia's $3 billion. It takes $2.7 or so to export 1 kg flowers from India, whereas for the same consignment from Africa, one needs to spend only $1.5. Hence, India can have the competitive edge once freight charges are slashed. Chart below gives export trends of floriculture.

Category-wise Exports

As can be seen from Table 7.10 India's exports of floricultural products during the period 2001-02 to 2003-04 continued to grow, having risen from ₹ 127.43 crore in

2001-02 to ₹ 180.79 crore in the subsequent year and then to ₹ 250.45 crore in 2003-04, registering a steep growth of 38.53 per cent over the previous year.

Table 7.10: Category-wise India's Exports of Floricultural Products during the Period 2001-02 TO 2004-05 (₹ Crore)

Item	2001-02	2002-03	2003-04	2004-05	Per cent Growth in 2004-05 Over 2003-04
Foliage branches and other plant parts, grasses, moses and lichens for bouquets/ornamental purposes (fresh, dried, dyed bleached, impregnated or otherwise prepared)	89.95	101.38	114.14	92.33	(-)19.11
Cut flowers and flower buds suitable for bouquets/ornamental purposes (fresh, dried, dyed, bleached, impregnated or otherwise prepared)	19.84	59.55	111.07	94.96	(-)14.50
Other live plant (incl. roots) cuttings and slips, mushroom spawn	14.02	16.40	21.94	22.08	0.64
Bulbs, tubers, tuberous roots, corms, crowns and rhizomes, dormant in growth/in flower, chicory plants and roots excluding roots of HS Code 1212	3.62	3.46	3.30	3.33	0.91
Total (All India)	127.43	180.79	250.45	212.70	**(-) 15.07**

Source: DGCI and S, Monthly Statistics of India's Foreign Trade : Exports, and Re-Exports, March 2002, 2003, 2004 and 2005 Issues, Kolkata.

Detailed data on India's exports of floricultural products during the period 2001-02 to 2003-04 is shown in **Table 7.11.**

Table 7.11: Item-wise India's Exports of Floriculture and Floricultural Products During the Period 2001-02 To 2004-05 (₹ Crore)

Item	2001-02	2002-03	2003-04	2004-05	Per cent Change in 2004-05 Over 2003-04
Foliage branches, etc., not fresh without flower buds, and grasses suitable for bouquets/ornamental purposes excluding fresh.	19.38	58.08	108.71	91.60	(-)15.74
Other cut flower and flower buds suitable for bouquets for ornamental purposes	51.96	58.15	68.40	59.61	(-)12.85
Cut flowers and flower buds suitable for bouquets or ornamental purposes, fresh	37.99	43.23	45.74	38.35	(-)16.16
Plant for tissue culture	7.63	11.51	17.59	13.77	(-)21.72

Item	2001-02	2002-03	2003-04	2004-05	Per cent Change in 2004-05 Over 2003-04
Other foliage branches, etc., fresh without flowers/buds, and grasses for bouquets/ornamental purposes, fresh	0.45	--	2.36	0.42	(-)82.20
Other live plants	4.31	2.22	2.21	4.67	111.31
Bulbs, tubers, tuberous roots, corms, crown and rhizomes, dormant	2.56	2.17	1.28	1.54	20.31
Bulbs horticultural	0.57	0.15	0.90	0.40	(-) 55.56
Chicory plants	--	--	0.57	0.50	(-)12.28
Flowering plants excluding roses and rhodondrn	0.09	1.03	0.56	-	-
Other bulbs, tubers, tuberous roots, etc.	0.49	0.40	0.53	0.80	50.94
Mushroom spawn	0.01	0.15	0.40	1.00	150.00
Uprooted cuttings and slips of live plants	1.25	1.13	0.38	0.68	78.95
Roses, grafted or not	0.69	0.27	0.36	0.20	(-)44.44
Other trees, shrubs and bush, grafted or not	--	--	0.33	1.11	236.36
Edible fruit/nut trees grafted or not	0.03	--	0.09	-	-
Cactus	0.01	--	0.03	0.11	266.67
Chicory roots	--	--	0.02	0.09	350.00
Total (All India)	**127.43**	**180.79**	**250.45**	**212.70**	**(-) 15.07**

Source: Compiled from the data of DGCI and S, Monthly Statistics of India's Foreign Trade, March 2002, 2003, 2004 and 2005 Issues, Kolkata.

Foliage of plans and cut flowers account for about 90 per cent of the total exports of floricultural products. These are followed by other plant cuttings and tubers, *etc.* Exports of cut flowers and flower buds, *etc.*, continuously grew from mere ₹19.84 crore in 2001-02 to ₹59.55 crore in the subsequent year and ₹111.07 crore in 2003-04, registering a record growth of 86.52 per cent over the previous year. Other live plant cuttings also showed a significant growth of 33.78 per cent in 2003-04 over the previous year when the same reached a level of ₹ 21.94 crore as against ₹16.40.

Country-wise

Table 7.12 shows that USA continues to be the largest market, having a share of 21.65 per cent in 2003-04. Exports to this market during the period 2001-02 to 2003-04 made a quantum jump from ₹25.81 crore in 2001-02 to ₹ 47.40 in the subsequent year and then to ₹71.10 crore in 2003-04, registering phenomenal growth of 54.21 per cent over the previous year. The other countries witnessing a similar trend during the period include Italy (175.67 per cent), Australia (145.71 per cent), Germany (153.63

per cent) and France (21.86 per cent). On the contrary, UAE showed a negative growth of 38.53 per cent.

Table 7.12: India's Exports of Floricultural Products to the Major Markets during the Period 2001-02 To 2004-05

Country	2001-02	2001-03	2003-04	2004-05	Per cent Growth in 2004-05 Over 2003-04
USA	25.81	47.40	73.39	48.75	(-)33.57
UK	13.37	14.14	28.10	25.12	(-)10.61
Netherlands	23.35	25.70	27.11	26.38	(-)2.69
Japan	12.78	23.16	25.48	12.31	(-)51.69
Germany	12.41	13.91	21.38	20.57	(-)3.79
Italy	6.01	4.48	11.57	8.59	(-)25.76
France	7.74	6.68	8.15	5.31	(-)34.85
Australia	2.26	2.10	5.16	4.15	(-)19.57
UAE	2.70	4.58	3.54	4.90	38.42
Total (all India)	127.43	180.79	250.45	212.70	(-)15.07

Source: Compiled from the data of DGCI and S, Foreign Trade Statistics of India (Principal Commodities and Countries), March 2005, Kolkata.

Detailed data on India's exports of select floricultural products to major markets from the period 2001-02 to 2003-04 is shown in Table 7.13.

Table 7.13: India's Exports of Major Floricultural Products to Important Markets during the Period 2001-02 To 2004-05

Country	2001-02	2002-03	2003-04	2004-05	Percent Growth in 2004-05 Over 2003-04
Foliage branches, etc., not fresh without flower buds, and grasses suitable for bouquets/ ornamental purposes excluding fresh					
USA	7.52	24.28	36.06	27.69	(-)22.46
UK	2.72	6.62	15.59	13.67	(-)12.32
Germany	3.33	8.13	1.66	12.50	653.01
Italy	0.75	2.72	6.96	3.71	(-)46.70
Netherlands	1.39	3.05	6.94	7.73	11.38
Poland	0.55	2.03	5.00	2.62	(-)47.60
France	0.44	1.48	3.73	2.79	(-)25.20
Spain	0.74	1.13	2.53	2.16	(-)14.62
Belgium	0.03	1.53	1.71	2.99	(-)74.85
Austria	0.18	0.43	1.55	3.10	100.00
Canada	--	0.64	1.37	1.24	(-)9.49
Sweden	0.12	0.45	1.33	1.22	(-)8.27
Japan	--	0.41	1.21	0.48	(-)60.33
Israel	0.11	0.32	1.12	0.29	(-)74.11

Country	2001-02	2002-03	2003-04	2004-05	Percent Growth in 2004-05 Over 2003-04
Portugal	0.06	0.14	1.07	1.11	3.74
Total (incl. others)	**19.38**	**58.08**	**108.71**	**91.61**	**(-)15.73**
Other cut flowers and flower buds suitable for bouquets or ornamental purposes					
USA	14.03	20.57	28.32	17.06	(-)39.76
UK	5.43	6.09	10.26	8.84	(-)13.84
Netherlands	8.46	11.41	8.48	6.86	(-)19.10
Germany	6.77	5.20	4.03	6.61	64.02
Italy	3.11	1.51	3.56	3.48	(-)2.25
France	2.38	1.81	2.42	1.93	(-)20.25
Poland	0.25	0.53	1.61	0.07	(-)95.66
Total (incl. others)	**51.96**	**58.15**	**68.40**	**59.61**	**(-)12.85**
Cut flowers and flower buds suitable for bouquets or ornamental purposes, fresh					
Japan	11.05	21.37	23.41	4.18	(-)82.14
Netherlands	7.30	3.16	3.85	6.76	75.58
Austria	1.14	1.17	2.82	1.81	(-)35.82
UAE	1.70	2.92	2.60	1.84	(-)29.23
Singapore	2.19	2.29	2.15	1.49	(-)30.70
France	2.34	2.91	1.73	0.39	(-)77.46
Switzerland	2.49	1.23	1.38	0.91	(-)34.06
UK	--	1.13	1.34	1.15	(-)14.18
Total (incl. Others)	**37.99**	**43.23**	**45.74**	**35.35**	**(-)22.72**
Plants for tissue culture					
Netherlands	3.99	5.72	5.90	5.41	(-)8.31
UK	0.79	--	6.95	1.24	(-)82.16
USA	1.02	1.72	--	2.36	(-)66.04
Total (incl. others)	**7.63**	**11.51**	**17.59**	**13.77**	**(-)21.72**
Other live plants					
Austria	0.13	--	0.14	0.38	171.43
China P. Rep.	0.04	--	0.22	1.44	554.55
Germany	0.19	0.10	0.32	0.06	(-)81.25
Netherlands	0.22	0.24	0.24	0.13	(-)45.83
Singapore	0.05	0.17	0.13	0.03	(-)78.92
UAE	0.33	--	0.18	0.29	61.11
UK	0.03	--	0.20	-	-
USA	0.23	0.17	0.24	0.18	(-)25.00
Total (incl. others)	**4.31**	**2.22**	**2.21**	**4.67**	**111.31**
Bulbs, tubers, tuberous roots, corms, crown and rhizomes, dormant					
Japan	0.26	0.54	0.16	0.31	93.75
Netherlands	1.03	0.74	0.56	0.40	(-)28.57

Country	2001-02	2002-03	2003-04	2004-05	Percent Growth in 2004-05 Over 2003-04
Sri Lanka	--	--	0.24	-	-
UAE	0.06	--	0.23	-	-
Total (incl. others)	**2.56**	**2.17**	**1.28**	**1.54**	**20.31**
Total (All India)	**127.43**	**180.79**	**250.45**	**212.70**	**(-)15.07**

Source: Compiled from the data of DGCI and S, Monthly Statistics of India's Foreign Trade : Exports, and Re-Exports, March 2002, 2003, 2004 and 2005 issues, Kolkata.

Recent Picture

During the latest 3years 2013-14 to 2015-16 (Tables 7.14 and 7.14A) shows that during the previous 11 years, exports of floriculture has increased from ₹127.4 crores in 2001-02. ₹ 479.4 crores by 2015-16. The main import was USA during all this period maintaining its share from 20 per cent in 2001-02 23 per cent in 2015-16. Top importers did not change much. Germany has actually come to occupy the 2nd position other major importers, U.K. Netherland continuing to remain at the top. It is Japan which has shifted its position among the top four. Other major importers remain the same. Reading importers reaming the same. With the export advantages which India has over other countries, it should be possible for India to give a push to export and floriculture should be in a position to occupy its place of price.

Table 7.14: Exports of Floriculture From India

Value in ₹ Lacs, Quantity in MT

Country	2013-14		2014-15		2015-16	
	Qty	Value	Qty	Value	Qty	Value
Product: Floriculture						
United States	5,158.70	8,459.38	5,490.00	9,813.61	5,185.09	9,679.11
Germany	2,841.16	5,928.94	2,240.04	5,547.21	2,336.20	5,692.88
United Kingdom	2,583.87	5,512.57	2,557.24	5,947.56	2,197.52	5,597.00
Netherland	1,983.51	6,615.28	2,060.74	5,125.10	1,883.90	5,567.55
United Arab Emirates	1,026.05	1,701.07	1,582.65	2,204.17	1,499.63	2,699.31
Canada	567.78	1,365.31	856.16	1,538.45	945.88	1,736.13
Japan	727.07	1,621.35	608.91	1,467.35	421.97	1,596.52
Singapore	817.98	910.51	916.94	1,067.92	1,092.47	1,337.17
Australia	270.35	996.34	474.72	1,458.86	380.96	1,330.67
Italy	682.75	1,332.93	561.65	1,207.84	444.91	1,135.73
China P. Rep.	515.43	951.67	379.42	851.95	334.52	1,082.84
Malaysia	155.46	540.1	297.28	736.72	392.55	810
Poland	383.69	582.43	280.12	606.16	382.85	807.92
Saudi Arabia	165.26	291.81	220.08	407.93	407.02	641.9
Belgium	281.93	751.28	241.28	551.9	180.05	490.45
Austria	203.01	424.38	139.76	298.45	194.66	479.26
New Zealand	155.76	558.34	104	522.67	114.69	477.53

Country	2013-14		2014-15		2015-16	
	Qty	Value	Qty	Value	Qty	Value
Lebanon	72.25	252.16	91.18	357.88	114.78	454.7
South Africa	227.3	509.9	396.83	280.03	421.63	438.45
Chile	174.48	272.1	173.5	282.72	326.66	409.32
Spain	310.38	456.43	397.41	522.42	175.03	395.41
France	350.01	438.58	377.38	495.5	255.47	377.62
Qatar	138.23	293.24	139.81	255.89	132.68	335.86
Latvia	197.82	363.94	121.97	428.17	114.14	303.5
Thailand	155.82	195.09	86.76	94.91	127.36	292.54
Korea Republic	81.68	304	25.92	209.76	19.14	226.78
Oman	59.6	147.81	111.45	156.14	277.31	212.5
Hungary	172.85	172.57	144.37	181.17	136.86	191.6
Kuwait	293.99	213.26	69.59	217.78	109.5	190.48
Bahrain	66.4	148.34	114.47	227.7	79.75	183.29
Switzerland	110.18	127.58	131.12	180.37	114.53	177.77
Bangladesh	225.04	53.92	272.08	52.8	299	155.25
Greece	99.15	133.37	78.26	168.06	95.18	138.18
Morocco	31.72	79.89	16.18	53.2	8.54	119.53
Czech Republic	16.26	32.64	49.57	85.07	70.98	113.33
Nepal	29.06	47.57	20.89	29.59	67.65	109.02
Mexico	37.83	82.05	40.49	121.28	29.72	105.56
Mauritius	13.28	85.83	12.52	97.13	26.49	103.3
Portugal	9.39	39.1	18.25	91.45	228.52	102.54
Slovenia	37.85	60.4	77.13	106.05	52.2	100.84
Sri Lanka	48.92	95.79	109.56	81.41	56.82	81.12
Maldives	27.58	96.8	53.79	64.5	57.51	70.59
Ecuador	13.52	130.26	28.44	47.69	1.55	67.72
Philippines	22.16	63.93	27.48	62.27	22.08	62.44
Jordan	22.69	43	12.4	49.05	13.88	61.78
Kenya	0.54	25.41	2.5	60.39	1.2	61.58
Ireland	6.12	6.6	37.59	57.3	27.59	59.65
Egypt Arab Republic	141.06	186.58	86.97	91.49	42.35	56.56
Panama Republic	11.85	35.13	26.86	48.73	26.83	54.25
Sweden	38.24	64.78	19.58	45.7	20.66	47.36
Pakistan	4.51	40.45	0.51	5.02	43.24	46.68
Russia	63.5	97.35	2.68	8.13	21.43	45.14
Lithuania	15.51	31.31	31.2	26.72	50.33	44.77
Hong Kong	99.49	89.53	18.41	30.99	20.69	43.9
Peru	18.92	18.36	3.94	11.09	16.6	43.51
Israel	4.06	34.69	17.18	46.83	21.57	43.28
Colombia	19.51	65.57	15.6	73.43	13.94	42.44

Country	2013-14		2014-15		2015-16	
	Qty	*Value*	*Qty*	*Value*	*Qty*	*Value*
Croatia	29.34	53.27	28.76	38.22	38.49	39.76
Turkey	48.31	112.58	9.98	27.85	15.96	39.6
Brazil	0.47	5.95	20.39	55.84	64.73	39.46
Argentina	35.31	78.02	15.83	50.04	11.5	37.79
Vietnam Social Republic	9.87	54.96	8.69	42.7	8.02	35.34
Denmark	12.1	37.04	14.24	34.26	39.42	33.43
Zambia	4.99	10.59	1.09	6.41	7.92	30.66
Slovak Republic	2.29	2.93	58.09	57.46	44.97	29.05
Iran	16.4	15.28	6.15	11.37	3	26.19
Bulgaria	26.23	71.18	43.45	274.79	12.5	26.09
Guatemala	10.42	17.58	6.8	17.44	15.7	23.11
Taiwan	35.07	68.58	6.37	45.15	0.87	21.46
Eritrea	0	0	10.25	50.85	2.95	19.45
Ethiopia	3.8	333.74	0.4	33.79	0.8	18.65
Uruguay	10.42	14.54	24.88	34.94	24.06	18.18
Estonia	28.18	65.46	0	0	6.02	13.2
Norway	4.81	7.18	18.08	12.45	12.85	12.01
Iraq	2.35	4.19	0.41	3.09	2.11	10.89
Dominic Republic	8.5	8.53	11.88	25.84	9.75	10.51
Indonesia	15.9	78.12	1.17	3.14	3.58	9.7
Azerbaijan	0	0	0.4	4.18	2.17	9.67
Cyprus	4.6	15.31	6.53	38.4	2.92	8.62
Ghana	1	2.58	0.16	6.67	2.98	8.45
Tunisia	4	4.43	8	9.92	4	7.95
Malawi	5.19	49.5	6.24	73.69	0.18	7.28
Reunion	0.15	0.27	0	0	2.87	6.86
Costa Rica	52.73	111.31	91.95	111.95	0.13	6.71
Bhutan	0	0	22	38.05	3	5.71
Congo P Republic	0	0	0	0	5.67	5.7
Cameroon	3.4	4.6	0	0	4.97	5.05
Cote D Ivoire	1.9	4.6	0	0	4.97	5.05
Senegal	1.9	4.6	0	0	4.97	5.05
Unspecified	6.41	3.4	0.9	0.52	2.72	4.83
Tanzania Republic	2.08	5.41	0.18	28.17	0.24	4.32
Angola	0	0	0	0	3.22	3.74
Afghanistan	3.64	4.17	4.32	6.13	1.35	3.44
Tajikistan	5.68	10.82	1.72	2.78	2.08	3.18
Seychelles	0	0	4.36	3.87	1	3.07
Burkina Faso	0	0	0	0	2.83	2.85
Kyrghyzstan	0	0	1.3	1.77	1.3	2.03

Country	2013-14		2014-15		2015-16	
	Qty	Value	Qty	Value	Qty	Value
Uganda	0.02	0.1	0.02	0.09	0.51	1.33
Cape Varde	0	0	0	0	1.02	1.05
Netherlandantil	0	0	0	0	0.8	0.83
Sudan	0.06	0.24	0.08	0.89	0.1	0.76
Ukraine	2.08	3.75	1.6	5.26	0.1	0.41
Kazakhstan	0.27	1.12	0	0	0.41	0.4
St Helena	0.5	0.27	0.87	1.15	0.24	0.25
Mali	0.08	0.37	0.01	1.64	0.08	0.19
Namibia	0	0	0	0	0.02	0.16
Albania	0	0	0	0	0.01	0.1
Brunei	0	0	0	0	0.01	0.1
Congo D. Republic	0	0	0	0	0.05	0.05
Dominica	0	0	0	0	0.08	0.05
Monaco	0.02	0.01	0	0	0.08	0.05
Georgia	10.85	22.6	0	0	0.01	0.04
Trinidad and Tobago	0.1	0.17	0.14	0.05	0.02	0.03
Belarus	8.09	6.44	0	0	0	0
Bosnia-Hrzgovina	0.01	0	1.63	3.46	0	0
Botswana	0	0	0.02	0.02	0	0
Cambodia	0	0	4.3	20.08	0	0
Finland	2.8	12.77	0	0	0	0
Guadeloupe	0	0	0.45	1.6	0	0
Jamaica	0.35	1.07	0	0	0	0
Liberia	0.02	0.18	0	0	0	0
Libya	0	0	1.62	2.05	0	0
Macao	0	0	0	0	0	0
Madagascar	1.58	25.06	0	0	0	0
Montenegro	0	0	7.44	6.27	0	0
Mozambique	0.19	0.08	0	0	0	0
Myanmar	2.68	5.86	0	0	0	0
Nigeria	8.72	15.95	5.49	7.17	0	0
Puerto Rico	47.54	33.05	0	0	0	0
Romania	10.11	12.72	10.03	10.42	0	0
Swaziland	0.01	0.01	0	0	0	0
Syria	2	6.38	0	0	0	0
Yemen Republc	1.18	2.71	5.7	5.51	0	0
Zimbabwe	0	0	0.05	0.18	0	0
Total	**22,485.21**	**45,590.63**	**22,947.23**	**46,077.23**	**22,518.57**	**47,942.00**

Source: DGCIS Annual Export.

Table 7.14A: India's Exports of Floricultural Products to the Major Markets during the Period 2001-02 To 2004-05

Country	2001-02	2001-03	2003-04	2004-05	Per cent Growth in 2004-05 Over 2003-04
USA	25.81	47.40	73.39	48.75	(-)33.57
UK	13.37	14.14	28.10	25.12	(-)10.61
Netherlands	23.35	25.70	27.11	26.38	(-)2.69
Japan	12.78	23.16	25.48	12.31	(-)51.69
Germany	12.41	13.91	21.38	20.57	(-)3.79
Italy	6.01	4.48	11.57	8.59	(-)25.76
France	7.74	6.68	8.15	5.31	(-)34.85
Australia	2.26	2.10	5.16	4.15	(-)19.57
UAE	2.70	4.58	3.54	4.90	38.42
Total (all India)	127.43	180.79	250.45	212.70	(-)15.07

Source: Compiled from the data of DGCI and S, Foreign Trade Statistics of India (Principal Commodities and Countries), March 2005, Kolkata.

Recent Export Promotion Measures

1. Ever since the setting up of an expert group on floricultural development by the government of India in 1989, several steps have been initiated for the speedy development of the sector. Some of the major steps taken in this direction are :

2. New Foreign Trade Policy (FTP): Under the new FTP (2004-09) announced by the Ministry of Commerce and Industry, government of India, a host of incentives have been taken to boost agri-exports. These inter alia include duty –free import of capital goods under the Export Promotion of Capital Goods (EPCG) scheme, duty credit scrip equivalent to 5 per cent of the free-on-board (FOB) value of exports, and launch of Vishesh Krishi Upaj Yojana which is aimed at promoting agri-exports, viz, flowers, vegetables, fruits, minor forest produce, *etc.*,

3. Identification of floral varieties. The expert group on floriculture, besides giving several recommendations, has identified a host of flowers, plants and bulbs suitable for export development on long-term as well as on short-term basis.

4. Exemption from import duty: The government has fully exempted import duty on ornamental plants, tubers and bulbs of flowers, and cuttings of saplings of flower plants used for the purpose of sowing and planting.

5. Entry of large units: To meet the burgeoning demand for floricultural products both in the domestic and overseas markets, a host of major Indian companies have entered this sector. Prominent among them are RPG Group, Harrison Malayalam and Oriental Floriculture. These companies

have started commercial production of roses with technical and marketing tie-ups mainly with Dutch companies.

6. Liberal import policy : According to the present Exim policy, imports of plants, seeds and other plants material are permitted against a licence on the recommendations of the Department of Agriculture and Cooperation, government of India, subject to meeting of the provisions of 'Plants, Fruits and Seeds Regulation of Import into India Order'.

7. APEDA : APEDA has been designated as the nodal agency for export promotion and development activities relating to floriculture. It has recently taken several steps to boost export of floriculture products so as to achieve an export target of ₹1000 crore within five years.

8. Commissioning of the Indian Institute of Packaging for standardizing the packing used for floriculture export items:

   a. Visit of Indian delegations to the Netherlands

   b. Finalization of a host of joint ventures with prominent foreign companies

   c. Visit of Dutch delegations to India for finalizing joint ventures on protected cultivation techniques, *i.e.*, the practice of growing plants in greenhouses as is done in European countries, and in the area of refrigeration marketing and propagation of planting.

   d. Setting up of integrated facilities for handling and storage of exportable perishable products like floricultural products cargo at international airports in Mumbai, Delhi, Chennai, Bangalore, Hyderabad and Thiruvananthapuram.

   e. Setting up of market-cum-auction centres for exports at Bangalore, Mumbai and Noida

   f. Setting up of a market facilitation centre in Aalsameer, Netherlands

   g. Setting up of export processing zones for floriculture products in Tamil Nadu, Uttarakhand, Maharashtra, Karnataka and Sikkim.

9. Air freight subsidy : To make export of floriculture more competitive in the international market, the government had decided to grant air freight subsidy on selected flowers. The scheme is being implemented by APEDA.

10. R and D efforts : With the recent settings up of scientific committee by the Ministry of Commerce, Floriculture exports are set to surge in the near future. The committee will suggest measures

   The exporters desiring to participate in specialized international trade fairs in floriculture may get in touch with the APEDA for getting details of these fairs.

necessary to take advantage of the emerging trade opportunities following the establishment of World Trade Organisation (WTO) and signing of the WTO Agreement in agriculture.

11. Setting up of model floriculture centers : The government of India has been sanctioning funds from time to time for development of floriculture in the states of UP, Maharashtra, Karnataka, Kerala, Sikkim, West Bengal, Jammu and Kashmir and Tamil Nadu. These centers will provide facilities for modern techniques in postharvest handling and also impart training to the floriculture.

12. Foreign Collaboration : The growing interest of the private sector has resulted in a spate of joint ventures for setting up production facilities in India, mainly with Europe – based companies. Through these joint ventures, the European companies transfer their production from Europe to India, while many local partners have access to techno-logical assistance for standardizing their production technologies. In other cases, however, Indian companies merely as production units with full technical assistance from their foreign partners and a 100 per cent buy –back arrangement.

13. Rehabilitation of floriculture units : Many of the floriculture units are reportedly facing problems for their survival. These units, which were set up in the initial phase, had to face the brunt of ignorance, which had a direct impact on the economic viability of these units. With the concerted efforts made by APEDA forgetting financial assistance, many of the units were rehabilitated.

14. Setting up of wholesale market-cum-flower auction centre: Realising the enormous export potential of the floriculture sector, the government has approved setting up of market-cum-flower auction centers for export of flowers at Bangalore, Mumbai and Noida.

15. Centre for perishable cargo: A centre for perishable cargo at Mumbai has been set up to handle the postharvest management of many fresh produce like floricultural products at Shivaji International Airport, Sahar Mumbai, by APEDA. The centre has been functioning since March 2003 and provided a vital platform for quick disposal of perishable cargo. Similar centers have already been approved by APEDA at Ahmedabad, Kolkata and Amritsar.

16. **Liberalized government policies to promote floriculture products**. The liberalized policies of the government both at the centre and state levels have resulted in mushrooming of floricultural units in the country over the years. Some of the recent policy initiatives include setting up of an exclusive agri-export zone for floricultural units in Bangalore, and approval for setting up of 100 per cent export-oriented units (EoUs) by the government of India to about 191 units. Of this, 70 EoUs are operational.

Besides, a full-fledged Department of Floriculture in the Ministry of Agriculture has been making all-out efforts to increase floriculture area and production in India. About 34,000 hectares are reportedly under the cultivation of floriculture in the country. The bulk of the area lies in the Southern states like Tamil Nadu, Karnataka and Andhra Pradesh.

Karuturi Sets up Floriculture Unit in Ethiopia

Karuturi Networks, a Bangalore-based company, has set up a 50-hectare floriculture unit at Holeta near Addis Ababa and is in the process of adding 50 hectares more. A few more Indian companies including Pushpam Florabase are in the process of setting up their floriculture operations in Ethiopia.

The Ethiopian government offers large tracts of land for floriculture companies on a perpetual long-lease at very attractive rentals. Ethiopia offers ideal climatic conditions that help in producing premium-grade roses and commands a significant premium compared to the Indian roses. Though it offers no fiscal incentives, growers can avail themselves of funds at low rate of interest from the World Bank, which has earmarked around $350 million for the development of floriculture in Ethiopia.

Global Floriculture Expo in Bangalore

A 3 day international floriculture exhibition called *FLORA 2005* was organized in Bangalore from July 3-5, 2005. While inaugurating the event, the Union Commerce and Industry Minister Kamal Nath said that exorbitant air cargo rate was one of the main bottleneck in the export of floriculture products, and added that he proposed to hold discussions with the international cargo movers to reduce the rates.

Though India is blessed with favourable climatic conditions for floriculture, our share of exports in the global market is only about 0.5 per cent. However, India's proximity to growing economies in East Asia and the Gulf offers good opportunity. A game plan should be evolved to boost the country's floriculture exports to achieve an export target of ₹1000 crore in the next five years.

The event was organized by APEDA and supported by National Horticultural Board and other industry associations. A large number of floriculture companies and delegations from the Netherlands, Japan, Israel, France, Germany, Italy, Singapore, Dubai, Pakistan, Sri Lanka and Bangladesh participated in the event. The event provided a platform to showcase the country's flora and expose the stakeholders to newer technologies.

Besides, a large number of Indian companies growing roses, carnations, gerberas, anthurium and flower seeds showcased their products, with the special feature being dry flowers and plants sector. Encouraged by the success of the Bangalore Flora, APEDA plans to organize FLORA 2006 at Delhi.

Entry of Large Units

To meet the burgeoning demand for floricultural products both in the domestic and overseas markets, a host of major Indian companies have entered this sector. Prominent among them and RPG Group, Harrison Malayalam and Oriental Floriculture. These companies have started commercial production of roses with technical and marketing tie-ups mainly with most of the Dutch companies.

Second Flora Expo

Asia's biggest floriculture event, the second Flora Expo and Landscape Expo 2006 opened at Delhi's Pragati Maidan on September 8, 2006. The three day this was attended by 24 countries including the Netherlands, Australia, Britain, Russia, Germany, New Zealand, Japan, Malaysia, Taiwan and several European nations. The whole idea of organizing Flora Expo is to showcase India's capability the variety that we have to offer and our ability to supply quality products to the international market. The mega event will make the world realize that our country has the potential to become one of the foremost suppliers of quality floriculture material to the international market. A whole segment of traditional flowers including prepared bouquets, garlands, chotis and gajaras have been added. The display will focus on India's diverse agro climatic spectrum and the product range it can offer like liliums, zantedesias, liatris from Uttarakhand, orchids from Sikkim, tuberose and marigold from West Bengal and foliage like ferns and dry arrangements from the North-East.

Export Strategy

Besides these developments, there is a need to adopt a coherent export strategy by initiating suitable measures, especially providing better and efficient infrastructural facilities. More cold storage facilities at the airports and a fleet of refrigerated vans for taking flowers from production units to the airports are required.

Lack of exposure to the major developments in international trade has resulted in the slow growth of the floriculture industry in India. There is a need to participate actively and regularly in the specialized international trade fairs in floriculture. The exporters desiring to participate in such fairs may get in touch with the APEDA for getting details of such fairs in regard to the dates, venue, and terms and conditions of participation.

Export Prospects

Floriculture offers immense potential for export. For tapping this potential, there is a need to chalk out a suitable export strategy by giving due impetus on:

1. Diversifying the product range of flowers; presently, roses constitute the main item of exports.
2. Improvements in the quality of the flowers.
3. Better supply chain.
4. Setting up of more flower auction centers.

With these policy initiatives taken by the government, exports of floricultural products are set to achieve a level of ₹ 500 crore by 2006-07.

Export Potential from West Bengal

West Bengal as a major producer and exporter of flower, has further geared up to focus on the development of floriculture, especially exports. By 2005-06 the flower exports from the state are expected to quadruple from the level attained in 2003-04. The state is stated to have been receiving encouraging enquiries from the prospective buyers in the USA, Holland, the U.K., the U.A.E. and Japan, which have evinced interest. Export to Japan may increase sizeably in the near future.

It is reported that the State Government has drawn up plans for export of flowers. An effort is being made to cultivate the right variety of flowers. Steps are being taken to set up facilities for preservation of flowers in areas such as Barasat and Englishbazar in Malda which would help facilitate quality exports. A pilot project for exports is being implemented in Birbhum district. The State Government feels that the prospects are bright because the local floriculturists are showing an interest in the project. About 3000 flower bunches are being exported every week in the first phase of the project.

Northern Hilly State Set to become Floriculture Hub

Himachal Pradesh and Jammu and Kashmir along with Uttarakhand and some North-eastern states are fast becoming hubs for floriculture exports. Presently in the global floriculture market of around $11 billion, though India's share is a miniscule, yet the area under its cultivation is fast growing especially in the foothills of Himalaya. A large number of commercial floriculture farms have come up in the recent past in these states thanks to the growing domestic demand and exports growing at over 20 per cent rate annually.

The floriculture exports have grown to ₹305 crore in 2005-06 from ₹18.83 crore in 1993-94. This year also the growth in expected to be around 20-25 per cent implying exports to reach ₹375 crore. In recent years, India has been recognized as one of the potential suppliers of floriculture to meet the growing international demand that is expected to reach $16 billion by 2010. Indian exporters are taking consignments through direct cargo flights leaving for global consumption centres in USA, Netherlands, UK and Japan from Bangalore, Mumbai and Delhi.

Parliamentary Standing Committee on Floriculture Report, 2005

One of the reasons why flowers grown in India do not reach European markets is because peak season for flowers clashes with the peak season for garment exports. And flowers lose out in the struggle for limited cargo space. These and many other ills that plague the floriculture industry have been highlighted by the parliamentary standing committee on commerce that tabled its report on floriculture in Rajya Sabha.

To enable floriculturists function more efficiently and competitively, the report recommends that the government treat it as an agricultural activity and not as an industry. This will help units get income tax benefits, and free them many procedural hassles and regulatory provisions, says the report. For an industry that started with a lot of expectations, the investments – adding up to ₹500 crores have shown little results. There are 70 sophisticated floriculture industry units in India, of which only eight have performed satisfactorily. The rest are financially sick, "Though the industry has grown significantly since 1994, export – oriented units are not doing well. The focus has shifted to domestic markets which grew about 150 per cent since 1994-95, from ₹200 crore to ₹500 crore. Metros like Delhi and Mumbai have markets worth more than ₹50 crores each, says the report. Highlighting the problems, it says that the development of infrastructure has not kept pace with growth in production. First, thee was the high capital costs of units. Then came the high interest rates on loans. Then the international market crashed following increased supplies from Africa and other developing countries.

The small size of floriculture holdings resulted in high operational costs. Recently, some measures have been introduced for the uplift of the floriculture industry, including air freight subsidy, cold storage and cargo handling facilities at airports, a marketing center in Holland and flower auction centers in Bangalore, Mumbai and Noida. Rehabilitation of sick units, too, is in the offing. However, these measures are clearly not enough. The committee had identified the following bottlenecks and recommended that a special task force be set up to clear up the glitches :

☆ The cultivation of flowers in greenhouse is dependent on foreign tech, in the absence of homegrown equivalents, raising the initial investments.

☆ There is death of warehousing space at airports.

☆ The certification process is cumbersome

☆ Air freight is expensive and cargo space inadequate.

☆ Growers do not get sufficient information about market trends on demands, prices and consumer preferences.

Taking the case of Karnataka, the committee has found that 200 units were registered and only 65 commissioned. Part of the reason could have been the high interest rates on loans : 18 per cent. The Indian industry also bears the impact of entering the business when international prices were falling and tariff barriers, such as import duties imposed by the EU, are being levied.

The committee believes that banks should take a sympathetic view of the problems faced by the floriculture industry and re-design credit delivery mechanisms. It has endorsed the concept of setting up Agriculture Export Zones (AEZ) for export promotion of floriculture and recommendable that the AEZ be set up within a definite time frame.

Export Performance in 2004

During the calendar year 2004, India made a significant stride in the export of flowers, though her share in the global market worth around $50 billion is less than 2 per cent. In fact during the year 2004, the country exported 1500 tonnes of flowers as against only 600 tonnes during the whole of 2003. It is reported that the growth in the international market, resulting in a shortage of flowers, and the high quality of Indian flowers has led to a sudden spurt in export. In fact, according to the floriculturists, over the years, they have learnt the technique of growing high quality flowers. Unlike in the past, India now produces roses with longer stems, bigger buds and with a longer shelf life up to two weeks. Higher air freight tariff

GLOBAL PRESENCE

1,500 tonnes of flowers exported during 2004 calendar year, the highest ever

Nearly three-quarters of flower exports are from Bangalore

In the global flower market worth around $50 billion, India's share is less than 2 per cent but is growing.

and stagnant prices presently, do not cause much worry to the floriculturists as they are complacent with the boom in volumes.

In this context, it may be stated that Europe is the largest producer and consumer of flowers. But its production levels are low during its freezing winters and India takes full advantage of the demand-supply gap there. Besides, countries like Kenya which thrive on the European market, have low productions during the peak demand season of December-February because of the warm weather. India has also made significant inroads into the Gulf, earlier fed largely by South American countries like Mexico and Columbia.

While describing India's grand performance in the export of flowers, it may be pointed out that Bangalore, the state headquarters of Karnataka, is making a fragrant impact in the international market with its high quality roses. In fact, during the previous Christmas – New Year, India was set to export some 600 tonnes of flowers – mostly roses–to Britain, Japan, Australia, the Meddle-East and South-East Asia, and Bangalore, which accounted for three quarters of India's flower exports (90 per cent of them being roses) was set to have a record export during the season. It is stated that the city's cold nights and not-so-hot days almost round the year make its surroundings ideal for growing high quality roses.

Major Bottlenecks in Indian Exports

The single most important bottleneck experienced by the Indian exporters in markets abroad is the lack of efficient and effective marketing network. The Indian growers are exporting floriculture products to their agent in importing countries. Only a few directly deal with auctions in the Netherlands.

In this context, it may be highlighted that volumes of Indian roses handled in one of the largest flower auctions at the Netherlands – Bloemenveiling Aalsmeer – has doubled over the year 2003, though the numbers are still very small and India is at the 15th place as a source of flowers. The turnover of Indian roses – main flower which is exported from the country, had been estimated at 1.1 million euro at Bloemenveilling Aalsmeer in 2003, around 0.7 per cent of the auction house turnover of 1.6 billion euro in the year. There are 19 Indian growers registered in Aalsmeer, of which only 5-6 are active.

Roses, as a group of flowers, see the highest trade in auctions, and about half of these flowers put for bidding through a reverse clock are imported from all over the world. In fact, at Bloemenvelling Aalsmeer, the auction house has a separate hall for rose trading find their way into the markets of Western Europe. India exports red roses and it is reported that the auction house is getting the top quality flowers from the country. The auction house handles three grades of flowers – A1, A2 and B1, which A1 being the top quality. B1 is very high quality, but the produce may carry remarks which may pertain to very slight change in quality due to changes in conditions in which the flower may have been grown.

Bloemenveiling Aalsmeer has been handling Indian roses for 5-6 years now and officials say they have seen the quality of roses sourced from the country improve. They now get A1 grade of roses from India. This means Indian growers are meeting the exacting standards in the highly competitive world flower market. But the country has a long way to go before it can catch up with the African countries such as Kenya, Zimbabwe and Tanzania as source for flowers. Israel is today the top source for flowers in the global market.

Further, India also exports roses to Japan and the annual rose exports to the country are placed between 13 million and 14 million stems. However, the future of Indian floriculture in Japan, the biggest in Asia and more than the combined value of the Middle East and Singapore markets, now seems to be tantalizingly poised. Japan is estimated to have as many as 300 floriculture auctions in operation. The Japanese ministry for agriculture has recently issued a notification seeking royalty payment related information from international suppliers. The Japanese notification assumes significance since representatives of various international breeders have been demanding that Indian growers pay royalties. Indian growers in turn have been refusing to pay up, citing several constraints, including poor yields from these varieties.

Competitiveness Issues

Freight from India to Europe and USA is much higher than Israel and African countries, India also suffers from import duty disadvantage in comparison to almost all its competitors. While Indian flowers are subject to the import duty in Europe, most of its competitors (Israel, African countries except South Africa, Latin America) have total duty exemption for export to EU countries. In the US, Equador and Columbia are able to provide roses at 70 cents a kg, which is way below the Indian rate of $ 3 a kg.

Price Fluctuation

Most of the exporters from India flood the market on special occasions, such as, Christmas, Valentine's Day, Mother's Day, Russian Women Day, *etc.* disturbing the overall market demand and supply equilibrium. This practice is considered as 'dumping' by the auctioneers and other importers besides bringing down the realizations. In fact, the pattern of exports of roses dictated by the seasons and festivals in the importing countries – March and April to Japan, Europe and Australia; in May to Singapore and the Middle East; from June through September to Australia; August is for Japan and Europe. Each stem costs the grower ₹ 1 and the margins are significant, especially when exported – ₹30 during valentine, ₹10 during Christmas and ₹5 during much of year.

During the lean export months of May, June and July, when Europe grows its roses, the growing domestic market has been absorbing the produce though at a far reduced rate of Re.1.50 a stem. But this is also the period when African countries like Zimbabwe and Kenya, which do not have a significant domestic market, have started dumping their roses in India. These roses are available in the Indian market at less than Re.1 a stem, much cheaper than the Indian blooms. Apparently, they are willing to sell on cost basis to establish themselves in the market and are able to do so by circumventing the Indian tariff barriers allegedly manned by pliable officials.

Quality Problem

The consignments from India often suffer adverse quality remarks. Quality remarks (on defective consignments) by the auctioneers result in heavy discounts on realizable price. Main reasons for poor acceptance of India produce include:

☆ Damaged flowers on arrival

☆ Bunching is not uniform both in terms of stem length as well as opening of the buds

☆ Diseases like Botrytis, mildew, etc, and

☆ Opening of flowers is not uniform.

Liberlisation

The liberalisation of industrial and trade policy in July 1991 paved the way for the development of export oriented production of cut flowers. The plants, fruits and seeds (Regulation of import into India) Order, 1989 known as New Seed Policy had already made it feasible to import the planting material of international varieties. The floriculture was also included as one the areas eligible for setting up of EOUS in the Exim policy. The enterprises started evaluating floriculture as one of the possible areas for diversifying into agri business. The Government agencies like APEA identified the potential for export of cut flowers.

Growth and Export Status of Indian Floriculture: A Review

Floriculture is the branch of horticulture that deals with the cultivation of flowering and ornamental plants for sales or for use as raw materials in cosmetic industry. Demands for floricultural products are steadily increasing both in the domestic as well as export markets. India has made significant improvement in the production of flowers, particularly cut flowers, which have good potential for export. Floriculture is important from the economic perspective as well. Commercial floriculture has been steadily increasing with increased use of protected cultivation employing greenhouse, shade nets, polyhouse etc Commercial flowers cultivation in India provides an opportunity for rural development owing to its higher returns per unit area and the new employment opportunities. India has a scope to bridge the gap between demand and supply as global demand of floricultural products is growing at a faster rate. India is enriched with diverse agro-climatic conditions such as, fertile land, suitable climate, abundant water supply, low labour cost, availability of skilled manpower, *etc.* which are quite beneficial for growing a variety of flower plants throughout the year.

International Scenario

Globally more than 145 countries are involved in floriculture industry and the global floriculture trade is estimated to be at US$ 70 billion at present (ICAR Vision 2050). According to The International Association of Horticultural Producer s (AIPH 2010), 702,383 ha area was under flower production in different countries of the world, of which the total area in Europe was 48,705 ha, North America was 21,067 ha, Asia was 523,829 ha, the middle East was 4,026, Africa was 7,604 ha,

North America was 21,067 and central and South America was 97,152 ha. In that year, according to Indian Horticulture Database, India occupied a floriculture area of 183,000 ha, which was 26 per cent of the global area.

The global floriculture industry is experiencing rapid changes due to globalization and its effect on financial development in the different regions of the world. At the same time, competition is increasing worldwide. The Netherlands, USA, Columbia, Japan and Italy are well known as traditional growers of flowers. Some Asian countries like India, China, Bangladesh, Thailand, Vietnam, *etc.*, are also steadily improving horticultural production. Also in Latin America and Africa, production is increasing very rapidly. Major flower consuming countries in the world concentrate in the Western Europe and North America. Germany, USA, UK, the Netherlands, France and Switzerland together consume around 80 per cent of the total flower production. Of the world's ten largest domestic markets for cut flowers, six are in Europe, namely Germany, the UK, France, Italy, the Netherlands and Spain. Other important markets are the US and Japan, accounting for around 20 per cent each. Recently, Russia and the Middle East have also become important markets demonstrating rapid market growth.

World floriculture trade is mostly depending on the trade of cut flowers and buds, cut foliage, potted plants and bedding plants. Main cut flowers in world trade are rose, chrysanthemum, carnation, gerbera, and lily.

Figure 7.9 shows the worldwide trade of the major countries in cut flowers and buds export. It is seen that the Netherlands is the world leader. Total world export of floriculture products stands at USD 9,784,525,000 and Netherlands claims 47.7 per cent of total world exports. The other major countries are Colombia, Ecuador, Kenya, Ethiopia and Belgium. India is in 14th position in exporting floricultural products.

Figure 7.10 shows the major importing countries in the world in cut flowers and buds. It is clearly seen that Ger man y possesses lea din g position, UK a n d USA possessing second and third position respectively. Other European countries are the major destinations for cut flowers and buds export.

Indian scenario: India's, commercial floriculture has gained momentum in the 1990's. The development of Indian commercial floriculture has centered around the production of rose, marigold, gerbera, chrysanthemum, gladiolous, anthurium, carnation, orchid, tuberose, lilium, alstroemeria *etc.*

Figure 7.11 shows the area under production of floricultural crops from 1993-94 to 2013-14. Production area shows a continual increase since 1993-94. Only in the year 2002-03 the area under cultivation decreases. In India, during the year 1993-94, the area under flower cultivation was 53000 hectare and then area has been increasing with CAGR (Compounded Annual Growth Rate) of 7.76 per cent. It is seen that the growth rate has drastically changed during the year 1995-96, 2003-04 and 2011-12.

Figure 7.12 shows the trend of both loose flowers and cut flowers production starting from the year 1993-94. It is evident from the figure that there is a gradual rise of flowers production in the following years. Loose flowers production almost keeps uniformity in its incremental rate.

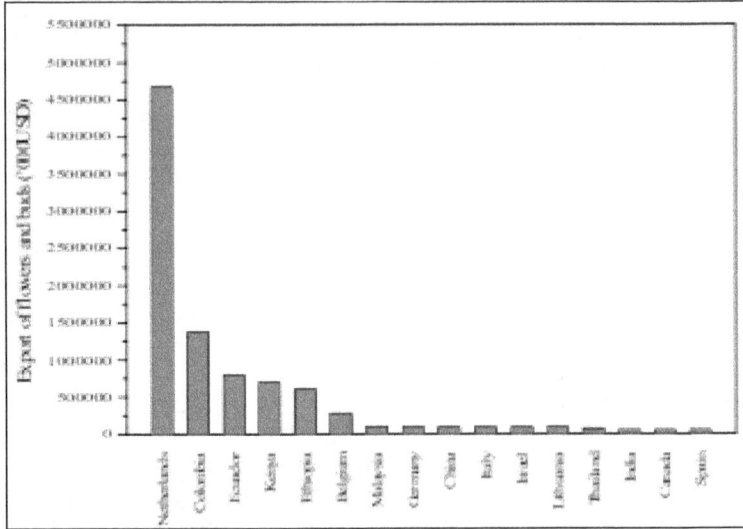

Figure 7.9: World's Leading Cut Flowers and Buds Exporting Countries.

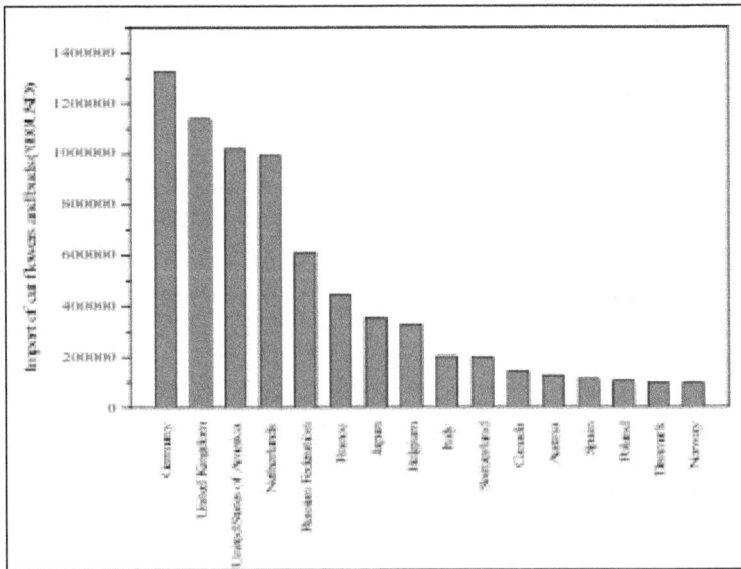

Figure 7.10: World's Leading Cut Flowers and Buds Importing Countries.
***Source*: ITC Trade Map- International Trade Statistics, 2014.**

It can be observed that cut flowers production slowly progresses up to the year 2006-07; thereafter its production tremendously rises at a CAGR of 40.5 per cent. From the graph, it is clear that the cultivation of cut flowers is in the process of accelerating the floriculture development in India. Since at present cut flowers are highly demanding particularly for export purpose and India has been shifting from traditional flowers to cut flowers production.

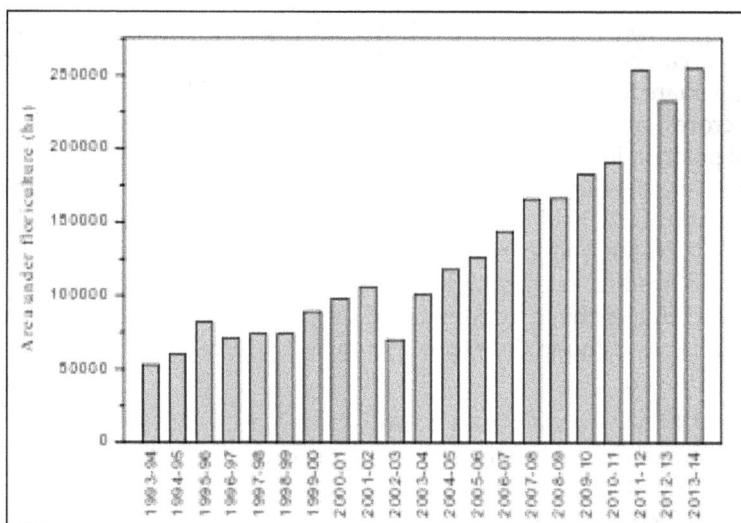

Figure 7.11: Area under Floriculture (ha) in India from 1993-94 to 2013-14.

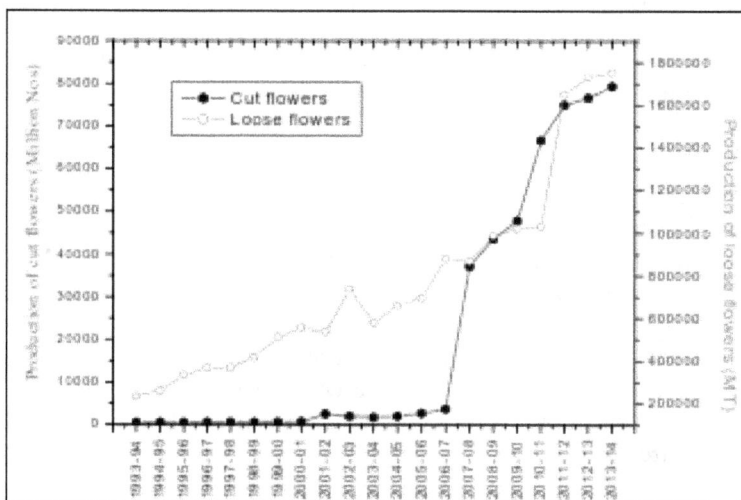

**Figure 7.12: Yearly Production Quantity of Important Indian
Loose Flowers (MT) and Cut Flowers (Million nos).**
Source: Indian Horticulture Database, 2008 and 2014.

Export Potential of Indian Floricultural Products

Indian floricultural products have been attaining to the world floriculture trade by increasing the productivity. The Agricultural and Processed Food Products Export Development Authority (APEDA) manages for developing and promoting agri-exports including flowers.

Figure 7.13 shows India's export of the floricultural products from 2000-01 to 2014-15. It can be observed that 2001-02 was the exceptional year. During that year India's share to the global market was just 18,803.67 MT with an export value of 115.33 crores, afterward both quantity as well as value increases and reaches maximum position in the year 2006-07. India's total export of floricultural products during that year was 42,545.29MT with an export value of 652.70 crores. It is also observed that from 2006-07 production quantity as well as value slowly decreases up to 2009-10. In the time periods 2009-10 to 2014-15 though the quantity of the total exported products varies slightly, export value has significantly increased. It has happened due to the increasing trend of global market price. Thus, India has an opportunity to increase the export potential by increasing the productivity of commercial flower.

Figure 7.14 shows the major countries, where India exports floricultural products, its quantity, and value. It is observed that the export to the different countries is quite uneven. In the year 2014-15 the floricultural products have been exported to 105 countries, of which 86.3 per cent are exported to the 18 countries shown in the figure. It is clearly seen that the biggest export market is the USA importing 5490 MT quantity by 98.13 crores. The USA imports 23.9 per cent of the total exported quantity followed by UK, Germany, the Netherlands, UAE, Canada, Japan, Australia, Italy and Singapore. These ten countries together import more than 70 per cent of the total exported quantity. The Netherlands, which is known as the leading exporter in the world trade of flowers, also imports a large amount of floricultural products from India.

Export of Cut Flowers from India

Floriculture is an important and upcoming trade with potential both in domestic as well as export markets. The world over, the flowers have gained an important place in one's life be it for religious purposes or personal decoration. The global floriculture industry with an investment of about US$ 40 billion is growing at an annual rate of 10 to 12 per cent. The USA, Japan, Western Europe and the major markets for the flowers. Besides, Eastern Europe, South Korea, Thailand and Indonesia are also coming up as large consuming countries. In the producing countries Netherlands alone enjoys 56 per cent followed by Columbia 11 per cent. Among the floriculture products major share is of "Cutflowers".

Total import/export trade of cutflowers in the world is estimated at US $ 4100 million. Main importing and exporting countries are given in Tables 7.1 and 7.2 respectively in value terms. Of the US $ 3716.8 millions of imports, the share of Germany was the highest (30.3 per cent) followed by USA (16.8 per cent), UK (9.7 per cent), France (9.7 per cent) and Netherlands (8.4 per cent). These five countries together shared nearly 75 per cent imports. Netherlands which exports large amount of flowers also imports cutflowers. Regarding exports (Table 7.15) Netherlands had the maximum share of US $ 2102.2 million (56.5 per cent) followed by Columbia (14.1 per cent), Israel (4.2 per cent) together constituting nearly 75 per cent of world's export. In fact, the share of Netherlands in the total exports has come down from 64 per cent in 1991 to 56.5 per cent in 1995. This is due to increasing shares of exports

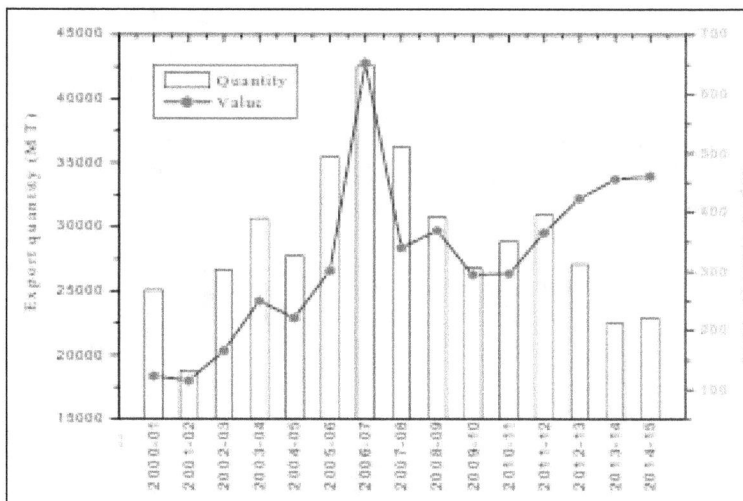

Figure 7.13: Export of the Floricultural Products from India in Quantity (MT) and Value (Crores).

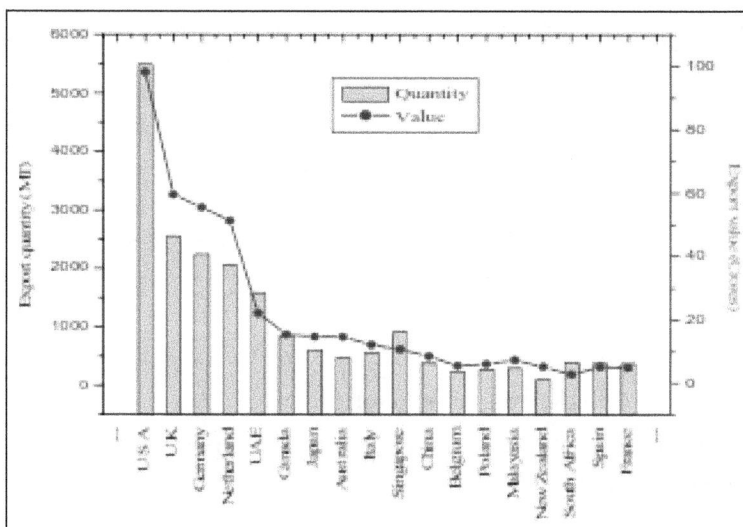

Figure 7.14: Floricultural Products Export from India to Major Importing Countries by Quantity (MT) and Value (Crores)
Source: APEDA 2000-01 to 2014-15.

from developing countries like Columbia, Kenya, Ecuador, Thailand, Zimbabwe *etc.* India has also appeared in the world cutflower trade with about 0.3 per cent share of exports.

Floriculture Exports from India

Floriculture is a very broad group consisting wide range of products such as flowers, cut flowers, flowering and ornamental plants, bulbs, tubers, corms, rhizomes, chicory, orchids, mosses *etc.* For export purposes, all floriculture products are grouped into four categories, which are given in Table 7.3 along with their qualities and values for the year 1998-99. It is seen that cutflowers in the largest group sharing nearly 72 per cent value of floriculture exports. This is followed by a group of foliage, branches mosses and plant parts (17.16 per cent), group of rooted cuttings, slips, mushroom spawn (9.92 per cent) and the last bulbs, tubes, tuberous roots, rhizomes, chicory *etc.* (1.05 per cent). Thus the cutflowers - fresh and dried dominate floriculture export trade of India.

Table 7.15: Main Import Markets of Cutflowers by Value 1995
(Value million US $ [CIF])

Important Markets	Value	Per cent
Total world imports of which	3716.8	100
Germany	1124.9	30.3
USA	622.8	16.8
United Kingdom	360.0	9.7
France	358.9	9.7
Netherlands	310.4	8.4
Japan	216.1	5.8
Switzerland	168.7	4.5
Italy	118.9	3.2
Belgium	95.8	2.6
Denmark	63.0	1.7
Canada	49.6	1.3
Other 15 countries	227.07	6.1

Cutflowers Exports

The cutflowers are exported in two forms:

a) Fresh form for bouquets and ornamental purposes.

b) Dried, dyed, bleached *etc.* for other purposes.

The information of export of fresh cutflowers in respect of quantities values and the countries to which exported is given in Table 7.4 for the 1998-99. Total quantity exported was 2722 MT valued at ₹25.12 crores. The biggest export market was Japan importing 19.65 per cent cutflower followed by USA 10.44 per cent, Germany (7.96 per cent), UK (7.40 per cent), Australia (5.43 per cent) and Netherlands (4.67 per cent). The fresh cutflowers are exported to 54 countries, but nearly 68 per cent are exported to 10 countries shown in the table. Unit value revealed that the highest price was paid by Netherlands (₹376.62/kg), followed by Japan (₹155.92/kg) and Singapore (₹141.49/kg). The overall average price received was ₹92.26/kg. The price offered by USA was quite low (₹ 51.33/kg). Information of export of other cutflowers is given in Table 7.16. The quality of cutflowers exported was 8295 MT valued at ₹51.04 crores. They were exported to in all 55 countries but the major export markets were USA sharing 35.25 per cent exports followed by Japan (27.98 per cent), Netherlands (9.50 per cent), Germany (5.41 per cent), U.K (4.91 per cent) and so on. These five countries shared about 83 per cent of total exports. Thus, the exports to different countries are unevenly spread. Unit values or prices received varied from ₹26/kg by Japan to ₹137.38 by Germany. The overall average unit value received was ₹61.52/kg. There was a very wide difference in the prices received from different importing countries. Although USA and Japan imported largest quantities, they offered low prices.

Table 7.16: Main Exporting Countries of Cutflowers by Value 1995
(Value US million dollars)

Exporting Countries	Value	Per cent
Total world exports of which	3716.8	100
Netherlands	2102.2	56. 5
Columbia	525.0	14.1
Israel	158.7	4.2
Kenya	103.5	2.7
Ecuador	102.2	2.7
Italy	92.3	2.4
Thailand	80.5	2.1
Spain	60.0	1.6
Zimbabwe	51.9	1.3
France	39.6	1.1
New Zealand	37.1	1.0
Other 13 countries	364.0	9.8

Table 7.17: Export of Floriculture Products from India (1998-99)

Product Group	Quality	Value	
		₹ in Crore	Per cent
Bulbs, tubers, tuberous root corms, rhizomes chicory dormant, growing, in flower etc.	1889.69 (000 No.)	1.11	1.05
Other live plants incl. Rooted cuttings, slips, mushroom spawn, etc.	2971.96 (MT)	10.51	9.92
Cutflowers flower bulbs, suitable for bouquets/ornamental purposes-fresh, dried, dyed, bleached, impregnated	11017.78 (MT)	76.16	71.87
Foliage, branches, other plants parts, mosses for bouquets/ ornamental purposes – Fresh, dried, dyed, bleached etc.	3076.19 (MT)	18.18	17.16
Total (000 No.) (MT)	**1889.69 17865.93**	**105.96**	**100.00**

Table 7.18: Export of Fresh Cutflowers from India-Direction of Trade (1998-99)

Country	Quantity		Value	Unit value
	MT	Per cent	₹ in crore	₹/kg
Japan	535	19.65	8.34	155.92
USA	284	10.44	1.46	51.33
German Fed. Rep	217	7.96	1.38	63.87
UK	202	7.40	1.40	69.32
Australia	148	5.43	1.52	102.95
Netherlands	127	4.67	4.78	376.62
UAE	121	4.46	1.00	82.28
Singapore	116	4.25	1.64	141.49
Italy	53	1.96	0.47	87.23
Denmark	45	1.64	0.23	51.75
Other 44 countries	875	32.14	2.90	33.15
Total	**2722**	**100.0**	**25.12**	**92.26**

Table 7.19: Export of other Cutflowers-Direction of Trade (1998-99)

Country	Quantity		Value	Unit value
	MT	Per cent	₹ in crores	₹/kg
USA	2924	35.25	12.74	43.57
Japan	2321	27.98	6.03	26.00
Netherlands	788	9.50	9.68	122.84
German Fed. Rep	449	5.41	6.16	137.38
UK	407	4.91	2.69	66.00
Sri Lanka	246	2.97	0.76	31.05
Italy	167	2.01	2.07	124.06
France	164	1.98	1.80	110.29
Spain	125	1.51	1.08	86.33
Other 46 countries	704	8.49	8.03	114.06
Total	**8295**	**100.00**	**51.04**	**61.52**

World Floriculture Scenario

Introduction

Floricultural production contains a wide variety of different types of plants and plant mate- rials. They can be divided into cut flowers and foliage, potted plants, garden plants, nursery stock (trees), flowering leafy, annuals and perennials and bulbs and tubers. Parental material like plant parts from tissue culture, cuttings is used as a start in production systems to grow the end products for the market.

The production of floricultural products has grown quite consistently over the last 35 years. Based on Dutch and American articles the production was estimated 11 billion dollars in 1985 growing to 24 billion in 1990, via 31 billion in 1996 to 44 billion dollars in 2000. We estimate that the production value was about 60 billion dollars in 2003, 26 billion EUR in 2016. This means an average yearly growth of 6 to 9 per cent. The highest growth has developed in Asia and the USA. And the future is promising.

Floriculture as an industry exists in almost every country of the world, but only few countries have emerged as major suppliers and buyers of flowers. The emergence of countries in Asia, Latin America and Africa as new production centers has increased the competition in the market; these production centers are able to produce quality products at lower costs. Their produce is favoured in the markets of Western Europe, America and Japan. Besides emergence of these new centers, the significant changes in the competitive relationships (internationally) and liberalization of world trade in the context of the WTO agreements are also expected to affect the international trade.

World Producing Areas

Area and Production

Flowers in the world are grown on are of 702 thousand hect. valued at 26.2 thousand EUR. value per hect is 0.03 Eur (Table 8.1) Asia with an area of 523.8 thousand hect is the leader, followed by China – 286.hect. Etheopia with 645 hect is at the bottom with just its value of 80 Euro per hect. All this will show that flowers do not play an important part in the world agricultural scenario.

Table 8.1: Production, Flowers and Pot Plants World

Region	Area of Land* (Ha.) (A)	Production Value*, min. EUR (B)	Productivity of Land (B/A)	Holdings*
Europe	48,705	10,843	0.22	37,319
Netherlands	7,560	3,780	0.50	5,372
Middle East	4,026	220	0.05	6,100
Africa	7,604	634	0.08	1,461
Kenya	2,180	299	0.14	140
Ethiopia	645	90	0.14	80
Asia	523,829	7,608	0.01	156,764
China	286,068	2,668	0.009	n/a
North America	21,067	5,450	0.25	9,319
Central and South America	97,152	1,441	0.01	12,494
World	702,383	26,196	0.03	223,457

* Based on latest data available in 2010.

Source: AIPH, International Statistics Flowers and Plants (2010).

The Production of Ornamentals Worldwide

Figures on the production of cut-flowers and pot-plants worldwide are tabulated in ISPF 2014, though there is no compilation for cut-flowers alone. Nevertheless, the information reveals much about the current state of ornamentals production. The statistics are shown in the Table 8.2.

The striking feature of these data is the enormous production area - though not necessarily value - and numbers of holdings in China and India. This is apparently mainly for national use - little is destined for western markets, as later export/import figures will show. Japan and the Philippines also have huge numbers of holdings, though they are also lacking in production areas and values.

In Europe, the largest areas of production and number of holdings are found in Italy, France, Germany, Spain and Poland, though, as expected, it is the UK that concentrates its production on relatively few, larger holdings, and the Netherlands which excels in production value. In Africa, South Africa has by far the largest production area, dwarfing that of new producer countries (Kenya and Ethiopia).

In the Americas, the USA has a large production area and number of holdings. Substantial areas of production in Brazil and Mexico are also linked to large numbers of holdings, whereas in Columbia the substantial area of production is associated with relatively few.

Table 8.2: Recent Production Areas, Production Values and Numbers of Holdings for Flowers and Pot-Plants

	Area (ha.)			Production Value (€ millions)	Number of Holdings
	Protected	Open	Total		
Europe					
Austria	211	197	408	195	1,320
Belgium	426	912	1,338	227	841
Czech Rep.	_2	-	612	129	1,000
Denmark	265	-	265	453	415
Finland	128	26	154	101	697
France	-	-	9,159	954	7,234
Germany	1,848	4,863	6,741	1,319	4,449
Greece	363	732	1,094	66	-
Hungary	280	680	960	42	850
Ireland	-	-	415	17	133
Italy	5,443	7,282	12,724	1,330	14,093
Netherlands	4,396	2,905	7,301	4,130	4,127
Norway	113	-	113	32	402
Poland	1,616	3,840	5,456	180	4,800
Portugal	610	1,090	1,700	258	1,415
Spain	1,911	4,611	6,522	880	3,969
Sweden	135	-	135	154	501
Switzerland	195	-	195	294	402
UK	545	5,163	5,708	430	304
Middle –East Israel	1,748	1,000	2,748	129	1,100
Turkey	-	-	1,192	57	-
Africa Ethiopia	700	1,300	2,000	470	300
Kenya	-	-	4,039	595	140
Morocco	113	52	165	10	-
South Africa	-	-	11,461	49	900
Tanzania	-	-	120	21	15
Uganda	-	-	205	42	20
Zambia	-	-	195	25	30
Asia/Pacific Australia	349	3,840	4,189	175	877
China	-	-	169,081	5,095	83,338

	Area (ha.)			Production Value (€ millions)	Number of Holdings
	Protected	Open	Total		
Hong Kong	-	-	153	5	-
India	-	-	242,000	-	-
Japan	10,190	9,869	16,840	2,512	77,980
Korea Rep.	3,132	-	3,132	598	10,383
Malaysia	-	-	2,000	102	600
Philippines	-	-	670	3	42,189
Singapore	-	-	312	27	<149
Taiwan China	-	-	4,929	199	-
Thailand	-	-	9,280	60	25,000
North America					
Canada	814	-	814	786	1,885
USA	21,294	8,113	29,407	4,434	26,884
Central/South America					
Brazil	-	-	13,800	1,747	8,000
Columbia	6,783	-	6,783	1,012	541
Costa Rica	-	-	850	116	-
Ecuador	5,377	1,292	6,669	630	-
Mexico	1,158	13,963	15,121	281	7,857
Regional Totals					
Europe			61,000	11,191	>46,952
Middle East			3,940	186	>1,100
Africa			18,185	1,212	>1,405
Asia/Pacific			452,586	>8,776	>240,367
North America			30,221	5,220	28,769
Central/South America			43,223	3,786	>16,398

Source: ISPF 2014, from a variety of worldwide sources (see original publication for details); naturally the information collated will be subject to many different collection protocols in the original countries, so it should be treated as a general guide.

2. Indicates data not available.

ISFP 2014 includes a similar dataset for bulb production, shown in the Table 8.3. The UK is not included in these figures, perhaps because statistics comparable with those of the countries shown were unavailable.

As expected, the Netherlands leads bulb growing, but China, the UK, the USA and France also have substantial production areas. It is interesting to note that in the Netherlands and Japan similar numbers of holdings are involved despite a 40-fold difference in production areas.

**Table 8.3: Recent Production Areas, Production Values and
Numbers of Holdings for Bulbs[1]**

	Area (ha.)	Production Value (€ million)	Number of Holdings
Belgium	171	[2]	46
France	1,115	-	-
Germany	270	-	161
Netherlands	18,528	570	1,551
Turkey	55	-	-
China	4,174	82	-
Japan	478	-	1,660
USA	2,521	55	193

1. Based on the most recent year available, mostly 2012 or 2013.

2. Indicates data not available

When it comes to the main production areas of ornamentals we see the following:

☆ In Europe the most ornamentals are produced. The Netherlands is known for cut flowers and potted plants, as well as bulbs, for annuals and perennials. Germany has a name in nursery stock and garden plants. In Italy lots of flowers and potted plants are produced. While Denmark is famous of their potted plants. France is a broad player when it comes to the different types of products. UK, Belgium en Spain are small players in this field.

☆ The total production value is about 10 billion dollars, and has stabilized. Countries with the largest share in cut flower production are Germany (11 per cent), Italy (18 per cent) and the Netherlands (35 per cent).

☆ In North and South America ornamental production mainly consist of flowers and cut-tings. In North America (USA and Canada), where 80 per cent of the flowers and potted plants are grown, California and Florida are the most important production regions. In South America, Colombia (6 per cent) and Mexico (3 per cent) together with Costa Rica and Ecuador have developed rapidly over the last decade as producers of flowers and parental material. Also Brazil (6 per cent), for a long time, has been a producer of parental material and cuttings of potted plants.

☆ Production in Africa has increased over the last decade, with Kenya in the frontline fol- lowed by Tanzania, South Africa and Uganda. Flower production, of roses in particular has made enormous progress. Production in countries like Zimbabwe and Ivory Coast has decreased because of the political situation in the country. For Africa we see rapid growth in a high risk environment. Besides an estimate of 0,2 billion dollar production there are hardly figures available about ornamental production in this continent.

☆ About Asian production we hardly have any figures from the developing countries. Japan is traditionally a producer of specialties in ornamentals. South Korea, India and Thailand are coming up strongly as producers of ornamentals, followed by Taiwan. The production of cut flowers in Israel has decreased because of the political situation. Local production for local markets seems to be big. Export is starting from countries like China, Vietnam, India, *etc.*

☆ In Oceania, Australia and New Zealand produce mainly cut flowers, but are relatively small producers.

Major flower and plant producing countries focus mainly on their respective domestic markets. The Netherlands is an exception, with its small home market and strong focus on exports. The production of flowers has increased manifold in the last two decades. The world production of floriculture products is estimated to be around USD 50 billion.

Traditionally countries like China, India Japan and USA, have been major growers of flowers. However, these countries have been producing floriculture products mainly for domestic consumption and have not been major suppliers of flowers internationally. Holland is the largest producer of cut flowers in the world for international supply, followed by Colombia and Italy. In all nearly 80 countries are active in the world floriculture trade. The total area under floriculture in the world (both under protected and open cultivation) is around 2,23,000 hectares (Table 8.4).

Table 8.4: Country-Wise Area Under Floriculture (Area in Hectares)

Country	Protected	Open Field	Total
Europe			
Austria	216	529	745
Belgium	542	1100	1642
Denmark	330	353	683
Finland			144
France	1747	2048	3795
Germany	273	3908	6621
Greece			882
Hungary			1050
Ireland	49	251	300
Italy	4402	3252	7654
Netherlands	5518	2499	8017
Norway			310
Poland			807
United Kingdom	859	5945	6804
Spain	2442	1788	4230
Sweden	27	89	116
Switzerland	272	373	644
Europe total			**44444**

Country	Protected	Open Field	Total
NORTH AMERICA			
Canada	988		988
Mexico	800	4200	5000
United States	7121	9279	16400
North America Total			**22388**
ASIA AND PACIFIC			
China			59527
Hong Kong			343
India			34000
Indonesia			1000
Japan	100488	11170	21218
Korea	2229	1718	3947
Malaysia			1286
Philippines			670
Singapore			162
Sri Lanka			200
Taiwan			4033
Thailand			7000
Asia and Pacific total			**133386**
MIDDLE EAST AND AFRICA			
Israel			1950
Kenya	350	850	1200
Morocco			450
Turkey	400	300	700
Uganda			40
Zimbabwe	245	997	942
Middle East and Africa total			**5282**
CENTRAL AND SOUTH AMERICA			
Argentina			800
Brazil		10285	10285
Colombia	4710	47	4757
Ecuador			1158
Guatemala			605
Central and South America Total			**17605**
WORLD TOTAL			**223105**

Source: Floriculture International, March 1997.

Japan, the US, Italy, Columbia and the Netherlands are the world's main producers of both cut flowers and pot plants. Brazil and China also have large areas under floriculture. While Brazil has specialized in cultivation of bulbs, China is yet to make a mark as a major exporter.

World Scenario of Floriculture

There has been a rapid growth in demand and consumption of floriculture products in recent decades. Cultivation and consumption of flowers have been part of tradition in world over. It indicates that Netherlands, Italy, Germany and Japan have strong tradition of growing and consuming flowers. The expansions in area and production of flowers in non-traditional regions have been one of the noticeable features. Recently, new production centres are developing in Latin America, Africa and Asia to meet the increasing demand of importing countries and to expand their domestic market. Columbia, Costa Rica Chile, Kenya, Rhodesia, Morocco, South Africa, Israel, India, China and Shrilanka these are the new floriculture centres. The floriculture market has concentrated in Western Europe, North America and Japan. Western Europe accounts for half of the world's cut flower production and consumption of the product. Flower council of Holland has projected a European consumption of cut flowers to the tune of US $ 16.6 billion in 2008. The new markets emerging in Europe are Poland, Hungary, Slovakia and Ireland. The mostly preferred cut flowers in the international market are roses, tulip, chrysanthemum, gerbera, orchids and gypsophilla. The world statistic of floriculture is presented in Table 8.5.

Table 8.5: Major Flower-Producing Countries

Sl.No.	Name of the Country	Share of Flower Production
1	Netherlands	33 per cent
2	Japan	24 per cent
3	U.S.A.	12 per cent
4	Italy	11 per cent
5	Thailand	10 per cent
6	Others	10 per cent

Source: S.K.Bhattacharjee, (2006) Advances in Ornamental Horticulture, Volume 6 pp. 20 to 32.

It is evident from the above (**Tables 8.5–8.8**) that Netherlands and Japan both are leading flower producing countries with 57 per cent. As far as concern to holding of area under greenhouse floriculture, it concentrates in Netherlands, U.S.A, Italy, Germany and Spain. Netherlands has the dominating contributor as area and production, export and import. Especially, Netherlands has obtained command position in export. Land area of Flowers and ornamental plants are given in Table 8.8A.

It is need to explain that the other exporters, those are Ecuador, Zimbabwe, Thailand, Belgium, France, Germany, Mexico, Costa-Rica, New Zealand, U.S.A., Turkey, Canada, Australia, South Africa, Zambia, Singapore, India, *etc.*

Table 8.6: Country-wise Area under Greenhouse Flowers

Sl.No.	Name of the Country	Area (ha)
1	Netherlands	5556
2	U.S.A.	4532
3	Italy	4402
4	Germany	2765
5	Spain	2369
6	France	1747
7	U.K.	999
8	Belgium	542
9	Denmark	330

Source: S.K.Bhattacharjee, (2006) Advances in Ornamental Horticulture, Volume 6 pp. 20 to 32.

Tale 8.7: Major Cut Flower Exporting Countries

Sl.No.	Name of the Country	Share of World Export
1	Netherlands	59 per cent
2	Columbia	10 per cent
3	Italy	6 per cent
4	Israel	4 per cent
5	Spain	2 per cent
6	Kenya	1 per cent
7	Others	18 per cent

Source: S.K.Bhattacharjee, (2006) Advances in Ornamental Horticulture, Volume 6 pp. 20 to 32.

Table 8.8: Major Cut Flower Importing Countries

Sl.No.	Name of the Country	Share of World Import
1	Germany	34 per cent
2	U.S.A.	18 per cent
3	France	12 per cent
4	U.K.	11 per cent
5	Netherlands	7 per cent
6	Japan	6 per cent
7	Switzerland	5 per cent
8	Others	7 per cent

Source: S.K.Bhattacharjee, (2006) Advances in Ornamental Horticulture, Volume 6 pp. 20 to 32.

Table 8.8A: Land Area (in thousand hectares)
Flowers and Ornamental Plants (including seets)

GEO/TIME	2001	2002	2003	2004	2005	2006	2007	2008	2009	2010	2011
Belgium	1,2	1,1	1,1	1,1	1,1	1,0	1,0	1,0	0,9	1,0	
Bulgaria	4,5	4,3	3,9	4,8	4,2	3,8	3,6	3,3			
Czech Republic	1,4	1,2	1,0	1,3	1,1	1,3	1,9	1,4	1,1	1,3	
Denmark	0,2	0,1	0,1	0,1	0,1	0,1	2,3	2,5	1,8	1,9	1,9
Germany	6,4	6,5	6,9	6,7	6,4	6,8	7,0	6,6	6,2	8,4	
Estonia	0,1	0,1		0,1			0,1				
Ireland	0,9	0,9	1,0	1,3	1,3						
Greece	0,8	0,8	1,1	1,1	1,1	1,2	1,2	1,1	1,1	0,8	
Spain	5,4	2,1	3,1	2,7	1,7	2,4	3,0	3,0	4,2	6,7	
France	8,4	8,2	8,2	7,9	7,8	7,9	7,9	7,8	8,8	8,1	
Italy	9,4	9,2	8,3	8,7	8,0		4,3	4,3			
Cyprus			0,1	0,1	0,1	0,1	0,2	0,2	0,2	0,1	0,1
Latvia		0,4	0,5	0,1	0,1	0,1	0,1	0,1		0,1	
Lithuania			0,1		0,1	0,1	0,1	0,1	0,1	0,1	
Luxembourg										0,0	
Hungary	1,0	1,0	1,3	1,4	1,4	1,1	0,4	1,5	0,9	0,5	
Malta										0,0	
Netherlands	26,1	28,2	28,4	27,4	26,8	27,7	27,5	28,1	27,4	26,2	
Austria	0,3	0,3	0,3	0,2	0,2	0,2	0,2	0,2	0,2	0,4	
Poland	10,9	3,2	7,4	7,5	2,8	3,3	2,7	1,7	2,6	4,1	
Portugal	0,6	1,3	1,4	1,4	1,5	1,5	1,6	1,6	1,1	1,2	1,2
Romania	0,2	0,2	0,2	0,2	0,2	0,3	0,2	0,2	0,2	0,2	
Slovenia	0,1	0,1	0,2	0,2	0,2	0,1	0,2	0,2	0,2	0,1	
Slovakia	0,5	0,1	0,2	0,2	0,2	0,2	0,3	0,3	0,2	0,2	
Finland	0,1		0,1	0,1							
Sweden		0,5	0,5	0,5	0,5	0,5	0,5	0,5	0,5	0,4	
United Kingdom							6,0	6,0	5,0	5,0	
Croatia	0,2	0,2	0,2	0,2	0,2	0,1	0,1	0,1	0,3	0,3	

Source: Eurostat.

Emerging Supply Centres

The emergence of new production centers in Asia, Latin America and Africa has increased the competitiveness in the international floriculture market. The countries of these regions derive their competitiveness from low cost, high volumes and rapid innovation. These countries produce highly qualitative floriculture products at low cost, and are favoured in the markets of Northern America and Japan.

Cut flower production in Africa particularly in Kenya, Zimbabwe, Ivory Coast, Morocco, South Africa and Tanzania is highly competitive. Professionally managed companies owned by foreign investors with technical experts from Europe are active in floriculture in these countries. These countries have benefited from developing infrastructure, knowledge, transport facilities and entry to European market at zero duty on the basis of trade agreements. Bulk of flowers produced in these countries, are intended for export to European markets. Efforts are also being made by these countries to enter Asian markets.

Roses are major cut flowers grown in these countries because of higher returns from exports. Growers in South Africa, especially small growers, have focused on production for domestic consumption. African countries make use of market opportunities, such as supplementing the European supply during the winter months and cultivating species which are difficult to grow in Europe or which require high inputs of energy due to prolonged growing time. The climate is favourable and there is adequate supply of cheap land, water and labour in these countries but they suffer from lack of knowledge and absence of developed domestic markets. However, due to good coordination on supply side, African countries would continue to be strong in products with high volume and low cost.

World Consumption

Consuming Pattern

In the past, the flowers were mainly consumed by way of gifts on special occasions and in meeting institutional requirements. With the increase in income levels, personal consumption of flowers is also rising, both in the developing and

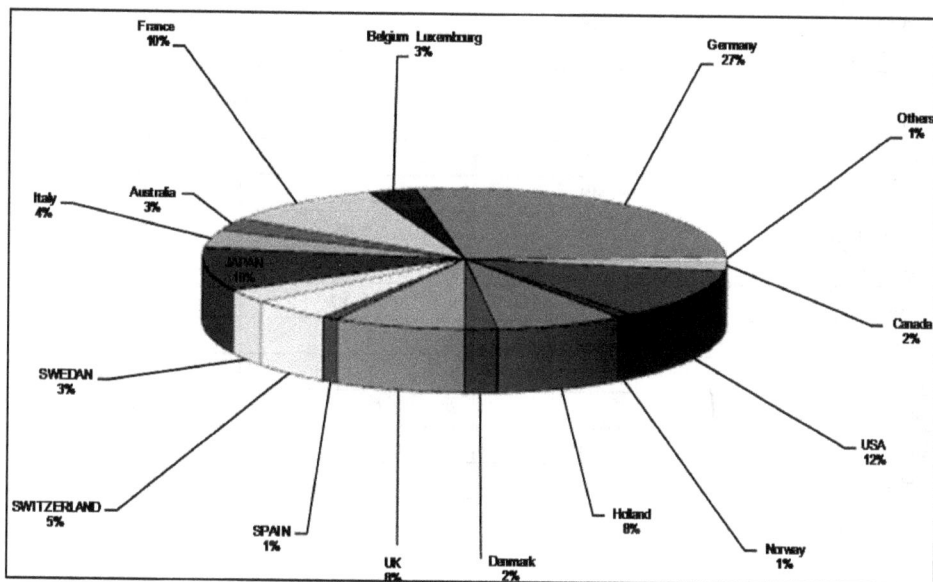

Figure 8.1
Source: Indian Horticulture Database – 2003.

developed countries. The consumption has increased almost four fold in terms of value in the past two decades *i.e.* 12.5 billion US$ in 1980 to 50 billion US $ in 1999. The increased consumption of flowers has mainly been concentrated in North America, Japan and Western Europe.

Western Europe consumes more than 50 per cent of the total cut flower production. Among the European countries, Germany continues to be the largest consumer of flower. Of late, consumption has also registered an increase in East European markets. The world consumption of flowers is shown in the following **Figure 8.1.**

According to the statistics available from the Flower Council of Holland, Switzerland, Norway, Austria and Germany, in that order, are the countries with highest per capita consumption of floriculture products. The details of World consumption, per capita consumption of flowers, market size and world trade are shown below (Table 8.9)

Table 8.9: World Consumption of Flowers and Plants per Capita and Market Size

Country	Per Capita Consumption ($)	Population (Millions)	Value of Total Market ($ millions)
Austria	105	7.99	836
Belgium/Luxembourg	66	10.49	690
China	00.40	1,203.10	488
Czech Republic	9	10.43	92
Denmark	80	5.2	416
Finland	66	5.09	335
France	66	58.11	3828
Germany	94	81.34	7607
Greece	31	10.65	332
Ireland	12	3.55	44
Italy	60	58.26	3496
Japan*	43	125.51	5397
The Netherlands	77	15.45	1191
Norway	160	4.33	693
Spain	24	39.40	950
Sweden	85	8.82	752
Switzerland	167	7.08	1183
United Kingdom	29	58.30	1680
United States	55	263.81	14586
Total of world floral market			**$44596**

Source: Floriculture International, March 1997.

*Cut flowers only.

There are nearly 80 countries which are active in the world floriculture trade. The total area under floriculture world over both under protected area as well as open cultivation is around 2,23,105 hectares. According to a study done by ITC (1995), the rate of increase in world trade has been around 10 per cent and developing countries share of total trade is consistently around 20 per cent. The leading world supplier is the Netherlands, followed-by Colombia and Italy.

Patterns behind the Consumption

There are a few trends that can help to explain the general conclusion out of Table 8.5, that the consumption per capita has increased strongly over 8 years after 1994. Because of a growing income people have started to buy more ornamentals for personal use as well as the gifts. Also the institutional use has risen; companies, events and recreation has started to use more flowers and plants for attractive decoration. In Europe and USA we see more small households occur of one or two persons, with a higher than average salary. So the amount of homes to be decorated per capita is rising. Another reason can be found in a trend called 'cocooning'. People in Europe and USA especially have started to spend more time at home, to get a feeling of peace and safety. Doing so they started to make the home cosier, with decorations like ornamentals. The trend of individualisation results in that people want to show their own life style. They also use ornamentals to do so. Finally there is a demand for quality products has grown, since there is more than enough produce; *i.e.* the English markets compete in who is selling the best quality products. In Europe there is also a trend in bringing nature back in the daily life, by using natural products. All these developments have contributed to a growth in the consumption of products per capita.

If we look at the developments in trade from 1994 to 2002, we see the following (Table 8.10): The trade within Europe, which was 6055 million dollar in 2002, has changed with 0 per cent in the eight years before. The extra products, that the growth in consumption per capita required, has obviously come from Africa (+74 per cent up until 242 million dollar) and Latin America (+45 per cent up until 274 dollars). Import from Asia seems to have declined by 2 per cent in that period.

With North America the trade has increased with 94 per cent up until 459 million dollar. To be able to supply the increased consumption per capita, further more the imports from Latin America has increased with 72 per cent until 908 million dollar, from Europe with 33 per cent until 361 million dollar and from Asia with 89 per cent. Also trade within Latin America has grown strongly (137 per cent up until 32 million dollar) in the eight year period. Next to this imports from Europe have increased with 52 per cent until 36 million dollar.

Internal Asian trade has been expanded with 28 per cent over the years until 148 million dollars. This might have resulted in a decrease in imports from Europe(-2 per cent until 250 million dollars) and Oceania (-2 per cent until 32 million dollar). Looking into the figures the size of countries and amount of inhabitants of continents has to be taken into account. While export from The Netherlands to Belgium adds up to the total, the trade between California and New York is unobserved.

Table 8.10: Floricultural Trade 2002

Export	Import Europe	North America	Asia	Latin America	Africa	Oceania
Europe	6055	361	250	36	34	15
North America	73	459	13	30	0	0
Asia	228	46	148	5	1	4
Latin America	274	908	23	32	0	0
Africa	242	6	9	0	8	2
Oceania	14	12	32	0	0	2
World Total	6886	1791	474	103	43	22

Global Demand

The world trade in floriculture increased steadily during the recent years to reach US $ 50 billion. The major import market in the world for all floriculture products is Germany. In 1995 it had a US $ 2000 million market, the world's biggest. Its share of total world imports in 1995 was around 30 per cent. The United States of America was the second largest importer (US $ 831 million in 1995), but there was a moderate decline as the share of the total declined from 14.5 per cent in 1987 to 12.0 per cent in 1995. France with 11 per cent of total world floriculture imports was the third largest importer followed by the Netherlands and United Kingdom whose share was 9 per cent each. The five countries, Germany, the United States, France, the Netherlands and United Kingdom together accounted for a little above 71 per cent of all imports in 1992. During the period 1991 to 1995, the fastest growing export market was Japan, with indexed growth of 162 per cent. The details of major markets for world floriculture are illustrated below (Table 8.11).

Table 8.11: Major Import Markets for World Floriculture

Country	Value ($ thousands)
Germany	2,012,726
United States	830,727
France	722,809
Netherlands	585,634
United Kingdom	586,809
Switzerland	367,253
Japan	313,865
Italy	284,524
Belgium-Luxembourg	234,085
Austria	192,497
Sweden	185,077

Country	Value ($ thousands)
Canada	144,179
Denmark	113,375
Spain	90,853
Norway	76039
Others	65,701
Total	**6,816,143**

Source: United Nations Comtrade data base,ITC/UNCTAD/WTO, October 1996.

Global Supply

The Netherlands is the largest producer of cut flowers in the world from around 8,107 ha. as far as area is concerned and Japan is nearly three times as big as the Netherlands with 21,218 ha. However, the productivity of the farms in the Netherlands is very high.

Bloom Book

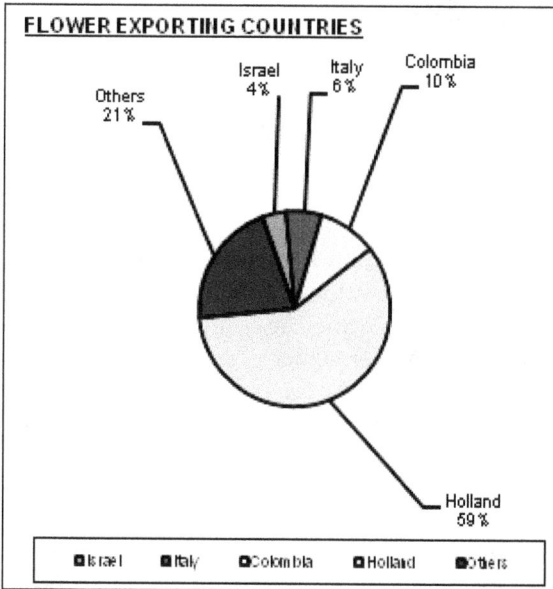

Figure 8.2

Greenhouse Cultivation

Global

50.000 hectares

Ecuador : 2,000 hec.

Colombia : 5,000 hec.

India

200 hectares

World Trade in Floriculture

In 2010 of the total production of flowers in the world around thousand tonnes, total quantity entering the trade (import/Export) was just 424 tonnes as imports and 466 tonnes export (Table 8.12). As regards imports the share of Kenya (100 tonnes was the highest at 100 tonnes and lowest among major countries was China (13.6 tonnes). Other countries which among the imports were Costa Rica (10.9 per cent), Ethiopia (9 per cent), Israel (6.5 per cent), USA (6.3 per cent), Ecuador(5.8 per cent), Colombia (5.2 per cent), Guatemala(3.2 per cent), Others (24.7 per cent) showed the balance. Cut flowers and cut foliage was height in the total at 315.7 tonnes. Priorities were not the same as in the case of total production of flowers. Which foe example Keya retained its first position at 96.5 tonnes, USA improved its position from 5 to 3. Other countries Costa Rica (33.9 per cent), China (19.6 per cent), Guatemala (7.8 per cent), Honduras (5.9 per cent), Sri Lanka (3.2 per cent), Thailand (2.8 per cent), Taiwan (2.9 per cent), Israel (2.6 per cent) and others (21.4 per cent) shared the balance 107 tonnes. Conifers and hardy perennial plants China(12.1 per cent), Israel (15.1 per cent), For JRep mac (22,5 per cent),Switzerland (5.6 per cent), Japan (4.6 per cent), Argentina (4.5 per cent) Egypt (3.8 per cent) Turkey (3.3 per cent) Others (28.5 per cent). Bulbs and corms Chile (25.4 per cent), Brazil (28.8 per cent), New Zealand (13.3 per cent) Taiwan (5.2 per cent), China (4 per cent) South Africa (4 per cent) Thailand (4 per cent), Others (15.3 per cent).

Other categories showed the balance. The order, of course changed in each case – But both the order of the quantity imported and the quantity imported was different in each case.

Trade Patterns

We have elaborated on the main streams of export of ornamental produce world wide. The figures show the value of export of ornamentals within a continent; one country exports to a market in another country in the same continent. Also the value of export streams of ornamental between continents will be shown. There is hardly any relation to production values of continents and countries, except that they are smaller a lot. Quite some countries for instance mainly produce for local markets. (See Italy which only produces for domestic use).

Cut Flowers

Based on figures of 2002, Europe is by far the largest market for cut flowers. The countries within together have an export value for Europe of 2660 million. US dollar. Further more Europe imports from main suppliers like South America (228 mln. USD), Africa (196 USD), Asia (174 mln USD) and North America (65 mln USD). So, the export in and outside of Europe totals about 3425 mln USD).

North America is the second largest market. USA and Canada have an export to each other of about 108 mln USD. The continent imports large amounts of cut flowers from the countries in South America (859 mln USD). Europe is with 141 mil USD the second supplier, followed by Asia with 26 mln dollar export value on cut flowers. So, the export in and outside of North America totals up to 1100 mln USD)

Table 8.12: Import of Live Plants and Products of Floriculture by Country (in tonnes)

Total

Partner/Period	2001	2002	2003	2004	2005	2006	2007	2008	2009	2010	2011	Share in 2011	2011 vs 2010
Kenya (23,5 per cent)	45.608	51.073	57.260	69.999	79.978	90.223	93.127	98.159	94.904	89.939	99.625	23,5 per cent	10,8 per cent
Costa Rica (10,9 per cent)	47.750	45.625	47.833	49.248	51.899	49.218	54.728	56.754	45.887	48.668	46.050	10,9 per cent	-5,4 per cent
Ethiopia (9 per cent)	233	256	673	1.382	3.729	8.441	16.129	22.686	29.707	34.065	38.188	9,0 per cent	12,1 per cent
Israel (6,5 per cent)	39.122	36.069	33.087	33.898	44.829	33.492	33.523	28.449	26.998	26.519	27.352	6,5 per cent	3,1 per cent
USA (6,3 per cent)	25.522	25.817	23.887	26.997	26.868	29.848	31.186	30.853	28.815	26.420	26.505	6,3 per cent	0,3 per cent
Ecuador (5,8 per cent)	14.188	13.003	14.160	15.909	17.253	18.486	22.228	24.674	24.345	23.562	24.659	5,8 per cent	4,7 per cent
Colombia (5,2 per cent)	18.526	19.213	22.726	21.119	23.779	24.523	25.483	25.945	24.888	24.372	21.916	5,2 per cent	-10,1 per cent
China (5,1 per cent)	10.669	15.242	21.085	26.233	29.504	34.504	35.771	30.984	26.489	23.914	21.427	5,1 per cent	-10,4 per cent
Guatemala (3,2 per cent)	16.981	18.429	20.372	20.245	19.833	18.702	19.936	20.399	15.352	12.775	13.609	3,2 per cent	6,5 per cent
Others (24,7 per cent)	110.069	125.003	149.000	152.548	145.681	136.880	134.269	121.071	113.115	110.888	104.644	24,7 per cent	-5,6 per cent
Total	328.668	349.729	390.082	417.579	443.350	444.317	466.379	459.973	430.500	421.122	423.976	100,0 per cent	0,7 per cent

a) Cut Flowers and Cut Foliage

Partner/Period	2001	2002	2003	2004	2005	2006	2007	2008	2009	2010	2011	Share in 2011	2011 vs 2010
Kenya (30,6 per cent)	44.093	49.398	55.265	68.000	77.992	88.187	91.031	95.971	92.769	87.938	96.468	30,6 per cent	9,7 per cent
Ethiopia (11,8 per cent)	233	256	673	1.382	3.555	7.922	15.291	21.961	28.841	33.079	37.104	11,8 per cent	12,2 per cent
USA (7,8 per cent)	23.611	23.790	22.043	24.779	24.043	28.118	29.035	28.550	26.435	24.940	24.779	7,8 per cent	-0,6 per cent
Ecuador (7,8 per cent)	14.129	12.987	14.134	15.890	17.238	18.429	22.126	24.602	24.296	23.535	24.619	7,8 per cent	4,6 per cent
Israel (7,4 per cent)	36.398	32.864	30.181	30.320	40.884	29.428	28.697	23.717	22.743	22.070	23.426	7,4 per cent	6,1 per cent
Colombia (6,9 per cent)	18.513	19.199	22.678	21.079	23.756	24.504	25.462	25.931	24.867	24.331	21.890	6,9 per cent	-10,0 per cent
Costa Rica (5,8 per cent)	24.598	22.124	21.173	20.291	21.142	22.510	23.533	22.175	20.886	20.331	18.209	5,8 per cent	-10,4 per cent
Turkey (3,1 per cent)	7.918	9.669	11.389	11.468	10.529	11.368	10.343	9.299	9.556	10.022	9.738	3,1 per cent	-2,8 per cent
Others (18,8 per cent)	72.027	76.360	80.068	79.339	79.104	73.771	77.069	74.270	66.822	63.341	59.438	18,8 per cent	-6,2 per cent
Total	241.520	246.648	257.605	272.548	298.242	304.236	322.587	326.475	317.214	309.586	315.671	100,0 per cent	2,0 per cent

b) Potted plants, bedding and balcony plants and nursery plants

Partner/Period	2001	2002	2003	2004	2005	2006	2007	2008	2009	2010	2011	Share in 2011	2011 vs 2010
Costa Rica (33,9 per cent)	23.049	23.383	26.485	28.771	30.656	26.532	31.039	34.264	24.577	27.995	27.681	33,9 per cent	-1,1 per cent
China (19,6 per cent)	8.185	11.830	16.261	20.908	23.131	26.487	27.490	23.358	19.848	17.294	16.045	19,6 per cent	-7,2 per cent
Guatemala (7,8 per cent)	7.903	8.016	9.082	8.945	9.268	8.851	10.331	10.539	6.991	6.192	6.394	7,8 per cent	3,3 per cent
Honduras (5,9 per cent)	3.802	3.849	4.616	5.153	5.379	5.492	5.130	5.920	6.342	6.710	4.825	5,9 per cent	-28,1 per cent
Sri Lanka (3,2 per cent)	979	1.017	1.212	1.464	1.572	1.833	1.956	2.047	2.416	2.825	2.612	3,2 per cent	-7,5 per cent
Thailand (2,8 per cent)	350	1.028	1.029	847	650	846	1.244	1.799	2.332	2.576	2.290	2,8 per cent	-11,1 per cent
Taiwan (2,9 per cent)	2.221	2.401	3.034	3.414	2.224	1.923	2.698	2.709	2.224	2.311	2.330	2,9 per cent	0,8 per cent
Israel (2,6 per cent)	1.797	2.108	1.952	2.001	2.190	2.165	2.391	2.624	2.040	2.394	2.111	2,6 per cent	-11,8 per cent
Others (21,4 per cent)	19.551	24.841	30.678	31.912	35.964	32.538	27.214	22.495	19.967	18.454	17.455	21,4 per cent	-5,4 per cent
Total	**67.837**	**78.473**	**94.348**	**103.414**	**111.033**	**106.665**	**109.493**	**105.754**	**86.736**	**86.750**	**81.742**	**100,0 per cent**	**-5,8 per cent**

c) Conifers and hardy perennial plants

Partner/Period	2001	2002	2003	2004	2005	2006	2007	2008	2009	2010	2011	Share in 2011	2011 vs 2010
China (12,1 per cent)	574	788	667	640	1.007	1.071	1.448	1.380	1.446	1.495	937	12,1 per cent	-37,3 per cent
Israel (15,1 per cent)	109	188	129	837	891	1.120	1.751	1.557	1.694	1.430	1.172	15,1 per cent	-18,0 per cent
For.J.Rep.Mac (22,5 per cent)	19	0	1			7			22	446	1.747	22,5 per cent	291,8 per cent
Switzerland (5,6 per cent)	465	608	659	450	291	274	356	295	348	361	431	5,6 per cent	19,4 per cent
Japan (4,6 per cent)	106	230	217	176	249	317	592	541	383	288	357	4,6 per cent	24,2 per cent
Argentina (4,5 per cent)	516	1.178	2.531	1.675	1.500	1.064	1.300	1.313	578	255	350	4,5 per cent	37,3 per cent
Egypt (3,8 per cent)	7.390	8.314	17.950	22.160	14.278	11.177	7.278	456	753	903	296	3,8 per cent	-67,3 per cent
Turkey (3,3 per cent)	40	92	131	44	40	682	89	113	328	370	254	3,3 per cent	-31,3 per cent
Others (28,5 per cent)	1.487	2.207	3.996	3.295	4.104	2.775	3.507	2.899	2.384	1.800	2.215	28,5 per cent	23,1 per cent
Total	10.705	13.605	26.280	29.278	22.359	18.485	16.321	8.552	7.936	7.347	7.759	100,0 per cent	5,6 per cent

d) Bulbs and corms

Partner/Period	2001	2002	2003	2004	2005	2006	2007	2008	2009	2010	2011	Share in 2011	2011 vs 2010
Chile (25,4 per cent)	1.457	2.005	2.727	2.816	2.423	3.532	4.510	4.785	4.432	4.759	4.779	25,4 per cent	0,4 per cent
Brazil (28,8 per cent)	2.213	2.971	2.806	2.973	2.556	4.051	4.827	5.900	6.256	4.667	5.414	28,8 per cent	16,0 per cent
New Zealand (13,3 per cent)	672	1.273	1.432	1.297	1.448	1.309	2.035	2.280	1.718	2.176	2.509	13,3 per cent	15,3 per cent
Taiwan (5,2 per cent)	727	665	610	862	909	520	724	734	676	844	970	5,2 per cent	15,0 per cent
China (4 per cent)	320	422	507	344	563	759	953	1.038	992	729	760	4,0 per cent	4,3 per cent
South Africa (4 per cent)	565	648	655	812	790	834	636	672	629	846	754	4,0 per cent	-10,9 per cent
Thailand (4 per cent)	121	122	181	298	324	543	749	726	640	617	747	4,0 per cent	21,1 per cent
Others (15,3 per cent)	2.532	2.897	2.932	2.937	2.703	3.382	3.544	3.058	3.272	2.801	2.871	15,3 per cent	2,5 per cent
Total	**8.606**	**11.003**	**11.850**	**12.339**	**11.716**	**14.930**	**17.979**	**19.192**	**18.614**	**17.439**	**18.803**	**100,0 per cent**	**7,8 per cent**

From our figures about Asia shows an internal export between countries of about 94mln USD. This is added by export of cut flowers of 53 mln USD out of European countries, 22 mln USD out of Oceania and 15 mln USD out of South America. This adds up to 184 mln. USD total. From Africa it is not known that they import cut flowers out of other continents or in between its countries.

Potted Plants

When it comes to potted plants, based on figures of 2002, we can see the patterns differ: Europe has an internal export between its countries with a value of 3013 million US dollar. In addition Europe imports from Asia for about 47 mln USD. This is followed by imports with a value of 42 mln USD from Africa and 39 USD from South America. In perspective of potted plants this mostly concern propagation material and cuttings for production in Europe. This totals 3140 USD value of imported product for Europe In North America the internal export between USA and Canada is 329 mln USD. This market imports from Europe mostly end product with a value of 54 mln USD. Also South America is a main supplier with 47 mln USD of products; propagation material as well as potted plants for the markets. This is the same with Asia with an export value of 13 mln USD. Together the trade of North America totals an import value of 440 mln. USD.

Asia shows an internal export of 'potted plants' between countries of about 52mln USD. This is added by export of potted plant of 76 mln. USD out of European countries. This adds up to 128 mln. USD total. From Africa it is not known that they import potted plants out of other continents or in between its countries.

Worlds Largest Exporting Countries

From the figures available about world wide trade in cut flowers it is established that the Netherlands is by far the biggest exporter. It is followed by Colombia. They are followed by Israel, Kenya, Belgium and Zimbabwe.

Their role is changing over the last 5 years. We see a negative growth of the export of The Netherlands (-2,7 per cent) and Israel (-6 per cent); the latter because of the political situation. The others show growth of export: Colombia (45 per cent), Kenya (182 per cent), Belgium (130 per cent) and Zimbabwe (46 per cent). Of Zimbabwe we assume that this growth has decreased or that export even has decreased because of the political situation in the country since 2002.

Incentive for Growing International Trade - IT, Communication and Distribution

There are a few global developments which will have their effect on international trade the years to come. There are rapid changes in the possibilities of information technology. Think of the possibilities that handheld computers will mean when it comes to management of the production process. Or what about the introduction of RFID-tags and micro sensors and their impact on tracking and tracing of products throughout the chain. Close to this is the development of communication technology.

If the hardware starts communicating with web-based applications all kind of stakeholders of the product chain can access the data and information. This also requires thinking about authorisation of the information access. Next to that more and more of the global population start using English, which make communication easier. There are also developments in in distribution. If the new IT-components en their communication aspects are linked to distribution and route planning, the distribution process can be optimised. Other developments are those in packing and transporting. Packaging called Modified Atmosphere can slow down the process of deterioration of products like fruits, vegetables and flowers. The product can be transported during a longer period of time and still have good quality. This process is also simulated with a ship. The whole shipload can be treated in way that fresh produce keeps its quality. This will increase possibilities of sending orders by ship instead of by plane, which makes transportation a lot cheaper.

International Economic Integration

There are different economic trends which will have their effect on the trade in fresh produce. For instances, countries more and more unite themselves in trade unions. And these unions are getting more powerful when more countries join. Different examples can be given. The European union existed out of 15 countries. But since may first 2004, another 10 Eastern European countries joined. Then there is ASEAN, a same kind of development in Asia. The USA, Canada and Mexico have there NAFTA-treaty. Trade within these unions is getting easier and cheaper by decreased regulations and trade barriers.

Another development we see is that (home) markets saturate. For producers, this saturation asks for the strategy to keep their share of wallet, either to enlarge the export activities to other continents. Finally we see that the activities of the World Trade Organisation (WTO-rounds) globally breaking down all kind of trade barriers. This means more and more the free market mechanism can do its work. This will have its effect on trade as well.

Shift in Market for Floricultural Products

The world market for floricultural product has been changing constantly. Going back to 1992 we see that, calculated in dollars, 57 per cent of the world market consisted of the Europe demand, 23 per cent was Japanese demand and 20 per cent was the demand from the USA. Over the years we see these shares changing. While the European market is loosing share, due to growth of market in the USA, Japan is keeping up its position. USA has proven to be a growing market the last decade up until 2002. The developments of the last two years, with dollar that is weakening, importing products into USA is getting more expensive and will strengthen the domestic production.

Overall we see an increase in the consumption of cut flowers and potted plants per capita in the different parts of the world, as the Table 8.13.

**Table 8.13: Development in Consumption of Ornamentals
per Capita in different Countries (dollar/year)**

Country	1994	2002	+
Germany	74	83	13 per cent
Denmark	63	83	31 per cent
Great Britain	23	52	130 per cent
Spain	19	30	55 per cent
The Netherlands	61	93	54 per cent
Japan	34	51	50 per cent
Russia	n.a.	4	-
China	n.a.	1	-
USA	44	64	45 per cent

The table shows us developments in Europe. In Europe we see that the erman market has grown only slightly compared to the others. It is known that the German market has become saturated. The Danish market shows quite a growth. The consumption has increased with even more than 50 per cent looking at the developments in The Netherlands and Spain from 1994 to 2002. The market Great Britain has been enlarged enormously. In 8 years time their people have more than doubled their consumption of ornamentals. The British Supermarkets have played a big role in this as a main outlet for quality products. And this will increase. Furthermore big volumes are needed to supply these parties.

In Japan the consumption in 2002 is one and a half time that of 1994, quite an increased. Com pared to that, with 45 per cent the increase of consumption in the USA is just a little less. We cannot compare the figures in the table directly to the trends of the shift in share of the global market. To do so, the figures in the table would have to be multiplied with the amount of inhabitants per country or continent.

World Trade Scenario

World trade in floriculture is estimated at $100 billion. It has reportedly been growing at the rate of 15 per cent annum. Developed countries account for more than 90 per cent of the total world trade in floriculture products. European countries, Japan and United States are the major importers of floricultural products. The Netherlands continues to be the world leader in the export of floricultural products. Italy, USA and Israel are the other major exporting countries. Since major part of the production goes into the respective markets of the producing countries, the world trade in terms of imports by the main markets, was only of the order of $7.9 billion, *i.e.* about 16 per cent of the value of world production. The rate of increase in the world imports has been around 10 per cent per annum in the recent years. The share of developing countries in the total trade has been around 20 per cent in the last few years 1.

Cut flowers constitute about 50 per cent of the world imports in floriculture products. In the total global import of floriculture products of 7.9 billion USD during 1999, the cut flowers segment accounted for USD 3.76 billion, followed by plants at USD 2.78 billion. The floriculture import figures of major markets during 1999 are given in the **Table 8.14**.

Table 8.14: Country-wise Imports of Floriculture Products (Year 1999) ('000 US$)

Sl.No.	Country	Bulbs	Plants	Flowers	Foliage	Total
1.	Germany	69183	673730	794251	141321	1678484
2.	USA	203280	257224	734804	85759	1281103
3.	France	69852	363353	414869	37208	885282
4.	U.K.	50704	264337	526637	26291	867969
5.	Netherlands	30989	192603	366418	157028	747035
6.	Italy	58840	168711	146278	15539	389368
7.	Japan	121523	70408	153680	38219	383830
8.	Switzerland	29493	141008	140949	32295	343745
9.	Belgium	11156	149300	108168	21775	290399
10.	Austria	11390	105673	90095	17079	224237
11.	Canada	31648	95216	62808	14178	203850
12.	Sweden	18210	93089	45657	4378	161333
13.	Denmark	10370	72532	64414	7049	154366
14.	Spain	16940	64884	38605	4815	125244
15.	Finland	6959	26553	15790	3864	53165
16.	Portugal	4746	22859	16940	1453	45121
17.	Ireland	778	10702	28089	849	40418
18.	Greece	3677	12839	21869	1294	39679
	Total	**749734**	**2785021**	**3769443**	**610430**	**7914629**

Source: www.pathfastpublishing.com

The country wise import figures indicated in the above table represent imports for consumption within the country. Imports into Netherlands are mainly for re-exports to other EU countries and the trade figures at the auction centers indicate imports of a much higher volume than indicated in the table.

The major importing countries for cut flowers are given in **Table 8.15.**

Table 8.15 : Major Importers of Cut Flowers

Country	Imports (Per cent)
Germany	39
U.S.A.	17
France	11
U.K.	10
Netherlands	6
Japan	6
Switzerland	5
Others	6

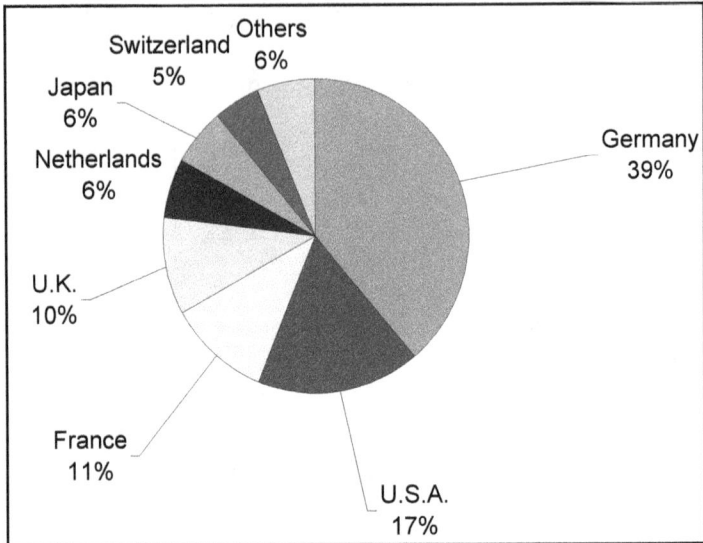

Figure 8.3

Germany is the largest importer and the total imports to West European countries including U.K. account for almost 70 per cent of the imports in international floriculture trade.

The main exporting countries *viz.* Holland, Colombia, Italy and Israel, accounts for about 80 per cent of the world exports. The share of exporting countries is given below. Other countries, which have emerged in recent years, are Korea and China. Both these countries have undertaken huge investments in production and infrastructure (particularly greenhouses) and are emerging as major competitors for supplies to Japan and South East Asia.

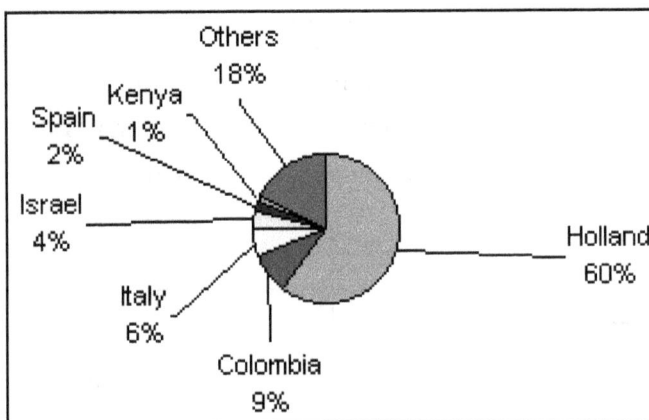

Figure 8.4

Composition

Rose is the most preferred flower with 51 per cent share in the international market. Carnations and chrysanthemums also have a substantial share. The demand for roses has been growing at a rate higher than that of other flowers. Market shares of different flowers in Europe are as under.

Flower	Share (per cent)
Rose	51
Carnations	19
Chrysanthemums	10
Gladioli and orchids	15
Other Species	5

The top ten cut flowers and pot plants traded at Dutch Auctions (an indicator of the world trade) are given in the **Table 8.16.**

Table 8.16: The Turnover of Top 10 Cut Flowers/Pot Plants at Dutch Auction (VBA, 1999)

Sl.No.	Cut Flowers	Million NLG	Pot Plants	Million NLG
1.	Rose	1672	Kalanchoe	27
2.	Tulip	569	Hedera	22
3.	Chrysanthemum	421	Ficus	17
4.	Gerbera	272	Saintpaulia	14
5.	Carnation	178	Pot Chrysanthemum	12
6.	Lily	143	Dracaena	11
7.	Freesia	141	Pot rose	11
8.	Alstroemeria	139	Hyacinth	10
9.	Iris	103	Primula	9
10.	Gypsophila	87	Begonia	3

Note: NLG- Dutch Guilders.

Source: APEDA, New Delhi.

EU Trade

Europe is the biggest buyer of cut flowers, and is also the major hub for world trade. Hence the trade scenario in Europe is described below in detail.

Demand

The demand for cut flowers and pot plants in Europe arises from both domestic consumption and trade. Western Europe accounts for more than half of the world's cut flower consumption. The Netherlands has a dual role as an exporter and as a major world distribution centre. Consequently the demand for floriculture products in the markets in the Netherlands reflects the demand of the consumers throughout

Western Europe. 38 per cent of all cut flowers and 21 per cent pot plants purchased by consumers in Europe, come from the Dutch trade. The maximum demand emanates from Germany (about 2 billion NLG at wholesale prices), which accounted for 38 per cent of Dutch exports of floriculture products in 1999.

Seasonality

The purchase of flowers by the Europeans for personal consumption is a common practice and, therefore, there is a consistent level of demand throughout the year. However, on special occasions like Mother's Day, Christmas, St. Valentine's Day, Easter *etc.*, the markets witness peak turnover of floriculture products.

Quality

Freshness and shelf life are the important requirements, which determine both value and volume of purchases of cut flowers and potted plants by the European consumers. Besides long shelf-life, the other important considerations, which influence consumer preferences are colour, shape and variety.

Supplying the Consumer-Retail Trade

There are a variety of outlets, which serve the consumers. Traditionally florists or flower shops, located near residential areas supply the requirements of the consumers. Supermarkets are now becoming popular as sellers of floriculture products and their market share is steadily increasing.

General profile of the retailing activities can be gauged from the share of various types of outlets in the Netherlands, (which is considered representative of the trend in other countries of Europe). There are over 12,000 outlets selling cut flowers and pot plants to consumers in Europe. For example in Netherlands, the specialist shops are the most important retail sales outlets along with street markets; the grocery trade/supermarkets also have a significant market share. The market share of traditional retail outlets like flower shops and garden shops has declined whereas that of supermarkets and flower stalls has shown an increasing trend, as indicated in the Table 8.17.

Table 8.17: Market Share of different Types of Retail Outlets in the Netherlands

Type of Outlets	1990	1993	1995
Flower Shops	53.0	55.0	50.4
Flower producers	2.0	1.9	1.4
Flower stalls	17.0	22.9	26.9
Super markets	8.0	11.9	14.1

Floriculture Trade Becoming more Turbulent and International

Rabobank's World Floriculture map 2015 illustrates that low-cost cut flower exporting countries close to the equator, such as Kenya, Ethiopia, Ecuador, Colombia and Malaysia, have increased their global market share in cut flower trade. These cost-efficient producers are strengthening their position in global production and trade, mainly driven by favourable growing circumstances, rising demand for competitively priced flowers in the main destination markets and improved logistics, including transportation by sea container. High-cost growers who do not want to sit on thorns need to differentiate themselves from low-cost competitors, or themselves become active in these low-cost regions.

Economic and financial turmoil—mainly in North America and Western Europe—has impacted global floriculture trade. The historically strong growth in global floriculture exports has taken a bumpier road from 2009 onwards (see Figure 8.5). In 2013, global exports of cut flowers, cut foliage, living plants and flower bulbs amounted to USD 20.6 billion as against USD 21.1 billion in 2011 and nearly USD 8.5 billion in 2001.

Cut flowers, which are traded worldwide, have always been the main group within global floriculture trade, followed by living plants, which are traded more regionally. As geographic expansion of cut flower production as well as further developments in logistics make long- haul transportation more viable, the share of cut flowers in floriculture trade will likely grow.

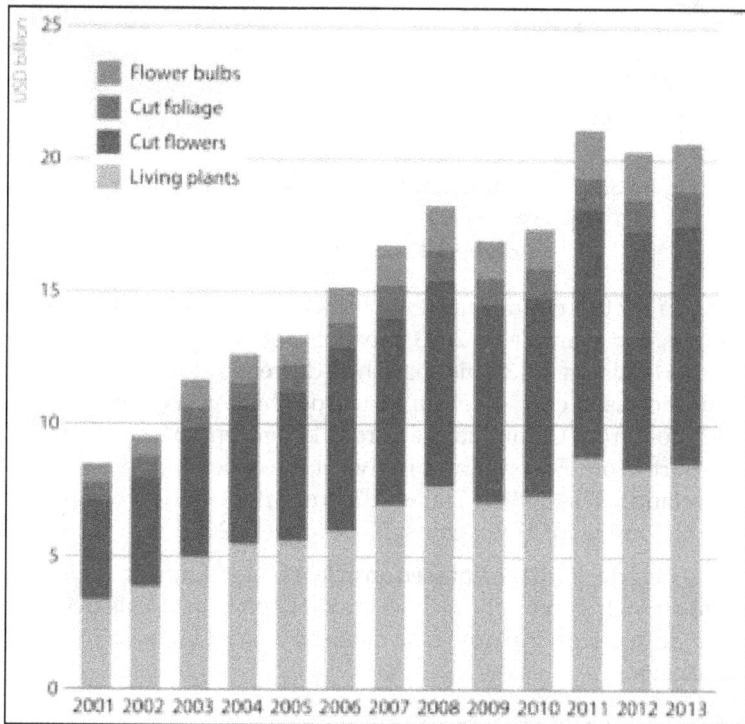

Figure 8.5: Rise in Floriculture Exports from 2001 to 2013 has Become more Volatile from 2009 Onwards
Source: UN Comtrade, 2014.

Competitive Playing Field in Cut Flowers is Changing

One of the main structural changes currently taking place in the world of floriculture is the increase in international competition, particularly for cut flowers. With a combination of locally produced flowers and imported flowers, the Netherlands is a dominant central market for global cut flower trade. However, the Dutch share in global cut flower exports is decreasing, declining from 58 per cent in 2003 to 52 per cent in 2013 (Figure 8.6). At the same time, Kenya, Ecuador, Ethiopia, Colombia and Malaysia have increased their share in global cut flower exports. Growers in these countries are able to achieve large-scale production of good-quality flowers for competitive prices.

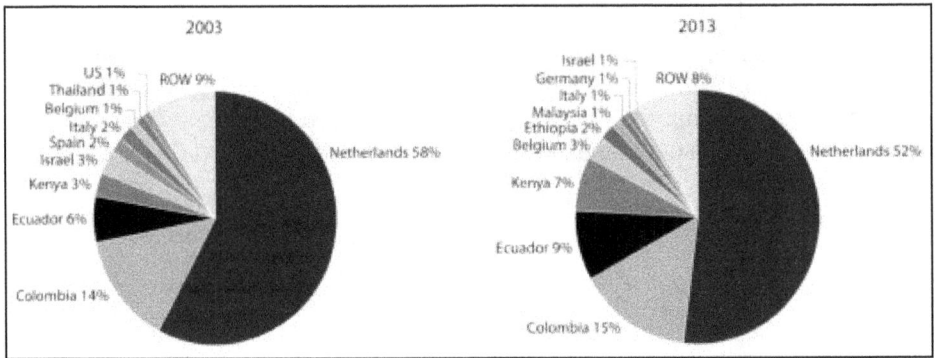

Figure 8.6: The Netherlands is World's Largest Cut Flower Export Nation, but its Share is Diminishing.
Source: **UN Comtrade, 2014.**

Zooming in on the main import markets exposes the shifts in international competitiveness. For example, in 2003, Japan imported 8 per cent of its cut flowers from the Netherlands. In 2013, this share had decreased to 2 per cent. In the same period, Japan increased cut flower imports from Malaysia from 10 per cent to 26 per cent, and those from Colombia grew from 14 per cent to 22 per cent. Russia has increased imports from Ecuador and Kenya at the expense of cut flowers from or via the Netherlands. The United States (US) imported about 65 per cent of all cut flowers from Colombia in 2013, against 55 per cent in 2003.

In the past decade, we have not seen any clear major trends in import and export flows in large emerging markets such as Brazil, China, India, Mexico and Turkey. However, we do see growing domestic demand and an increase in production of flowers and plants.

Breakthrough of Container Transport

Transport of cut flowers by sea container has become a major and an unstoppable development and has already become fairly substantial in a number of trade flows. About 15 per cent of total cut flower exports from Colombia are already shipped by sea. In 2013, Colombia shipped about 700 containers of mainly chrysanthemums to the United Kingdom (UK). One 40-ft container can be loaded with about 150,000 chrysanthemum stems. Other large container flows are from Vietnam to Japan and from Israel to Europe.

Transport by sea container is not new to floriculture, although it has been historically restricted to products with a relatively long shelf life, such as cut foliage, flower bulbs and young plants. Important foliage trade flows run from Costa Rica to the Netherlands and from China to Japan.

Expansion of container shipments in cut flowers is driven by various factors including the price difference between sea and air freight (the latter being roughly twice as expensive as the former), the ability to control conditions within containers, growing availability of port facilities and reefers, increasing knowledge of container transport and the best cut flower varieties to transport, and growing attention on sustainability issues.

Stagnation in the Main Floriculture Markets

In the US and Western Europe, growth in expenditure on floriculture products has come to a standstill. This is partly related to economic circumstances, in particular in the case of cut flowers that have a high correlation with disposable income. Consumer expenditure on bedding plants is very much linked with weather circumstances. But other factors such as availability and religious and cultural traditions also play an important role in flower and plant purchasing.

In certain markets, consumers seem to have moved to the low-value end of the market, which is mainly sold by supermarkets, discounters and DIY-stores. In the UK, the vast majority of cut flowers is sold by supermarkets, including discounters. In Germany, DIY stores are popular outlets for buying indoor plants. If organised retail chains professionalise and focus on their floriculture departments, they will be able to raise their market share. In Europe, the main growth in floriculture consumption has been in the Eastern part, including Russia. The size of the Russian market has already exceeded the Spanish market (Figure 8.7). However, the Russian market is a very unstable market.

Gearing up for the Next Decade

For the next ten years, Rabobank expects growth in global floriculture production and trade to continue, although it may be rather unsteady. The countries that are becoming more dominant in global supply and demand are also experiencing more instability compared to the current main production and consumption areas.

In the large traditional markets (*i.e.* Europe, the US, Japan), floriculture expenditure is not expected to increase, whereas in countries with rather strong income growth, floriculture consumption still has growth potential.

Although the US, Western Europe and Japan remain the world's largest floriculture markets in absolute terms, their dominance in the world of floriculture is slowly diminishing as both demand and supply grow relatively stronger in the rest of the world. Global trade in flowers and plants that are easy to grow and to transport (by container) will continue to expand and will benefit low-cost production regions.

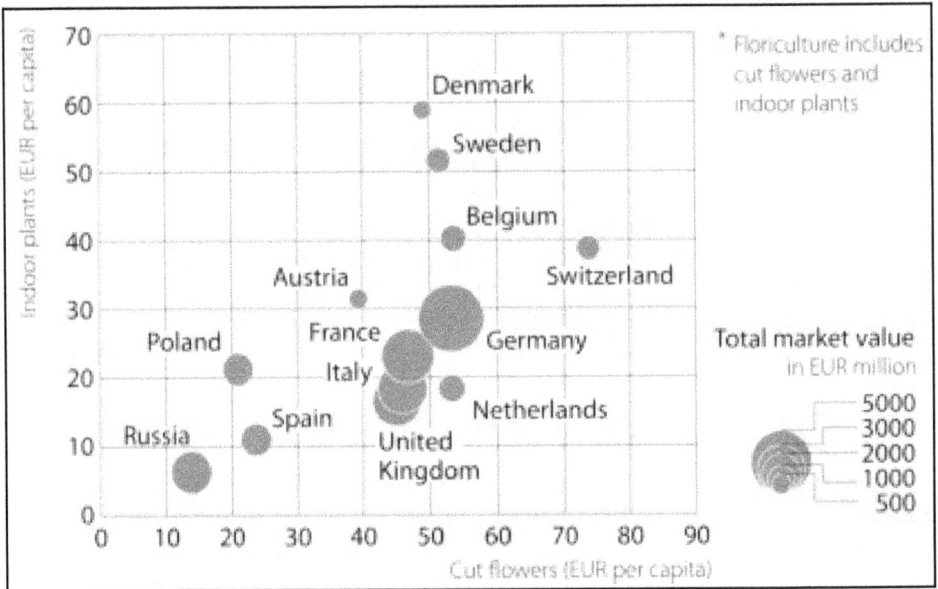

Figure 8.7: Large differences in per Capita Expenditure on Floriculture within Europe.

Growers operating in the high-cost regions of Europe, the US and Japan have to take action to remain competitive against increasing global competition and market stagnation. One option is to focus on products with a high weight/value ratio (*e.g.* potted plants), specialty products, typical local flowers and plants, and seasonal products. However, competition will remain fierce. Alternatively, these growers may use their knowledge and skills to produce or advise abroad in emerging flower-growing countries.

Dedicated supply chains will become more important for large-scale growers in competitive production regions. Within an ever more global and competitive business that is dealing with a high diversity of perishable and very delicate products, control over the supply chain will increasingly be a critical success factor. Only with long-term partnerships can growers as well as retailers control risks related to availability, quality, ethical issues and prices.

Within the next decade, dedicated floriculture supply chains will require a so-called 'facilitator' to make this value chain work smoothly. This facilitator should be able to match supply and demand in a very efficient way, take care of bulk break flows, handle quality control, arrange information exchange within the value chain and advise retailers on category management. Some floriculture value chains already work with one strong facilitator (*e.g.* distributor, marketer or importer/exporter) that provides services to its suppliers as well as retail customers. However, other value chains are still operating with many—sometimes varying—intermediaries. This model is outmoded.

Exports

Total world exports increased substantially from 272.6 tonnes to 466 tonnes in the from 2001 to 2011. Russia and Switzerland were the top exports at 140.8 and 121.6 tonnes exports respectively. Cut flowers formed the majority at 98.9 tonnes. They continued to be parties in other 4 categories also with changed positions. (Tables 8.18 and 8.19)

Flower Trade Across the World

World trade on floriculture produces like cut flowers, ornamental plants, flowering plants, flower seeds and plantlets gaining tremendous momentum. Many countries, particularly the developed ones, are importing flowers to meet their internal demand. It will be worthwhile to mention that the annual import figures of some of the largest importers on flowers - USA (232 crores US dollar) Japan (192 crores US $), Germany (180 crores US $) France (77 Crores Us Dollar), Italy (55.6 Crores US Dollar), Holland (50 Crores US Dollar). The other importers like Switzerland, Sweden, Denmark, Belgium, Middle-east countries *etc.* also import a sizable amount of cut flowers. In recent past, Israel has come up as the biggest grower of flowers, using modern agro-techniques like glass-house culture, drip irrigation, liquid pesticides and fertilisers application along with drip irrigation channels, Tissue Culture. It may be mentioned that the roses of Israel adjudged to be the best in the World. via-a-vis such a huge market potential of floriculture produce, India's contribution is not at all encouraging as its flower export amount to 30 lakh us dollar only, hence India has to do a lot to exploit this agro-business.

Present Status of Floriculture in Bangladesh

☆ The total area under floriculture in Bangladesh is estimated to range from 4250 to 4570 ha.

☆ Of which 3210 to 3450 ha of land is occupied by different flowers and approximate acreage under nursery producing ornamental plants was found to be 1040 to 1120 ha.

☆ Among the flowers grown in Bangladesh Rose, Tuberose, Gladiolus, Marigold, Gerbera and Jasmines is the main.

☆ Different ornamental plants are: Croton, Rangan, Thuja, Palms, Aurocaria, Musaenda, Cactus, Ferns, Dracaena, China box, Century plant, Bromeliads, Hydrangea *etc.*

☆ Dhaka, Chittagong, Cox's Bazar, Jessore, Chuadanga, Jhenidah, Bogra, Rangpur, Kushtia and Mymensingh are the major production belts of flowers.

☆ The other places like Barisal, Panchagarh, Gazipur, Khulna, Manikganj, Khagrachhari, Bandarban, Dinajpur, Rangamati and Tangail have started cultivating flowers.

☆ The major production belts of ornamental plants are Chittagong, Natore, Khulna, Pirojpur, Dhaka and Mymensingh regions. The other places such

Total

Table 8.18: Export of Live Plants and Products of Floriculture by Country (in tonnes)

Partner/Period	2001	2002	2003	2004	2005	2006	2007	2008	2009	2010	2011	Share in 2011	2011 vs 2010
Russia (23,2 per cent)	26.082	36.836	42.867	51.377	57.582	78.712	101.817	122.240	97.659	114.097	140.794	30,2 per cent	23,4 per cent
Switzerland (20 per cent)	70.206	80.284	80.758	80.995	81.191	96.869	102.643	105.401	105.452	115.897	121.588	26,1 per cent	4,9 per cent
USA (8 per cent)	55.490	66.325	62.136	55.701	58.661	65.289	63.030	53.369	52.934	52.770	48.601	10,4 per cent	-7,9 per cent
Norway (8 per cent)	30.826	32.708	30.441	31.460	35.471	38.970	42.788	43.037	41.623	45.688	48.591	10,4 per cent	6,4 per cent
Turkey (5 per cent)	9.606	9.678	8.797	11.560	18.551	28.954	26.792	30.097	16.035	22.110	30.169	6,5 per cent	36,4 per cent
Ukraine (3,9 per cent)	2.257	3.680	5.782	7.768	11.619	18.204	28.538	35.132	25.871	21.537	23.513	5,0 per cent	9,2 per cent
China (2,9 per cent)	4.465	6.407	8.296	8.609	9.334	9.869	12.687	14.768	16.683	17.721	17.859	3,8 per cent	0,8 per cent
Japan (2,9 per cent)	27.856	26.921	24.458	24.587	23.512	23.371	23.468	23.177	18.671	18.885	17.470	3,7 per cent	-7,5 per cent
Croatia (2,7 per cent)	7.533	8.357	9.219	10.925	12.839	14.726	16.668	18.623	18.193	15.875	16.492	3,5 per cent	3,9 per cent
Mexico (1,6 per cent)	2.400	3.282	4.827	4.763	5.036	7.164	8.531	8.924	8.905	10.477	9.906	2,1 per cent	-5,5 per cent
Others (21,7 per cent)	61.945	65.720	61.558	74.924	72.573	78.992	93.495	106.930	116.388	140.529	131.912	28,3 per cent	-6,1 per cent
Total	272.584	303.364	296.272	311.292	328.788	382.407	418.642	439.456	420.754	461.487	466.100	100,0 per cent	1,0 per cent

a) Cut flowers and cut foliage

Partner/Period	2001	2002	2003	2004	2005	2006	2007	2008	2009	2010	2011	Share in 2011	2011 vs 2010
Russia (36,9 per cent)	9.237	11.053	12.401	14.142	15.274	21.366	27.020	31.441	26.483	28.181	36.419	36,9 per cent	29,2 per cent
Switzerland (21,4 per cent)	17.210	21.030	19.915	18.935	17.723	18.644	19.893	18.258	19.499	21.983	21.125	21,4 per cent	-3,9 per cent
Norway (9,2 per cent)	7.836	7.902	7.616	8.385	8.462	9.473	9.648	9.013	8.482	8.872	9.061	9,2 per cent	2,1 per cent
USA (9,4 per cent)	17.007	17.067	13.298	12.308	10.933	11.311	10.146	7.947	7.814	7.474	9.281	9,4 per cent	24,2 per cent
Croatia (3,9 per cent)	1.759	2.088	2.264	2.709	2.821	3.250	3.419	4.025	3.986	3.686	3.826	3,9 per cent	3,8 per cent
Ukraine (3,4 per cent)	497	536	1.496	996	934	1.524	3.253	3.930	2.556	2.744	3.348	3,4 per cent	22,0 per cent
Serbia (2,3 per cent)					349	2.002	2.142	2.400	2.249	2.413	2.283	2,3 per cent	-5,4 per cent
Belarus (1,1 per cent)	455	460	534	689	1.119	1.279	1.454	1.412	1.419	1.308	1.080	1,1 per cent	-17,4 per cent
Others (12,6 per cent)	13.435	13.612	11.745	11.145	10.804	9.169	9.688	9.563	10.358	11.434	12.408	12,6 per cent	8,5 per cent
Total	**67.436**	**73.746**	**69.269**	**69.308**	**68.418**	**78.017**	**86.664**	**87.988**	**82.844**	**88.094**	**98.830**	**100,0 per cent**	**12,2 per cent**

b) Potted plants, bedding and balcony plants and nursery plants

Partner/Period	2001	2002	2003	2004	2005	2006	2007	2008	2009	2010	2011	Share in 2011	2011 vs 2010
Switzerland (33,6 per cent)	40.833	44.130	41.603	42.527	46.710	52.962	55.519	52.479	49.462	53.076	58.281	33,6 per cent	9,8 per cent
Russia (17,2 per cent)	10.761	15.988	18.779	21.599	22.591	24.325	29.253	33.545	23.925	27.238	29.788	17,2 per cent	9,4 per cent
Norway (11,5 per cent)	11.444	12.277	10.618	11.056	13.972	15.027	17.443	18.220	17.931	18.674	20.046	11,5 per cent	7,3 per cent
Turkey (5,1 per cent)	6.507	5.284	6.059	7.076	7.935	13.552	12.338	11.708	6.510	6.960	8.837	5,1 per cent	27,0 per cent
Ukraine (5 per cent)	1.007	2.209	3.452	4.454	6.856	9.607	13.593	16.105	12.790	9.538	8.655	5,0 per cent	-9,3 per cent
Croatia (4,3 per cent)	4.783	5.040	5.128	6.105	7.881	9.003	9.301	9.737	9.726	7.756	7.505	4,3 per cent	-3,2 per cent
Lebanon (3,3 per cent)	2.733	3.291	3.399	4.106	2.883	2.682	3.015	3.824	5.193	5.498	5.806	3,3 per cent	5,6 per cent
Albania (2 per cent)	2.364	2.237	1.205	1.867	1.832	1.822	2.226	3.441	6.069	4.507	3.517	2,0 per cent	-22,0 per cent
Others (18 per cent)	21.241	22.786	20.513	21.913	21.051	21.714	24.000	27.680	27.937	30.110	31.197	18,0 per cent	3,6 per cent
Total	**101.673**	**113.240**	**110.756**	**120.701**	**131.711**	**150.692**	**166.687**	**176.738**	**159.543**	**163.356**	**173.632**	**100,0 per cent**	**6,3 per cent**

c) Conifers and hardy perennial plants

Partner/Period	2001	2002	2003	2004	2005	2006	2007	2008	2009	2010	2011	Share in 2011	2011 vs 2010
Russia (34,2 per cent)	4.352	7.227	8.680	11.525	14.636	24.977	35.556	45.376	36.647	46.584	59.096	34,2 per cent	26,9 per cent
Switzerland (19,7 per cent)	9.000	10.458	13.891	14.843	11.995	17.633	21.296	27.519	28.565	32.569	34.056	19,7 per cent	4,6 per cent
Turkey (10,9 per cent)	2.659	3.699	2.295	3.920	9.501	13.744	12.389	15.855	7.794	12.790	18.817	10,9 per cent	47,1 per cent
Norway (7,6 per cent)	8.178	8.831	7.973	7.769	7.625	9.204	10.182	10.366	9.812	12.081	13.185	7,6 per cent	9,1 per cent
Ukraine (5 per cent)	612	742	640	2.018	3.418	6.381	10.470	12.516	8.659	7.284	8.556	5,0 per cent	17,5 per cent
Croatia (2,5 per cent)	660	871	1.249	1.593	1.776	2.074	3.501	4.284	3.779	3.679	4.368	2,5 per cent	18,7 per cent
Lebanon (1,8 per cent)	2.748	923	1.027	688	1.139	843	1.599	1.914	2.191	3.720	3.150	1,8 per cent	-15,3 per cent
Turkmenistan (1,1 per cent)	880	169	117	1.808	486	233	31	99	657	3.461	1.855	1,1 per cent	-46,4 per cent
Others (17,1 per cent)	3.908	7.009	5.800	7.700	6.654	7.799	11.899	16.485	18.571	31.112	29.482	17,1 per cent	-5,2 per cent
Total	32.996	39.929	41.670	51.864	57.229	82.888	106.921	134.413	116.674	153.279	172.564	100,0 per cent	12,6 per cent

d) Bulbs and corms

Partner/Period	2001	2002	2003	2004	2005	2006	2007	2008	2009	2010	2011	Share in 2011	2011 vs 2010
USA (22,9 per cent)	35.233	45.183	44.741	40.019	43.768	50.206	49.089	42.352	42.093	42.656	37.138	22,9 per cent	-12,9 per cent
Japan (9,8 per cent)	22.415	22.508	21.107	21.592	21.127	21.287	21.640	21.628	17.142	17.288	15.931	9,8 per cent	-7,9 per cent
China (9,8 per cent)	3.235	4.463	6.784	7.623	8.610	8.690	11.532	13.673	15.290	15.795	15.916	9,8 per cent	0,8 per cent
Russia (9,6 per cent)	1.733	2.568	3.007	4.112	5.081	8.044	9.988	11.879	10.605	12.095	15.492	9,6 per cent	28,1 per cent
Mexico (5,9 per cent)	2.191	3.026	4.571	4.487	4.765	6.857	8.214	8.616	8.682	10.299	9.559	5,9 per cent	-7,2 per cent
Canada (4,7 per cent)	5.767	6.386	6.546	7.143	6.417	7.215	8.090	7.501	8.774	9.291	7.630	4,7 per cent	-17,9 per cent
Switzerland (5 per cent)	3.162	4.666	5.349	4.690	4.764	7.631	5.935	7.145	7.926	8.268	8.126	5,0 per cent	-1,7 per cent
Norway (3,9 per cent)	3.368	3.699	4.235	4.251	5.413	5.267	5.516	5.437	5.398	6.062	6.300	3,9 per cent	3,9 per cent
Taiwan (3 per cent)	3.665	4.194	3.885	4.998	5.024	5.796	6.649	6.961	5.573	6.015	4.788	3,0 per cent	-20,4 per cent
Others (25,4 per cent)	15.793	16.591	17.219	21.882	24.044	28.529	33.534	37.367	37.870	44.899	41.185	25,4 per cent	-8,3 per cent
Total	96.562	113.285	117.444	120.797	129.013	149.522	160.188	162.557	159.352	172.669	162.065	100,0 per cent	-6,1 per cent

Source: Eurostat Comext.

Table 8.19: Import/Export Extra-EU of Live Plants and Products of Floriculture (in thousand euros, at current prices)

Import

	2001	2002	2003	2004	2005	2006	2007	2008	2009	2010	2011	2011 vs 2010	share in 2011
Cut flowers and cut foliage (76,3 per cent)	986.386	980.046	918.503	918.559	964.703	1.058.650	1.113.738	1.185.534	1.122.733	1.155.776	1.183.097	2,4 per cent	76,3 per cent
Potted plants (17,6 per cent)	193.221	212.064	222.136	225.206	231.079	243.786	261.203	262.298	250.469	264.942	272.437	2,8 per cent	17,6 per cent
Conifers and hardy perennial plants (0,8 per cent)	7.733	9.202	14.245	11.780	12.773	12.040	14.362	13.181	13.381	11.379	12.861	13,0 per cent	0,8 per cent
Bulbs and corms (5,3 per cent)	28.576	33.207	38.385	40.653	40.885	54.603	67.737	65.180	61.896	70.562	82.796	17,3 per cent	5,3 per cent
Total	1.215.916	1.234.519	1.193.270	1.196.198	1.249.441	1.369.080	1.457.040	1.526.193	1.448.479	1.502.659	1.551.190	3,2 per cent	100,0 per cent

Export

	2001	2002	2003	2004	2005	2006	2007	2008	2009	2010	2011	2011 vs 2010	share in 2011
Cut flowers and cut foliage (33,2 per cent)	428.897	464.202	433.221	418.056	424.186	493.962	553.497	551.215	476.070	535.275	579.413	8,2 per cent	33,2 per cent
Potted plants (28,5 per cent)	348.659	383.096	366.122	364.901	408.390	442.399	462.796	481.446	417.871	462.104	497.954	7,8 per cent	28,5 per cent
Conifers and hardy perennial plants (10,8 per cent)	52.531	62.763	57.944	73.638	77.889	99.937	131.696	163.036	151.756	188.750	188.093	-0,3 per cent	10,8 per cent
Bulbs and corms (27,5 per cent)	348.902	385.761	381.819	354.633	398.425	486.950	497.732	482.979	445.376	477.352	479.966	0,5 per cent	27,5 per cent
Total	**1.178.989**	**1.295.822**	**1.239.104**	**1.211.229**	**1.308.890**	**1.523.247**	**1.645.721**	**1.678.676**	**1.491.075**	**1.663.481**	**1.745.426**	**4,9 per cent**	**100,0 per cent**

Trade balance

	2001	2002	2003	2004	2005	2006	2007	2008	2009	2010	2011	2011 vs 2010
Cut flowers and cut foliage	557.489	515.844	485.283	500.503	540.517	564.688	560.241	634.318	646.663	620.501	603.683	-2,7 per cent
Potted plants	155.438	171.032	143.986	139.695	177.311	198.613	201.594	219.148	167.402	197.162	225.516	14,4 per cent
Conifers and hardy perennial plants	44.798	53.561	43.698	61.858	65.116	87.897	117.334	149.855	138.376	177.370	175.232	-1,2 per cent
Bulbs and corms	320.326	352.554	343.433	313.980	357.539	432.346	429.994	417.799	383.480	406.790	397.170	-2,4 per cent
Total	**- 36.928**	**61.303**	**45.834**	**15.030**	**59.449**	**154.168**	**188.681**	**152.483**	**42.596**	**160.822**	**194.236**	**20,8 per cent**

Source: Eurostat Comext.

as Chapai Nawabgonj, Rajshahi, Gazipur, Tangail, Bandarban, Rangamati and Khagrachhari are also considered to be prospective for producing ornamental plants.

☆ Approximately 1200 to 1500 traders are involved in flower business.

☆ Dhaka, Chittagong, Jessore, Feni, Rangpur, Natore, Bogra, and Mymensingh are the major markets of floriculture products.

☆ The number of flower shops has been increasing in different City/District towns of the country.

☆ Presently even at Upazilla level the flower shops are available for flower business With the increase in flower shops, the market size is also increasing.

☆ Presently the annual turn out in the whole sale market in Bangladesh is more than Tk. 1000 million.

☆ Roughly about 100,000 to 120,000 people are either directly or indirectly involved in floriculture industry.

Challenges Face by Floriculture Producers

Growers face many challenges including

☆ **Declining margins** - While prices have remained steady over the past several years, most input costs have risen steadily. To remain profitable, growers have had to become more efficient in production and management.

☆ **Environment** - Environmental issues are a major concern for growers. Growers have responded by re-using irrigation water, reducing pesticide and fertilizer use and reducing greenhouse runoff.

☆ **Pest control** - Concerns over pesticide use by the public and producers alike, along with pesticide resistance and the loss of approved pesticides, have prompted growers to adopt alternative pest control methods. Integrated pest management (IPM) is playing a larger role in greenhouse pest control. Many growers are now using biological or bio-rational control methods to supplement or replace existing pesticides.

☆ **Employment** - Labour is an important element in production. Bedding plant and cut flower growers face labour costs of up to one third of gross sales. Although increased mechanization is a necessary element of global competition, the industry continues to be a major agricultural employer.

☆ **Urban-rural conflicts** - Urban-rural conflicts are a fact of life for most agriculture in the Province. some municipalities look upon floriculture as more of a factory production industry rather than agriculture. Most municipalities have zoning regulations concerning the maximum site coverage for greenhouses.

☆ **Capital costs** - Modern, state-of-the-art greenhouse operations can cost up to $200 per square metre. This represents a barrier to entry for many potential growers. Field-grown cut flowers and bedding plant production have much lower capital costs, so they are often entry level crops.

☆ **Seasonal demand** - The demand for fresh floriculture products is seasonal and the product is very perishable. Large numbers of people want to buy flowers for special occasions or holidays like St. Valentine's Day, Easter, Mother's Day and Christmas. Growers must time their production to meet these periods of high demand. Some growers have 30 per cent of their annual sales in a three week period in spring.

Cut-Flower and Foliage Production in Key Countries

Some knowledge of changing production patterns for the various cut-flowers could be useful in informing future UK cropping plans. ISPF 2014 provides tables of production areas and values for numerous countries, though, as might be expected, the level of reporting varies widely from country to country. Information for the Netherlands is comprehensive and is summarised first. Data for some other countries follows in alphabetical order, though country statistics are included only when they include at least some breakdown to individual crops.

Netherlands

The striking feature of the Dutch production area statistics (see next figure) is the overall steady decline in the production of cut-flowers under glass, down by one-third over the 2005-2013 period, while over the same period the area of cut-flowers (and foliage) grown in the open has increased, albeit not fully compensating for the lost area under glass. In addition, the same statistics include an entry for 'shrubs for cutting in the open', down from 252ha in 2007 to 156ha in 2013. The much larger spring-flowering flower-bulb area remained relatively stable, with summer flower- bulbs suffering a small decline over the same period; both dipped in arean 2009-2012.

The area of cut-flowers under glass can be broken down to the main species, but unfortunately these statistics do not provide a breakdown of the flower and foliage crops grown in the open. Over the 2005-2013 period the decline in area of cut-flowers under glass was shared by virtually all crops – only the area of orchids remained stable. Rose was overtaken by chrysanthemum for first place in 2010. The major losses were in rose and carnation (each falling by about 50 per cent), freesia and 'other' cut-flowers (>40 per cent falls) and alstroemeria and anthurium (>30 per cent falls). Chrysanthemum, lily, gerbera and lisianthus suffered smaller – though still substantial – falls of 20 to 26 per cent. These figures reflect, no doubt, the production, of rose particularly, in the new producer countries of Africa and South America. Alstroemeria, lisianthus and 'other' crops' might provide opportunities for UK growers - promising results have already been achieved in the HDC-funded trials at CFC.

This sector is difficult to interpret in terms of cut-flower production because different crops will be grown for flowers, for dry bulbs, or for both. Many flower-bulb species, of course, are grown only for bulb sales, though for completeness they are all included.

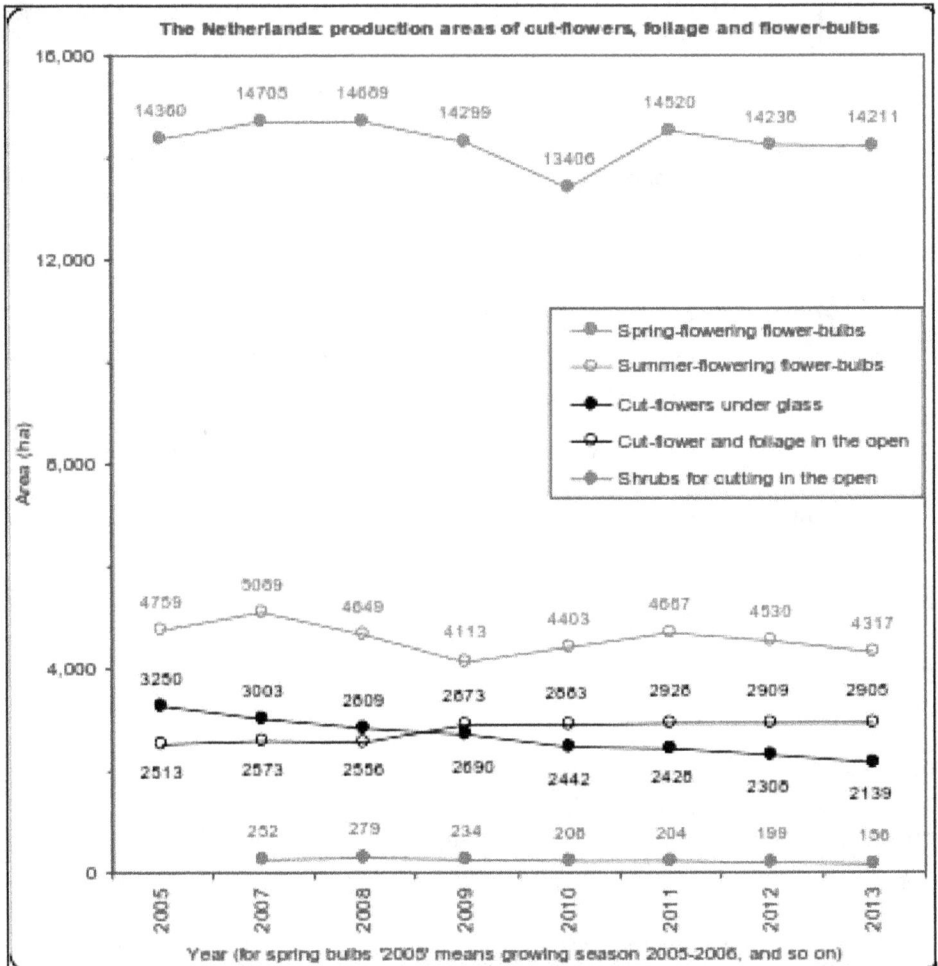

The Netherlands: production areas of cut-flowers, foliage and flower-bulbs

Legend:
- Spring-flowering flower-bulbs
- Summer-flowering flower-bulbs
- Cut-flowers under glass
- Cut-flower and foliage in the open
- Shrubs for cutting in the open

Area (ha) vs Year (for spring bulbs '2005' means growing season 2005-2006, and so on)

While the overall area of Dutch bulbs has remained more or less stable in recent years – down just 3 per cent between 2005 and 2013 – there have been some important changes for individual genera. Tulip and ornamental allium growing have increased during this period (by 5 and 16 per cent, respectively), the hyacinth area has remained stable, and the production of most other species has declined to some extent – iris by almost half and the 'minor' bulbs 25 by almost a third.

The Dutch tulip sector is notable for its continued flow of superior new cultivars that add to the tulip's enduring appeal as a cut-flower, garden plant and container plant. Growers of ornamental alliums have benefitted from the promotion of the crop as a trendy and architectural alternative to traditional flowers. The reason for the decline in many spring-flowering 'minor' flower-bulb genera, after considerable interest a few years ago, is unclear – perhaps difficulties in reliable production, which could indicate a niche opportunity for specialist growers.

USA

Due to increase in imports, production area under floriculture products in the USA is significantly decreasing. The total production area under floriculture crops declined by 2.5 per cent in 2007 compared to 2006. The flower industry in the United States is undergoing a period of consolidation. The large operations account for 96 per cent of the total value of floriculture crops. Annuals and perennials are the largest selling items accounting for 32 per cent of the total floricultural sales in the USA in 2007. The wholesale value of sales of floricultural products in 2007 was estimated at US$ 3.9 billion, registering an increase of 2 per cent over the year 2006.

Canada

The Table 8.20 indicates the total number of stems produced in Canada in 2013 was a 16 per cent increase over the 2009 figures. Notable increases over this period were recorded for chrysanthemum (74 per cent), tulip (40 per cent, though the time-course of the increase was irregular), gerbera (29 per cent) and antirrhinum (17 per cent). However, production of alstroemeria, rose, lily and daffodil fell over this period (as did the number of 'other' crops, which may be partly due to changes over time in the genera that have been recorded.

Table 8.20: Cut-Flower Production in Canada, 2009-2013

Category	Production (thousands of stems)				
	2009	2010	2011	2012	2013
All cut-flowers[1] of which:	285,037	310,656	302,624	321,099	331,900
Tulip	91,685	96,670	80,473	-[2]	128,743
Gerbera	60,939	69,029	66,046	75,086	78,560
Alstroemeria	19,504	15,617	15,924	-	13,180
Chrysanthemum	20,108	27,531	23,105	-	34,949
Antirrhinum	16,481	-	-	-	19,272
Rose	11,483	12,047	10,456	8,435	8,306
Lily	16,049	19,778	15,197	14,465	13,307
Daffodil	12,827	-	-	3,724	4,098
Lisianthus	-	-	2,344	-	-
Iris	-	-	-	2,723	2,408
'Other'	48,788	69,984	66,303	-	21,141

1. These figures do not appear to sum the values below.

2. Indicates data not available.

The Canadian data also records the area of flowers and pot-plants grown in glasshouses or other structures (polyfilm, rigid plastic and all other types of enclosed protection) over 2008-2013 (Table 8.21). While the overall area fell slightly over this time, an increase in glasshouse area was compensated by a fall in the area of rigid plastic structures, which may reflect the types of ornamentals now being grown.

Table 8.21: Production Area of Flowers and Pot-plants in different Greenhouse Types in Canada, 2008-2013

Structure Type	Area (ha)					
	2008	2009	2010	2011	2012	2013
Glass greenhouses	217	198	215	218	218	257
Fibreglass and other rigid greenhouses	92	65	62	41	54	42
Film plastic	523	565	561	556	544	515
Total	833	828	838	816	817	814

China

China's production of ornamental plants rose hugely over the period 2005-2013, with the production area for all cut-flowers and foliage increasing by 68 per cent over this period. All the recorded crops showed an increase – particularly gerbera (185 per cent), carnation (125 per cent) and lily (104 per cent). The small dried flower sector also showed a large increase, 121 per cent. However, the area of bulbs cultivated fell slightly (by 9 per cent) (Table 8.22).

Table 8.22: Production Area of Cut-Flowers and Foliage, Bulbs and Dry Flowers under Protection and in the Open in China, 2005-2013

	Production Area (ha.)								
	2005	2006	2007	2008	2009	2010	2011	2012	2013
Cut-flowers and foliage of which:	38,853	41,603	44,325	44,079	44,603	50,859	57,935	59,382	65,128
Rose	7,676	7,424	8,140	7,388	9,021	9,946	12,530	13,870	14,316
Lily	-[1]	4,908	4,717	5,373	5,827	7,485	8,831	9,105	9,989
Chrysanth.	7,486	4,279	4,722	4,500	4,122	4,927	5,719	7,185	8,475
Gerbera	-	2,180	2,531	2,764	4,563	5,156	5,444	5,378	6,204
Carnation	2,362	2,068	2,674	2,658	2,396	2,826	3,582	3,380	5,312
Gladiolus	2,523	2,140	2,365	2,386	2,447	2,891	3,414	3,327	4,896
Cut-foliage	4,926	4,720	5,835	5,362	6,037	5,609	6,805	7,409	8,592
Cut-branches	-	3,701	3,713	3,798	5,190	4,960	5,404	4,170	5,187
Bulbs	4,606	3,404	3,897	4,680	4,132	4,794	4,514	4,471	4,174
Dried Flowers	-	-	-	24	32	23	44	62	53

1. Indicates data not available.

Ecuador

Ecuador is an important player in the world floriculture trade. The country has around 3,500 ha under floriculture, of which, 2,500 ha is under rose cultivation. Other flowers grown in Ecuador are Chrysanthemum and Carnation. The sector employs around 60,000 people directly and over 100,000 people indirectly.

Ethiopia

The floriculture industry of Ethiopia is one of the fastest growing flower industries in the subSaharan Africa. The Ethiopian flower industry took-off in 2005, and currently, ranks 22nd amongst world exporters of floriculture products, with a share in world exports of around 0.51 per cent. There are over 70 flower farms in Ethiopia. Ethiopia grows Roses, Carnations, Carthamus and Satice. The sector generated employment for over 50,000 persons (permanent and temporary), and has become one of top five foreign exchange earning products of the country. The Ethiopian flower industry is an export-oriented industry and has been benefiting from Government support at large. About 40 per cent of the farms are fully foreign owned, 23 per cent are joint ventures, and 36 per cent are fully domestic owned.

Israel

Israel's cut-flower industry is regarded as one of the most advanced flower industries in the world by virtue of the technologies used in production of large varieties of flowers in the desert-like conditions. Flowers and ornamental plants account for 8.0 per cent of Israel's total agricultural production. Israel's flower sector is relatively small by international standards, but is highly profitable. Flowers are Israel's leading agricultural export item (29 per cent). Israel ranked 12th amongst world exporters of floriculture products and its share in world exports was around 1.5 per cent (US$ 259 million) in 2007.

Germany

Four-yearly cut-flower production figures are available for Germany, though the actual data collected appear to have varied from time to time, rendering them less useful than they may have been.

The area of cut-flowers and foliage grown in the open between 1992 and 2012. Over this period the area of chrysanthemum and of dried flowers each fell by nearly 80 per cent. In contrast the areas of annual and perennial flowers, although fluctuating widely, finished the period 34 per cent up, and shrubs for cutting increased by 40 per cent. The rose area remained relatively stable. Only the data for 2008 and 2012 include an entry for 'other' cut-flowers (including gladiolus and tulip), approaching 1,000ha by 2012: it is not clear how these were recorded previously, but the net effect is to increase the overall area of cut-flowers in the open considerably.

In contrast to open-ground growing, the production of cut-flowers and foliage under protection has declined dramatically over the same period, though data are difficult to interpret because of changes in the categories collected and missing data for some years. However, the figure below shows that the area of most subjects has fallen, drastically so in the case of chrysanthemum (85 per cent) and rose (48 per cent). The growing of 'other cut-flowers and foliage' also fell dramatically at the start of this period, though seems to have revived more recently (it reached 172ha by 2012). Overall, the figures may reflect the higher energy costs for glasshouse production, competition from new producer countries, and perhaps a fashion shift towards 'more natural looking' products.

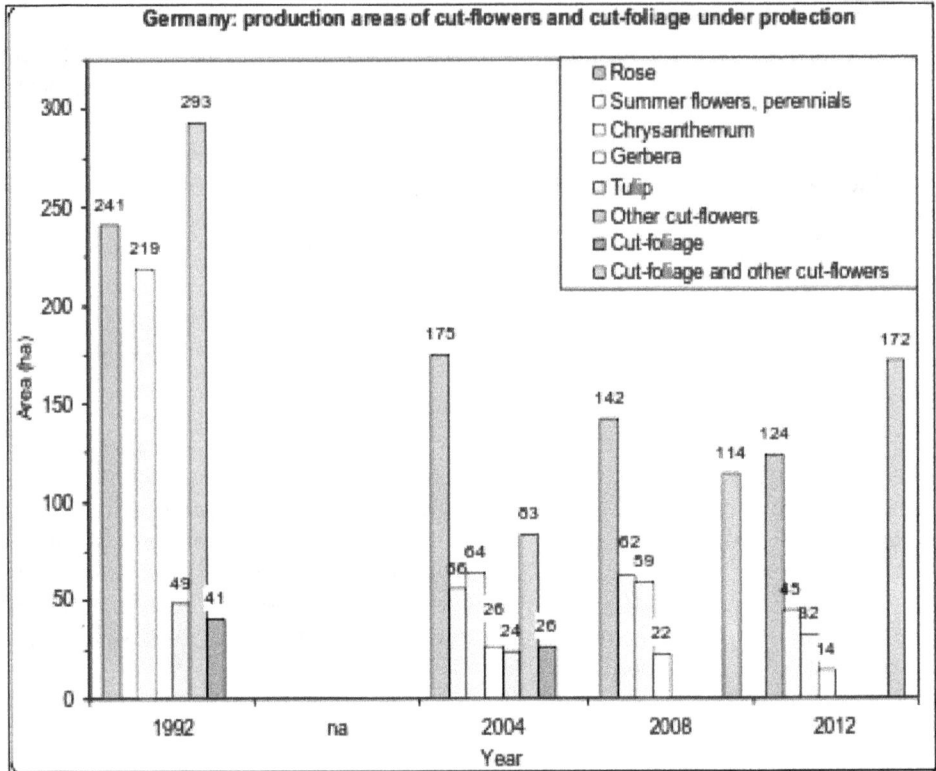

Germany: production areas of cut-flowers and cut-foliage under protection

Greece

Only relatively dated statistics are available, and the Table 8.23 shows that between 1990 and 1995 the overall production area was quite stable at around 400ha. The areas of the main crops – carnation, rose, chrysanthemum and gladiolus – were equally stable.

Table 8.23: Production of Cut-Flowers and Cut-Foliage under Glass and in the Open in Greece, 1990-1995

	Area (ha.)			
	1990	1993	1994	1995
Total of which:	402	421	380	420
Carnation	173	190	170	190
Rose	74	90	90	90
Gladiolus	95	80	56	80
Chrysanthemum	52	55	53	54
Tulip	3	1	6	3
Foliage	5	5	5	3

Japan

Japanese production areas are shown in the Table 8.24. They illustrate the importance of the traditional chrysanthemum crop that occupies almost one-third of the overall area – over 5,000ha of the about 17,000ha total. The production areas of alstroemeria, carnation, gerbera, gypsophila, lily, limonium, lisianthus, orchids and rose are small in comparison – but would be considered very substantial in other situations. The area of 'other' cut-flowers – over 8,000ha or nearly half of the total cut-flower area – suggests a very broad spectrum of species and probably a search for innovative products; unfortunately figures for 'others' are not available after 2008. Additional areas include >700ha for cut-foliage, >3,000ha for cut-tree branches, and nearly 500ha of flower-bulbs.

Table 8.24: Production Area of Cut-flowers, Cut-foliage and Bulbous Plants in Japan, 2006- 2012

	Area (ha)						
	2006	2007	2008	2009	2010	2011	2012
Cut-flowers of which:	17,450	17,230	16,840	-1	-	-	-
Chrysanthemum	5,700	5,650	5,530	5,420	5,331	5,233	5,230
Lily	880	860	860	869	855	833	805
Lisianthus	449	465	466	462	454	439	435
Rose	490	484	474	457	432	419	409
Carnation	430	410	410	405	390	378	367
Gypsophila	290	270	270	260	255	253	251
Limonium	250	210	210	212	204	214	212
Tropical rchids	189	185	178	172	167	161	157
Istroemeria	102	97	95	92	89	91	93
Gerbera	103	98	99	98	94	93	92
Other cuts	8,547	8,500	8,243	-	-	-	-
Cut-foliage Cut tree	672	681	744	768	763	742	713
Branches	4,024	4,100	3,996	3,739	3,813	3,707	3,703
Bulbous plants	575	564	567	538	505	478	472

1. Indicates data not given.

Apart from cut-foliage, all these crops and sub-sectors saw decreases in production areas between 2006 and 2012, though these were relatively modest compared with those in some European countries. These area reductions are reflected in the Table 8.25 of production amounts.

Thailand

Thailand is one of the leading producers of tropical floriculture crops. The development of ornamental industry in the country has been initiated by the private sector with little support from the government. Dendrobium orchid, which has been developed solely from introduced species, is the most commercially important floriculture crop of the country. It is a home of diverse ornamental plant genera; such as Aglaonema, Curcuma, Dracaena, Homalomena, Impatient, Ixora and many Orchid genera. The core competency of the industry is rapid propagation as well as breeding a new cultivar. Currently, the country is known to be the world center for breeding Aglaonema, Curcuma and tropical Orchids. The Orchid planting area in Thailand covers about 6,000 acres, mainly in Bangkok and nearby provinces, namely Nakhon Pathom, Samut Sakhon, Ratchaburi, Pathum Thani, Ayutthaya, Suphan Buri, and Chon Buri. The flowering season in Thailand usually takes place during the rainy season, from July to October. There are more than 1,000 species of Orchids in the country.Besides Orchids; over 30 types of flowers are commercially cultivated in the country, which include Roses, Carnations and Chrysanthemums.

Table 8.25: Production Amount of Selected Cut-Flowers in Japan, 2000-2008

	2000	2006	2007	2008
Cut-flowers of which:	5,593	4,934	4,829	4,734
Chrysanthemum	2,028	1,857	1,814	1,792
Lily	204	174	170	171
Rose	459	371	355	347
Carnation	495	412	387	388
Gypsophila	87	69	60	61
Limonium	138	127	118	120
Other cuts	2,182	1,924	1,925	1,855

Kenya

The Kenyan cut-flower production area is summarised in the next Table 8.26. Between 2010 and 2011 the total area fell, but it more than recovered in the following year, to >4,000ha. Rose is the predominant crop, making up >2,000ha in 2012, but this had fallen from the 2011 figure. The areas of Easter lily and eryngium showed major rises in 2012, the former from 24ha in 2011 to >1,000ha, and eryngium from 56 ha to 176ha. Asclepias, carnation and 'other' cut-flowers showed smaller increases over the same period, while in contrast the area of 'arabicum' (ornithogalum?) was halved and that of alstroemeria was down by one-third. Kenya looks like another emerging producer seeking its best options.

Kenya's economy largely relies on the agriculture sector. Horticulture sub-sector is one of the top foreign exchange earners for the country generating approximately US $ 1 billion annually. In 2015, the sub-sector contributed 1.45 per cent to the national GDP while flower exports contributed 1.01 per cent was from the flower industry. It has grown in significance to a vibrant flower industry worldwide.

It has recorded growth in volume and value of cut flowers exported every year from 10,946 tons in 1988 compared to 86,480 tons in 2006, 120,220 tons in 2010 136, 601 tons in 2014 and 122,825 in 2015. According to Horticultural Crop Directorate (HCD) in 2015, the floriculture industry earned Kenya Shillings 62.9 billion.

Table 8.26: Production Area of Cut-Flowers in Kenya, 2010-2012

	Area (ha)		
	2010	2011	2012
Total of which:	3,419	3,213	4,039
Rose	2,674	2,597	2,164
Carnation	160	157	252
Hypericum	-1	78	80
Arabicum	286	92	140
Easter lily	152	24	1,073
Asclepias	25	52	55
Alstroemeria	55	55	38
Eryngium	20	56	176
Anthurium	-	7	3
'Other'	-	26	155

1. Indicates data not given.

Kenya is the lead exporter of rose cut flowers to the European Union (EU) with a market share of 38 per cent direct sales are growing approximately 50 per cent of exported flowers are sold through the Dutch Auctions. Although In the United Kingdom, supermarkets are the main outlets. Over 25 per cent of exported flowers are delivered directly to these multiples, providing an opportunity for value addition at source through sleeving, labelling and bouquet production. Kenya flowers are sold in more than 60 countries.

It is estimated that in Kenya, over 500,000 people, including over 100,000 flower farm employees depend on the floriculture industry creating over 2 million livelihoods. The main production areas are around Lake Naivasha, Mt. Kenya, Nairobi, Thika, Kiambu, Athi River, Kitale, Nakuru, Kericho, Nyandarua, Trans Nzoia, UasinGishu and Eastern Kenya. The main cut flowers grown in Kenya are roses, carnations, and Alstromeria. Other flowers cultivated include, Gypsophilla, Lilies Eryngiums, arabicum, hypericum, Statice, a range of summer flowers amongst many others.

The industry continues to attract investors due to solid infrastructure, favourable climate, global-positioning of Kenya and a productive workforce. It comprises large, medium and small scale producers who have attained high management standards and have invested heavily in value addition through adoption of modern technology in production, precision farming and marketing. The farmers utilize technologies some of which includes drip irrigation, fertigation systems, greenhouse ventilation systems, net shading, pre-cooling, cold storage facilities, grading, bouqueting, fertilizer recycling systems to prevent wastage, wetlands for waste water treatment, artificial lighting to increase day length, grading/packaging sheds, and refrigerated trucks.

On the global front, a growth of 5 per cent is anticipated every year over the next five years. To be a part of this opportunity, Kenya is going to continue investing and expansion in growth rate than previous years. Going forward, the industry will continue benchmarking itself against other producer countries. Information thus gathered will be useful in formulating strategies, to capture prospects of expansion in current and emerging markets.

Flower Growing Areas in Kenya

The main production areas are around Lake Naivasha, Mt. Kenya, Nairobi, Thika, Kiambu, Athi River, Kitale, Nakuru, Kericho, Nyandarua, Transzoia, Uasin Gichu and Eastern Kenya.

Conclusions about Production, Consumption and Trade Patterns

If we overview the facts about the development of production, trade, and markets from the we can conclude the following:

- ☆ Worldwide the production has expanded. We see this in these countries where entre-
- ☆ preneurs see opportunities to enter markets combined with good growing conditions;
- ☆ Worldwide consumption has grown in these countries where use of ornamentals could increase and in countries where the level of prosperity has risen;
- ☆ With rising income consumption per capita will go up. A higher level of prosperity in countries will increase consumption of ornamentals personal use (life style), gifts and institutional segment (decorations);
- ☆ The international trade of ornamental products has increased, following the development in consumption of ornamentals in the different continents and countries.
- ☆ Regional supply will be leading. The transportation costs will be the lowest and the product will have the highest possible physiological quality;
- ☆ Trade liberalisation will be (or is) followed by non-trade barriers. When the regulations about quota and tariffs are decrease, other regulations increase *i.e.* on quality levels and phytosanitary condition.

Lessons Learned about Production, Consumption and Trade Patterns

- ☆ With higher income in countries and continents, the demand will rise;
- ☆ High volume 'bulky' products will be between North-South pattern or interregional because low product cost are essential;
- ☆ There are niche markets for high quality product all over the world; when product is special of exceptional, product price is not so much an issue.

But constant innovation is essential to be able to 'surprise' them with new products

☆ Strategies for supplying niche markets should depend on a supply chain concept of

☆ 'Quality'; from 'seed to vase'; the supply of product for this market should not be incidental. Structure and year-round supply are prerequisites.

☆ There is a demand for co-operation for all actors in international chains.

Market Developments

People from all over the world use floricultural products for culture and traditions. In the more prosperous countries they are also used to express themselves in line with their life style. Furthermore businesses and organisations also use these products for decorations to create a certain atmosphere contributing to their image.

To a consumers and institutions a floricultural product is more than a plant with foliage and flowers (core product). Plant go together with a pot, it has labels with treatment information and attributes *i.e.* for Valentines day, etc (tangible product). Then the flowers or plants can be combined with others *i.e.* in a bouquet, with certain packaging information and services *i.e.* home delivery (extended product). All these features belong to a product concept of a certain brand of product, with a certain exclusiveness (or not), with certain guarantees *i.e.* on vase life. Consumers buy products from concepts of 'basic' with low price until 'extend' with all kinds of added values. When they repeatedly buy a product from a certain concept they expect it to be of a constant quality and price; these are prerequisites to the suppliers.

Consumers and institutions buy their product at different kind of outlets. In Europe, USA and in the big cities all over the world there are local florists or even artist selling ornamentals and decorations. Adapting to developments sometimes they form chains, *i.e.* in Holland a Business Florist Association was established for the business segment. Then floricultural products can be bought at local garden centres and warehouses. But more and more chains of these, with a franchise element, occur. Furthermore in Europe more and more floricultural products are sold in supermarkets. In the U.K. for instance, the main part of flowers and potted plants are sold thru *i.e.* Tesco, Sainsbury's, Waitrose supermarkets, *etc.*

These different outlets get their products from different types of suppliers. Local florists and garden centres may buy their product at the auction directly or at the cash and carry service. In Germany, France and England product can also be acquired at wholesale markets. The chains of florists, garden centres, warehouse and supermarkets, more and more have their (preferred) supplier. It concerns large quantities.

More and more these supermarket chains have, lead by the choices consumers make, a clear market position in either the discount segment or service retailing with added value product. Because of fast growth of these chains, their economies of scale increase and every now and than price battles are the result. The supply chain therefore is forced to further reduce costs. We also see the growth of private label. Ornamentals in supermarkets *i.e.* in the U.K, France Germany *etc.* occur under private label. Supermarkets see that to make money, fresh product requires professional approach. The trend is to outsource category management to partners in the chain. For image building the supermarkets use all kind of quality assurance schemes, which have to be met by the supply chain.

These developments are the reason that wholesalers/traders specialise, professionalize and grow rapidly, to be of sufficient scale for their clients. More and more they operate as a logistical service provider to their clients. To be able to meet the demand they start working more and more close with growers instead of solely buying at the auction. This, to be sure, their demand is met. The marketing departments of these wholesalers develop new product concepts with their chain partners.

Supply Chain Management

While product, from all over the world solely used to be sold at the auction to all kind of buyers, more and more product is sold directly to supermarkets and specialised wholesalers. Dependent on their market focus and desired image they have product specifications on quality, diversity and quality assurances schemes to be met. The trend we see is that supermarkets, garden centres and warehouses want to be sure the products are produced in a sustainable way. This is also called production in the triple P-concept; a way of producing combining making money (Profit), with care for the environment (Planet) and a social responsible way (People). The word Profit contains items like assets, return on investment, partnerships, mergers *etc.* Planet refers to entities as soil, air, water and the effects of the use of fertilizer, chemicals for pest control, use of energy and the amount of waste. 'People' requires that attention is paid to human rights, poverty, workers, health and security.

These are schemes like, Eurep-GAP with demands on producing in a environmental and social responsible way. A scheme like this can also be covered with the MPS-label for environmental inputs, the MPS-scheme for socially qualified holdings, and product specs from a label called Florimark, combined in MPS-GAP.

The supply chain will have to combine commercial quality (branding) of the product, together with environmental quality (labels/certificates) and social quality (labels/certificates) in their product proposition towards their customers. Additional of managing the logistics of product to the outlets, from the chain partners this requires registrations, certificates, quality audits, information supply in the chain.

Important Aspects of the Flower Supply Chain

In this article it is established that the world wide trade of ornamentals has increased over the last decade. This also means there is quite a bit of competition in supplying the markets. The markets and different type of outlets set the standard. Further more, their hosting countries and continents have their regulations. So if exporting product to other countries and continents is considered, the following has to be taken in to account:

☆ Disease free propagation material; to deliver floricultural products free of disease, for Asian producers it is important to start with healthy propagation material. Because southern countries have good climate conditions for producing propagation materials like seeds, cuttings it is interesting for European and USA growers to import this material. Because of phytosanitary measures in their countries and also their own growing results, it is important that this material is guaranteed disease free.

☆ High level crop management: To be able to meet the demand and the arrangements made about amount of product, product quality, moment of delivery, it is important that crop management is of high level to assure this.

☆ Grading according to international standards is important to make the delivery accepted.

☆ Independent quality inspection: When propagation material or end products are inspected at borders or at the customers', the process will take shorter when it can be shown that it has been inspected by a independent quality inspection organisation in the exporting country. If these are organisation working world wide, *i.e.* S.G.S., this will build trust. This kind of inspection also will assure the exporting party the product won't returned or destroyed.

☆ Flexible marketing system: Exporters will have to try to find different (types of) outlets for their produce. Each customer will have a different demand and different product specifications. The more the exporting company is able to play the customers with more different type of specifications, the higher the chance of being able to sell the produce.

☆ Specialised outlets: Outlets try to build an image and to play target groups by marketing.

☆ They fill this in with their outlet buildings and locations, their product assortment and price setting contribute to this. When potential suppliers are able to strengthen this formula with their product and services, they will be welcome as a supplier. Next to that they can distinguish themselves from other suppliers to invest in this relation.

☆ Climate controlled storage and transport: 'Due diligence' or 'do as you say' is extremely important when it comes to meeting the quality level agreed on. During the transport, the quality of the product will decrease. With a climate control focused on the physiological needs of the product, the process of deterioration will be slowed down. This will rise the chances that the product will arrive with the quality of the right level

Expectations Nearby Future

☆ Looking at the facts presented about the developments in production, trade and consumption.

☆ in the last 10 to 15 years, combined with the actual trends and lessons learned, we come to the following expectations for the nearby future:

☆ Higher demand: The demand will rise because of higher income and because certain markets have not saturated yet. Rising income and prosperity will increase the demand in the Asian, Latin American and Eastern Europe;

☆ Higher quality products: The rise of the average income, the competition in production, market strategies of professional supermarkets will lead to a demand for higher quality product;

☆ Volume/Logistics: For their competitive edge for supermarket chains it will be more and more important to be delivering year-round, to cut in the amount of suppliers and to constantly have the story full with product. Suppliers will have to be able to help solve these problems by being able to supply big volumes, and to have the logistic well arranged to be able to meet the arrangements made;

☆ Professionals/providers (category management/vendor management inventory).

☆ Because supermarkets sell a whole range of products they could use help in the treatment of the fresh products category. Sometimes this category has problems *i.e.* with a lot of waste. They have to improve this because of the competition. If suppliers are able to offer category management for the supermarkets in a way that there is more turn over and less waste, it will benefit both parties;

☆ Long term commitment: This offers possibilities to invest in the relation and to optimise the fitting of the primary process of the enterprises in the supply chain. This also will benefit the parties in the chain;

☆ International alliances: Continents, countries and unions have their own regulations, and business culture. For exporting producers it is important to understand and cope with this. International alliances will make market access easier.

Market Information needs to be fed.

Watching trends in production, trade and consumption on markets world wide makes it possible to draw ones conclusion about their own position. In order to do so, there is a (constant) need for good statistics. With the help of a network of specialists and local sources throughout the world this could be achieved.

Annexure 1: EU Production of Flowers and Plants (in millions euros, constant prices 2005=100)

Member State	2001	2002	2003	2004	2005	2006	2007	2008	2009	2010	2011	2011 vs 2010	Share in 2011
Netherlands	5.572	5.468	5.583	5.751	5.636	5.628	5.632	5.663	5.593	5.627	5.655	0,5%	30,8%
Italy	2.902	2.852	2.898	2.847	2.846	2.940	2.972	2.975	2.725	2.663	2.658	-0,2%	14,5%
France	2.489	2.460	2.393	2.341	2.334	2.148	2.115	2.159	2.155	2.261	2.261	0,0%	12,3%
Germany	2.503	2.437	2.484	2.665	2.626	2.626	2.626	2.626	2.393	2.288	2.127	-7,0%	11,6%
Spain	2.200	2.215	3.613	2.531	1.800	1.202	1.887	2.282	1.925	1.925	1.925	0,0%	10,5%
United Kingdom	1.096	1.206	1.146	1.187	1.137	1.021	1.065	1.082	980	980	957	-2,3%	5,2%
Belgium	263	289	304	426	564	734	763	780	806	810	791	-2,3%	4,3%
Denmark	240	285	340	411	380	464	473	442	385	416	394	-5,4%	2,1%
Portugal	552	528	499	499	460	487	484	478	456	409	394	-3,7%	2,1%
Austria	265	268	259	260	256	242	235	233	231	227	227	-0,4%	1,2%
Sweden	198	189	174	155	161	174	187	184	194	202	199	-1,7%	1,1%
Greece	105	144	138	166	157	155	167	161	162	169	171	0,9%	0,9%
Finland	128	130	129	124	120	121	127	122	123	124	126	2,0%	0,7%
Romania	54	51	65	42	52	128	84	99	74	374	108	-71,1%	0,6%
Hungary	158	187	165	127	118	84	106	108	95	107	107	0,0%	0,6%
Czech Republic	66	54	66	55	93	93	86	95	95	97	96	-0,6%	0,5%
Poland	58	57	67	67	74	73	78	89	83	84	84	0,0%	0,5%
Slovenia	24	31	23	27	30	25	22	22	25	19	19	-1,7%	0,1%
Lithuania	6	12	9	9	17	14	9	7	9	13	13	0,0%	0,1%
Cyprus			26	23	21	14	18	16	13	13	12	-4,1%	0,1%
Slovakia	13	12	13	12	12	12	12	12	12	12	12	-0,1%	0,1%
Latvia	3	4	7	8	6	9	7	7	5	4	3	-5,7%	0,0%
Estonia	6	7	5	4	3	5	7	6	4	3	3	-2,2%	0,0%

Member State	2001	2002	2003	2004	2005	2006	2007	2008	2009	2010	2011	2011 vs 2010	Share in 2011
Luxembourg	3	3	4	4	5	3	5	4	5	3	3	0,0%	0,0%
Bulgaria	10	109	80	90	57	72	57	109	22	13	1	-91,2%	0,0%
Malta	11	10	11	6	2	1	0	0	0	0	0	0,0%	0,0%
Ireland													
UE-27	18.922	19.008	20.500	19.834	18.966	18.475	19.224	19.758	18.569	18.840	18.344	-2,6%	100,0%
UE-15 (OMS)	18.514	18.395	19.899	19.285	18.412	17.953	18.695	19.167	18.084	18.353	17.868	-2,6%	97,4%
UE-12 (NMS)	408	613	601	550	554	522	528	591	485	488	476	-2,4%	2,6%

Source : EUROSTAT ([aact_eaa03] code 04200) - extraction 19/09/2012. Data for Ireland (and Croatia) are not available all data are EUROSTAT estimates.

Total

Annexure 2: Main EU Exporters to EU and Third Countries (in tonnes) in 2011

	Nether-lands	Germany	Italy	Belgium	Denmark	Spain	France	Poland	Utd. Kingdom	Other MS	Total EU	%of Imports
Netherlands	-	23,5%	22,6%	11,2%	14,2%	38,2%	9,0%	10,4%	29,4%	6,8%	238.540	7,0%
Germany	28,6%	-	7,6%	21,1%	6,5%	7,9%	43,7%	15,4%	1,7%	5,0%	767.381	22,5%
Belgium	5,2%	3,5%	-	4,2%	3,5%	6,3%	2,6%	0,3%	0,0%	0,8%	143.648	4,2%
Italy	9,0%	3,1%	7,9%	-	13,1%	6,6%	1,2%	0,2%	0,3%	6,2%	236.565	6,9%
Spain	1,8%	0,6%	1,5%	3,8%	-	14,2%	0,1%	0,5%	0,8%	17,2%	92.239	2,7%
France	10,2%	11,0%	39,4%	23,2%	40,2%	-	4,6%	0,9%	1,5%	1,8%	471.028	13,8%
Denmark	1,8%	3,6%	0,3%	1,0%	0,4%	1,0%	-	4,6%	3,6%	1,9%	59.139	1,7%
Poland	3,1%	3,4%	0,3%	0,5%	0,5%	0,4%	2,4%	-	0,6%	0,4%	78.653	2,3%
Utd. Kingdom	12,6%	3,3%	8,6%	5,8%	5,9%	8,6%	6,7%	2,0%	-	9,9%	337.405	9,9%
Ireland	1,2%	0,3%	0,2%	0,4%	0,0%	0,2%	0,3%	0,0%	47,8%	0,0%	35.908	1,1%
Greece	0,4%	0,2%	0,4%	1,4%	0,1%	0,2%	0,0%	0,0%	0,0%	0,1%	15.178	0,4%
Portugal	0,6%	0,1%	0,2%	0,5%	7,8%	0,2%	0,0%	0,1%	0,0%	0,0%	27.770	0,8%
Luxembourg	0,1%	1,1%	1,5%	0,1%	0,0%	0,3%	0,1%	0,0%	0,0%	0,0%	8.577	0,3%
Sweden	2,2%	4,4%	0,1%	0,2%	0,4%	0,0%	12,1%	0,3%	3,3%	0,6%	79.345	2,3%
Finland	0,9%	2,2%	0,3%	0,0%	0,0%	0,0%	2,4%	0,8%	0,2%	1,0%	30.469	0,9%
Austria	1,7%	10,3%	0,5%	2,3%	0,3%	0,8%	1,6%	1,0%	0,1%	3,4%	82.670	2,4%
Malta	0,0%	0,0%	0,0%	0,3%	0,0%	0,0%	0,0%	0,0%	0,0%	0,0%	1.400	0,0%
Estonia	0,1%	0,3%	0,5%	0,0%	0,0%	0,0%	0,1%	0,2%	0,0%	0,2%	4.852	0,1%
Latvia	0,2%	1,6%	0,2%	0,4%	0,0%	0,1%	0,0%	2,2%	0,0%	0,3%	12.897	0,4%

Exporter

	Exporter										Total EU	%of Imports
	Nether-lands	Germany	Italy	Belgium	Denmark	Spain	France	Poland	Utd. Kingdom	Other MS		
Lithuania	0,2%	0,5%	0,1%	0,0%	0,0%	0,0%	0,0%	0,7%	0,0%	0,1%	5.981	0,2%
Czech Rep.	1,3%	3,7%	0,2%	0,9%	0,1%	0,1%	0,8%	0,9%	0,0%	1,8%	44.689	1,3%
Slovakia	0,4%	1,9%	0,3%	0,7%	0,0%	0,1%	0,2%	3,6%	0,7%	2,0%	22.117	0,6%
Hungary	0,7%	0,9%	0,1%	1,0%	0,0%	0,1%	0,6%	0,3%	0,2%	1,6%	24.397	0,7%
Romania	0,5%	0,7%	0,1%	2,1%	0,3%	0,2%	0,3%	0,1%	0,0%	2,7%	24.832	0,7%
Bulgaria	0,1%	0,1%	0,0%	0,2%	0,0%	1,0%	0,0%	0,0%	0,5%	0,8%	5.278	0,2%
Slovenia	0,3%	0,5%	0,2%	0,5%	0,0%	0,1%	0,0%	0,0%	0,0%	1,8%	12.484	0,4%
Cyprus	0,1%	0,8%	0,1%	0,3%	0,2%	0,0%	0,0%	1,9%	0,0%	0,3%	7.461	0,2%
Third countries	16,9%	18,6%	6,7%	18,2%	6,5%	13,4%	11,1%	53,8%	9,2%	33,4%	544.281	15,9%
Total exports	1.957.161	294.119	172.083	405.234	164.103	59.808	170.867	64.734	20.061	160.576	3.415.183	100,0%
%of exports	57,3%	8,6%	5,0%	11,9%	4,8%	1,8%	5,0%	1,9%	0,6%	4,7%	100,0%	100,0%

Source: Eurostat Comext.

Annexure 3: Main EU Exporters to EU and Third Countries (in thousand euros) in 2011

Total

	Exporter										Total EU	%of Imports
	Nether-lands	Germany	Italy	Belgium	Denmark	Spain	France	Poland	Portugal	Other MS		
Netherlands	-	28,1%	12,9%	40,7%	9,7%	17,3%	22,3%	20,7%	28,4%	10,6%	691.134	7,1%
Germany	26,3%	-	22,8%	5,5%	35,9%	10,8%	10,2%	22,2%	0,9%	7,5%	2.153.568	22,2%
Italy	6,4%	3,1%	-	2,1%	2,1%	8,4%	13,2%	0,4%	10,0%	2,3%	511.974	5,3%
Belgium	4,5%	3,2%	3,2%	-	2,4%	3,0%	5,8%	0,4%	1,5%	2,6%	368.663	3,8%
Denmark	2,1%	3,3%	1,0%	0,2%	-	0,8%	0,9%	8,5%	0,1%	2,4%	183.303	1,9%
Spain	1,6%	0,8%	3,5%	0,6%	0,2%	-	5,7%	0,9%	41,8%	0,4%	174.546	1,8%
France	12,8%	9,7%	20,4%	35,5%	4,4%	31,7%	-	1,9%	10,5%	1,4%	1.349.453	13,9%
Poland	2,4%	3,6%	0,9%	0,2%	2,0%	0,5%	0,7%	-	0,0%	1,1%	205.609	2,1%
Portugal	0,6%	0,2%	0,7%	0,1%	0,0%	8,5%	0,9%	0,3%	-	0,0%	62.027	0,6%
Utd. Kingdom	12,1%	3,1%	4,8%	8,1%	6,5%	4,7%	8,8%	3,7%	5,7%	6,7%	975.123	10,0%
Ireland	1,4%	0,4%	0,3%	0,2%	0,3%	0,0%	0,3%	0,0%	0,0%	11,9%	129.104	1,3%
Greece	0,4%	0,3%	1,5%	0,2%	0,0%	0,3%	0,6%	0,0%	0,0%	0,1%	43.779	0,5%
Luxembourg	0,1%	0,8%	0,1%	1,2%	0,2%	0,0%	0,4%	0,0%	0,0%	0,0%	20.440	0,2%
Sweden	2,3%	4,0%	0,3%	0,1%	15,3%	0,8%	0,1%	0,6%	0,0%	1,1%	255.603	2,6%
Finland	1,1%	1,9%	0,2%	0,2%	3,3%	0,0%	0,1%	0,6%	0,0%	1,3%	108.335	1,1%
Austria	2,4%	12,6%	3,4%	0,4%	1,4%	0,5%	1,1%	1,4%	0,0%	2,8%	282.899	2,9%
Malta	0,0%	0,0%	0,2%	0,0%	0,0%	0,0%	0,0%	0,0%	0,0%	0,0%	2.288	0,0%
Estonia	0,1%	0,2%	0,0%	0,1%	0,1%	0,0%	0,0%	0,3%	0,0%	0,5%	12.425	0,1%
Latvia	0,2%	1,2%	0,2%	0,1%	0,0%	0,0%	0,3%	2,0%	0,0%	0,8%	30.471	0,3%

	Exporter										Total EU	%of Imports
	Netherlands	Germany	Italy	Belgium	Denmark	Spain	France	Poland	Portugal	Other MS		
Lithuania	0,2%	0,4%	0,0%	0,0%	0,0%	0,0%	0,1%	0,9%	0,0%	0,2%	17.221	0,2%
Czech Rep.	1,3%	3,0%	0,9%	0,1%	0,8%	0,1%	0,4%	2,6%	0,0%	2,9%	126.121	1,3%
Slovakia	0,4%	1,5%	0,6%	0,1%	0,3%	0,0%	0,1%	1,7%	0,0%	3,9%	53.025	0,5%
Hungary	0,7%	1,1%	0,9%	0,1%	0,4%	0,1%	0,1%	0,3%	0,0%	2,7%	70.203	0,7%
Romania	0,5%	0,7%	2,9%	0,0%	0,4%	0,7%	1,8%	0,2%	0,0%	4,0%	70.964	0,7%
Bulgaria	0,1%	0,1%	0,2%	0,0%	0,1%	0,0%	0,6%	0,1%	0,0%	0,5%	10.444	0,1%
Slovenia	0,4%	0,8%	0,6%	0,1%	0,1%	0,0%	0,2%	0,0%	0,0%	2,9%	42.307	0,4%
Cyprus	0,1%	0,4%	0,2%	0,1%	0,0%	0,0%	0,0%	1,1%	0,0%	0,2%	10.270	0,1%
Third Countries	19,5%	15,7%	17,2%	3,9%	14,1%	11,7%	25,1%	29,4%	0,9%	29,0%	1.745.426	18,0%
Total exports	6.612.511	677.388	626.215	621.127	454.664	187.272	112.205	93.796	68.137	253.411	9.706.725	100,0%
%of exports	68,1%	7,0%	6,5%	6,4%	4,7%	1,9%	1,2%	1,0%	0,7%	2,6%	100,0%	100,0%

Source: Eurostat Comext.

Index

Carnations 12

Chrysanthemum 3, 12, 45, 110, 201

Climate controlled storage 409

Climatic conditions 3, 8, 30, 33, 36, 44, 50, 60, 64, 81, 82, 88, 96, 97, 100, 127, 146, 187, 235, 285, 292, 307, 340, 346

Climbing and rambling 172

Commercial commodity 9

Commercial floriculture 35, 37, 96, 131, 132

Commercial flower production 9

Cosmeceuticals 28

Cretaceous period 11

Crop management 409

Cut cultivated greens 3, 120

Cut flowers 2-4, 45, 51, 82, 83, 88, 89, 90, 117, 123, 127, 329, 332, 366, 370, 378, 383, 387, 393-395

Cymbidium aloifolium 128

D

Daffodil 38

Dahlia 12

Dahlia hortensis 62

Dahlias 275

Daisy 12

Damask 161

Delphinium 13

Dendranthema 49, 91, 202

Dendrobium orchids 286

Designer houses 33

Dicot plant 24

Disease management 115

Diseases 9

Domestic floriculture industry 49

Dry flowers 62

Dutch Roses 181

E

Economic importance 8

Employment 37

English/David Austin 173

Eria andamanica 128

Export commodities 1

Export Promotion of Capital Goods (EPCG) scheme 337

Exports 90

F

Fertilization 267

Financing 59

Floribunda 170

Floriculture export 308, 312

Floriculture products 2, 4, 51, 52, 130, 308-316, 353, 379

Flower cultivation 9, 61, 124

Flower fascination 183

Flower food 182

Flowering garden 3

Flower market 107, 108

Flower supply chain 408

Flower trade 53

Foliage production 397

Forget-me-not 39

French marigold 192

G

Gallica 160

Garden centres 4, 52

Garden roses 157

Gerbera 5, 110

Gerbera daisies 268

Gladiolus 5, 39, 111, 236

Government incentives 53

Grandiflora 171

Greenhouse 4, 36, 37, 52, 77, 97, 363, 400

Gross domestic product (GDP) 317

Growers 4

Gymnosperms 23